Comparative Anatomy of the Mouse and the Rat

A Color Atlas and Text

Printed in the USA.

Library of Congress Control Number: 2011900825

ISBN 978-0-9789207-2-2

American Association for Laboratory Animal Science • 9190 Crestwyn Hills Drive • Memphis, TN 38125-8538 - 901.754.8620 • fax 901.753.0046 • http://www.aalas.org

Comparative Anatomy of the Mouse and the Rat
A Color Atlas and Text

Illustrations and Text
Gheorghe M Constantinescu, DVM, PhD, mult. DrHC

Scientific Editor and Contributor
Nicole Duffee, DVM, PhD

Contributor
Cynthia Besch-Williford, DVM, PhD, DACLAM

Reviewers
Lynda Lanning, DVM, DABT
Tracy Gluckman, MS, DVM, DACLAM
Mark Hoenerhoff, DVM, PhD, DACVP
Julie Watson, MA, VetMB, DACLAM

Graphic Design and Layout
Amy B Tippett, BFA

Editorial Assistance
Nicole Brown, MA
Melissa Bagaglio, BS
Pamela Grabeel, MA

Table of Contents

Acknowledgments viii
Preface ix
Nomenclature and Abbreviations xi

Body Regions 1
Figure 1. Body regions (lateral aspect), shown in the male rat. 2
Figure 2. Body regions (ventral aspect), shown in the male rat. 3
Figure 3. Female reproductive landmarks (ventral aspect), shown in the rat. 4
Figure 4. Facial regions of the head (lateral aspect), shown in the rat. 5
Figure 5. Landmarks for access to cervicothoracic vessels (ventral aspect), shown in the rat. 6
Juvenile Features and Sex Differentiation 9
Figure 6A. Mouse. Neonate, albino (lateral aspect), age less than 24 hours, after feeding. 10
Figure 6B. Rat. Neonate, albino (lateral aspect), age less than 24 hours, after feeding. 11
Figure 7A. Mouse. Sexing juveniles, albino (perineal aspect), 5th day of age. 14
Figure 7B. Rat. Sexing juveniles, albino (perineal aspect), 5th day of age. 15
Figure 8A. Mouse. Sexing juveniles, agouti (perineal aspect), 11th day of age. 18
Figure 8B. Rat. Sexing juveniles, albino (positioned with head down, perineal aspect), 11th day of age. 19
Figure 8C. Rat. Sexing juveniles, albino (positioned with head up, perineal aspect), 11th day of age. 20
Figure 9A. Mouse. Sexing juveniles (perineal aspect), 21st day of age. 24
Figure 9B. Rat. Sexing juveniles (perineal aspect), 21st day of age. 25
Figure 10A. Mouse. Sexing adults (perineal aspect). 28
Figure 10B. Rat. Sexing adults (perineal aspect). 29
External Features 33
Figure 11A. Mouse. Male adult (dorsal aspect). 34
Figure 11B. Rat. Male adult (dorsal aspect) 35
Figure 12A. Mouse. Male adult (ventral aspect). 38
Figure 12B. Rat. Male adult (ventral aspect) 39
Figure 13A. Mouse. Female adult (ventral aspect) 42
Figure 13B. Rat. Female adult (ventral aspect) 43
Figure 14A. Mouse. Feet (palmar and plantar aspects) 46
Figure 14B. Rat. Feet (palmar and plantar aspects). 47
Mammary Glands 51
Figure 15A. Mouse. Mammary glands in lateral, ventral, and dorsal aspects. 52
Figure 15B. Rat. Mammary glands in lateral and ventral aspects. 53
Structures of the Head and Neck 57
Figure 16A. Mouse. Head, facial (lateral aspect), showing vibrissae. 58
Figure 16B. Rat. Head, facial (lateral aspect), showing vibrissae. 59
Figure 17A. Mouse. Head, lacrimal and salivary glands (lateral aspect). 62
Figure 17B. Rat. Head, lacrimal and salivary glands (lateral aspect). 63
Figure 18A. Mouse. Orbital veins and venous plexus (lateral aspect). 66
Figure 18B. Rat. Orbital veins and venous plexus (lateral aspect). 67
Figure 19A. Mouse. Brain with arterial and venous vasculature (dorsal aspect). 70
Figure 19B. Rat. Brain with arterial and venous vasculature (dorsal aspect). 71
Figure 20A. Mouse. Brain with arterial vasculature (ventral aspect). 74
Figure 20B. Rat. Brain with arterial vasculature (ventral aspect). 75
Figure 21A. Mouse. Brain with arterial vasculature (left lateral aspect). 78
Figure 21B. Rat. Brain with arterial vasculature (left lateral aspect). 79

Figure 22A. Mouse. Brain (median section). 82
Figure 22B. Rat. Brain (median section). 83
Figure 23A. Mouse. Head (median section). 86
Figure 23B. Rat. Head (median section). 87
Figure 24A. Mouse. Pharynx (rostral aspect from the open mouth). 90
Figure 24B. Rat. Pharynx (rostral aspect from the open mouth). 91
Figure 25A. Mouse. Head and neck to upper thorax, including the heart and great vessels (ventral aspect). 94
Figure 25B. Rat. Head and neck to upper thorax, including the heart and great vessels (ventral aspect). 95
Heart, Vascular Tree, and Respiratory Tract 99
Figure 26A. Mouse. Projection of the thoracic viscera (left aspect). 100
Figure 26B. Rat. Projection of the thoracic viscera (left aspect). 101
Figure 27A. Mouse. Projection of the rib cage and the thoracic viscera (left aspect). 104
Figure 27B. Rat. Projection of the rib cage and the thoracic viscera (left aspect). 105
Figure 28A. Mouse. Projection of the thoracic viscera (right aspect). 108
Figure 28B. Rat. Projection of the thoracic viscera (right aspect). 109
Figure 29A. Mouse. Projection of the rib cage and the thoracic viscera (right aspect). 112
Figure 29B. Rat. Projection of the rib cage and the thoracic viscera (right aspect). 113
Figure 30A. Mouse. Projection of the thoracic viscera (ventral aspect). 116
Figure 30B. Rat. Projection of the thoracic viscera (ventral aspect). 117
Figure 31A. Mouse. Projection of the rib cage and thoracic viscera (ventral aspect). 120
Figure 31B. Rat. Projection of the rib cage and thoracic viscera (ventral aspect). 121
Figure 32A. Mouse. Topography of the heart in situ (left aspect). 124
Figure 32B. Rat. Topography of the heart in situ (left aspect). 125
Figure 33A. Mouse. Topography of the heart in situ (right aspect). 128
Figure 33B. Rat. Topography of the heart in situ (right aspect). 129
Figure 34A. Mouse. Heart (atrial and auricular aspects). 132
Figure 34B. Rat. Heart (atrial and auricular aspects). 133
Figure 35A. Mouse. Heart in situ (ventral aspect) and reflected cranially (dorsal aspect). 136
Figure 35B. Rat. Heart in situ (ventral aspect) and reflected cranially (dorsal aspect). 137
Figure 36. Longitudinal section through the heart (atrial aspect), semi-schematic 140
Figure 37A. Mouse. Cervicothoracic organs (ventral aspect). 142
Figure 37B. Rat. Cervicothoracic organs (ventral aspect). 143
Figure 38A. Mouse. Heart and vascular tree (ventral aspect). 146
Figure 38B. Rat. Heart and vascular tree (ventral aspect). 147
Abdominal Structures 151
Figure 39A. Mouse, male. Abdominal topography (left aspect). 152
Figure 39B. Rat, female. Abdominal topography (left aspect). 153
Figure 40A. Mouse, male. Abdominal topography (right aspect). 156
Figure 40B. Rat, female. Abdominal topography (right aspect). 157
Figure 41A. Mouse, female. Abdominal topography (ventral aspect). 160
Figure 41B. Rat, female. Abdominal topography (ventral aspect). 161
Figure 42A. Mouse, female. Liver, stomach, and intestines (ventral aspect), intestines displaced. 164
Figure 42B. Rat, male. Liver, stomach, and intestines (ventral aspect), intestines displaced. 165
Figure 43A. Mouse. Upper abdominal structures (ventral aspect), liver reflected cranially. 168
Figure 43B. Rat. Upper abdominal structures (ventral aspect), liver reflected cranially. 169
Figure 44A. Mouse. Stomach, distal esophagus, and proximal duodenum (internal aspect). Left, median section through the long axis. Right, section through the major curvature. 172

Figure 44B. Rat. Stomach, distal esophagus, and proximal duodenum (internal aspect). Left, median section through the long axis. Right, section through the major curvature. 173
Figure 45A. Mouse. Parietal lymph nodes of the roof of the abdominal cavity (ventral aspect). 176
Figure 45B. Rat. Parietal lymph nodes of the roof of the abdominal cavity (ventral aspect). 177
Male Urogenital Apparatus 181
Figure 46A. Mouse. Male caudal abdominal and pelvic viscera including the testicle (left lateral aspect). 182
Figure 46B. Rat. Male caudal abdominal and pelvic viscera including the testicle (left lateral aspect). 183
Figure 47A. Mouse. Male reproductive and urinary apparatus (ventral aspect). 186
Figure 47B. Rat. Male reproductive and urinary apparatus (ventral aspect). 187
Figure 48A. Mouse. Testicle, epididymis, and spermatic cord (left lateral aspect). 190
Figure 48B. Rat. Testicle, epididymis, and spermatic cord (left lateral aspect). 191
Figure 49A. Mouse. Testicle, epididymis, and spermatic cord (right medial aspect). 194
Figure 49B. Rat. Testicle, epididymis, and spermatic cord (right medial aspect). 195
Figure 50A. Mouse. Penis, reflected caudally (dorsal aspect). 198
Figure 50B. Rat. Penis, reflected caudally (dorsal aspect). 199
Figure 51A. Mouse. Proximal urethra and urethral recess (dorsal aspect, positioned with head up), median section through urethral recess. 202
Figure 51B. Mouse. Latex cast of pelvic urethra (lateral aspect, positioned with head to the left). 203
Figure 51C. Mouse. Latex cast of pelvic urethra (dorsal aspect, positioned with head to the left). 204
Figure 51D. Rat. Proximal urethra and related structures (median section, positioned with head to the left). 205
Figure 52A. Mouse. Penis (median section), stained by hematoxylin and eosin. 208
Figure 52B. Rat. Penis (median section), stained by hematoxylin and eosin. 209
Female Urogenital Apparatus 213
Figure 53A. Mouse. Female caudal abdominal and pelvic viscera (left lateral aspect). 214
Figure 53B. Rat. Female caudal abdominal and pelvic viscera (left lateral aspect). 215
Figure 54A. Mouse. Female reproductive and urinary apparatus (ventral aspect). 218
Figure 54B. Rat. Female reproductive and urinary apparatus (ventral aspect). 219
Figure 55A. Mouse. Round ligament of the uterus (ventral aspect). 222
Figure 55B. Rat. Round ligament of the uterus (ventral aspect). 223
Pelvic Limb Vessels and Nerves 227
Figure 56A. Mouse. Pelvic limb, left (lateral aspect). 228
Figure 56B. Rat. Pelvic limb, left (lateral aspect). 229
Figure 57A. Mouse. Pelvic limb, left (medial aspect). 232
Figure 57B. Rat. Pelvic limb, left (medial aspect). 233
Structures of the Tail 237
Figure 58A. Mouse. Base of the tail (cross section), stained by hematoxylin and eosin. 238
Figure 58B. Rat. Base of the tail (cross section), stained by hematoxylin and eosin. 239
Skeletal Structures 243
Figure 59A. Mouse. Skeleton (left lateral aspect). 244
Figure 59B. Rat. Skeleton (left lateral aspect). 245
Figure 60A. Mouse. Skull and details of teeth (left lateral aspect). 248
Figure 60B. Rat. Skull and details of teeth (left lateral aspect). 249

Bibliography 253
References 255
Index 257

Acknowledgments

The **Color Atlas and Text of Comparative Anatomy of the Mouse and the Rat** could not have been printed without the contribution of many specialists in the field.

Many thanks go to Dr. Beth Bauer, clinical assistant professor, for her contribution to identification of different strains of mice and rats; to senior research laboratory technician Bonita (Bo) Cowan, laboratory supervisor and histotechnologist Jill Gruenkemeyer, and histotechnologist Jan Adair for their contribution to processing the histological sections; and to Laurie Wisdom, laboratory supervisor, for providing hundreds of mouse and rat specimens from the necropsy room for dissection.

High appreciation and respect for her continuous help in identification of anatomical structures and in providing various information about the anatomy of these species go to my friend and colleague Dr. Cynthia (Cindy) Besch-Williford, associate professor, who was in constant contact with the author throughout the dissection of specimens and illustration of the anatomical structures of mice and rats as well as during the editing stage. Cindy's exceptional professional knowledge and expertise in the field of laboratory animals was very helpful.

All of the contributors listed above belong to the Research Animal Diagnostic and Investigative Laboratory (RADIL) of the Veterinary Pathobiology Department of the University of Missouri College of Veterinary Medicine.

Special thanks go to my friends and day-by-day contributors: medical photographer and senior multimedia specialist Howard Wilson and senior multimedia specialist Don Connor from the College of Veterinary Medicine of Missouri University in Columbia.

The author thanks wholeheartedly the review subcommittee: Dr. Lynda Lanning (also chair of the Educational Resources Committee), Dr. Tracy Gluckman, Dr. Mark Hoenerhoff, and Dr. Julie Watson, who spent hours in teleconferences coordinated by Dr. Nicole Duffee reviewing and suggesting improvements of the text accompanying the illustrations. As a result of these discussions, several more anatomical structures have been investigated and illustrated for both species.

Thanks also to the Educational Resources Committee for their continuous support and encouragements.

Sincere thanks and gratitude go to Dr. Ann Tourigny Turner, executive director of AALAS, for remaining in contact with us and following step by step the development of the atlas.

Many thanks to the AALAS staff for their editorial support: education manager Pam Grabeel, communications resource editor Nicole Brown, and editorial specialist Melissa Bagaglio.

Special thanks to Amy Tippett, freelance graphic designer, for her professional hard work and dedication to this project. She meticulously followed our directions, sometimes more than once for the same label, until all labels were updated and correctly added to the illustrations.

I am deeply grateful to, and have the highest appreciation for, Dr. Nicole Duffee, director of education and scientific affairs at AALAS. She is a respected and dear colleague and friend of mine, who orchestrated this atlas on the basis of two previous posters: Anatomy of the Mouse and Anatomy of the Rat. During a phone conversation, Nicole and I envisaged the opportunity to build up a comparative atlas of mouse and rat anatomy, extending and deepening the two posters and adding explanatory text for each figure. Through hundreds of email messages, and phone calls between Nicole and me, and teleconferences with the review subcommittee, we built the foundation of the atlas.

All my contributors merit a big round of applause.

And finally, I thank warmly my lovely wife, Dr. Ileana A. Constantinescu, for her unlimited support, understanding, encouragements, and above all her sacrifice throughout the development of this atlas.

The author is open for any suggestions, comments, or criticism, and kindly asks readers to send them to the AALAS office in care of Dr. Nicole Duffee.

Gheorghe M. Constantinescu
Professor of Veterinary Anatomy and Medical Illustrator
Professional Member of the Association of Medical Illustrators
College of Veterinary Medicine
University of Missouri-Columbia

Preface

Comparative Anatomy of the Mouse and Rat: a Color Atlas and Text was developed to provide detailed comparative anatomical information for those who work with mice and rats in animal research, mainly researchers and laboratory veterinary professionals. These individuals require information on the anatomical features and landmarks for conducting a physical examination, collecting biological samples, making injections of therapeutic and experimental materials, using imaging modalities, and performing surgeries. This atlas compares in these species the structures of the skeleton; the skull and teeth; the pharynx; the digestive, respiratory, and genitourinary organs; the heart and major blood vessels; the brain; the tail; and the major vessels and nerves of the pelvic limb. The nomenclature for regions of the body and the topography of thoracic and abdominal structures is shown in lateral and ventral views.

An important aspect of this atlas is the use of the veterinary anatomical nomenclature from the Nomina Anatomica Veterinaria (NAV), 5th edition, 2005.[7] Anatomical terms are referenced in the Index, in which the page numbers refer to the labeled figures only, not the text corresponding to the figures.

In each set of illustrations, the same view is depicted in the mouse and the rat. Text is provided with all illustrations to draw attention to the anatomical features which are important for supporting the care and use of these animals in research. This work departs from a classical atlas illustrating all body systems and structures because its purpose is to provide the reader with essential information for research and clinical purposes and to describe structures that are not shown in any other anatomy atlas. For example, the muscles of the pelvic limb are deemphasized in transparency to depict the vessels and nerves used for common procedures such as injections and blood collection.

In the development of the figures, considerable emphasis was given to revealing the structures of the urogenital apparatus, particularly in the male. In both species, a median section of the penis, stained with hematoxylin and eosin, was prepared to show the corpus spongiosum glandis and the os penis (penis bone) in detail. This atlas includes a dorsal view of the internal genitalia, which is an unusual view in anatomical publications but quite valuable for developing an understanding of the glandular structures associated with the urethra, such as the coagulating gland, the vesicular gland, and the prostate gland. In the mouse, the urethra is incised to show the urethral fold over the fibro-cartilaginous plate in the urethral floor. A latex cast was prepared of the mouse pelvic urethra and the initial part of the penile urethra to highlight the urethral recess, the bulbourethral diverticulum, and the spatial relationship of the urethra with associated glands and ducts in the lateral and dorsal aspects of the urethral epithelium. In the rat, the proximal urethra and related structures are shown in a saggital section to best illustrate the presence of a urethral recess, the opening of the bulbourethral gland, and the absence of a bulbourethral diverticulum. These differences add a new dimension to the knowledge of the comparative anatomy of these two species.

Multiple specimens were dissected to generalize the normal anatomical findings in each species. For example, abdominal organs can vary greatly in their location due to the animals' state of feeding and fasting. Therefore, body landmarks approximate, at best, organ positions in the abdomen. The illustrations provide a typical position for the organs, as determined over the multiple specimens used to prepare each figure.

In each set of illustrations, mice and rats are presented in the same size, so that comparative details in anatomy can be best appreciated. Anatomical information is generalized to the species, without regard to animal strain or stock. Albino animals were most commonly dissected: CD-1 or Swiss Webster mice and CD/SD or Wistar rats. Pigmented animals were used for some figures: C57BL/6 mice, agouti mice of a mixed background (F2 generation from B6C3F1 or B6D2F1 cross), and rats of a mixed background (cross of Sprague Dawley with either Long Evans or ACI).

Nomenclature and Abbreviations

Nomenclature, synonyms (in parentheses), and abbreviations used in this atlas are from the Nomina Anatomica Veterinaria.[7] Individuals accustomed to the human medical literature should note that the directional terms "anterior" and "posterior" are generally not relevant to rodents, except for the structures of the head (eye, brain, etc.).

The figures illustrating regions of the body show only the rat; the nomenclature in these figures also applies to the mouse. These figures will assist the reader to describe approaches to the body for veterinary or experimental purposes, such as physical examination and surgery. Landmarks for palpation are provided to assist in locating vessels and related structures useful in common procedures, such as injections and blood collection.

The abbreviations and veterinary directional terms used in this atlas are listed below:

Abbreviations

A. .... Artery
Aa. .... Arteries
Br. .... Branch
CN .... Cranial Nerve (I through XII)
Lig. .... Ligament
Ligg. .... Ligaments
Ln. .... Lymph node
Lnn. .... Lymph nodes
M. .... Muscle
Mm. .... Muscles
N. .... Nerve
Nn. .... Nerves
R. .... Region
V. .... Vein
V.A. .... Vein and Artery
V.A.N. .... Vein, Artery, and Nerve
Vv. .... Veins

Directional Terms

Caudal .... Closer to the tail; used on the head, neck, trunk, and limbs proximal to the carpus and tarsus.
Cranial .... Closer to the cranium; used on the neck, trunk, and limbs proximal to the carpus and tarsus.
Distal .... Toward the periphery, away from the origin of the limbs and tail, the free end of them.
Dorsal .... Closer to the dorsum. This term refers to the back or dorsum of the neck, trunk, and tail, to the corresponding surface of the head, and to dorsum of manus and pes (corresponding to the hand and foot, respectively, in humans).
Lateral .... More distant or away from the median plane.
Medial .... Closer to the median plane.
Palmar .... Refers to the inner (caudal) surface of the palm (of manus).
Plantar .... Refers to the inner (caudal) surface of the planta (of pes).
Proximal .... Toward the trunk, near the origin of the limbs and tail.
Rostral .... Closer to the tip of the nose or the orifice of the mouth; used on the head.
Ventral .... Closer to the ground in the standing position of a quadruped.

Planes

Dorsal .... Parallel to the dorsal surface of the body or part, and perpendicular to the median and transverse planes (see below).
Median .... Divides the head, body, or limbs longitudinally into equal halves.
Sagittal .... Passes through the head, body, and limbs parallel to the median plane.
Transverse .... Across the head, body, or limbs perpendicular to the long axis of any structure.

Body Regions

Figure 1. Body regions (lateral aspect), shown in the male rat.

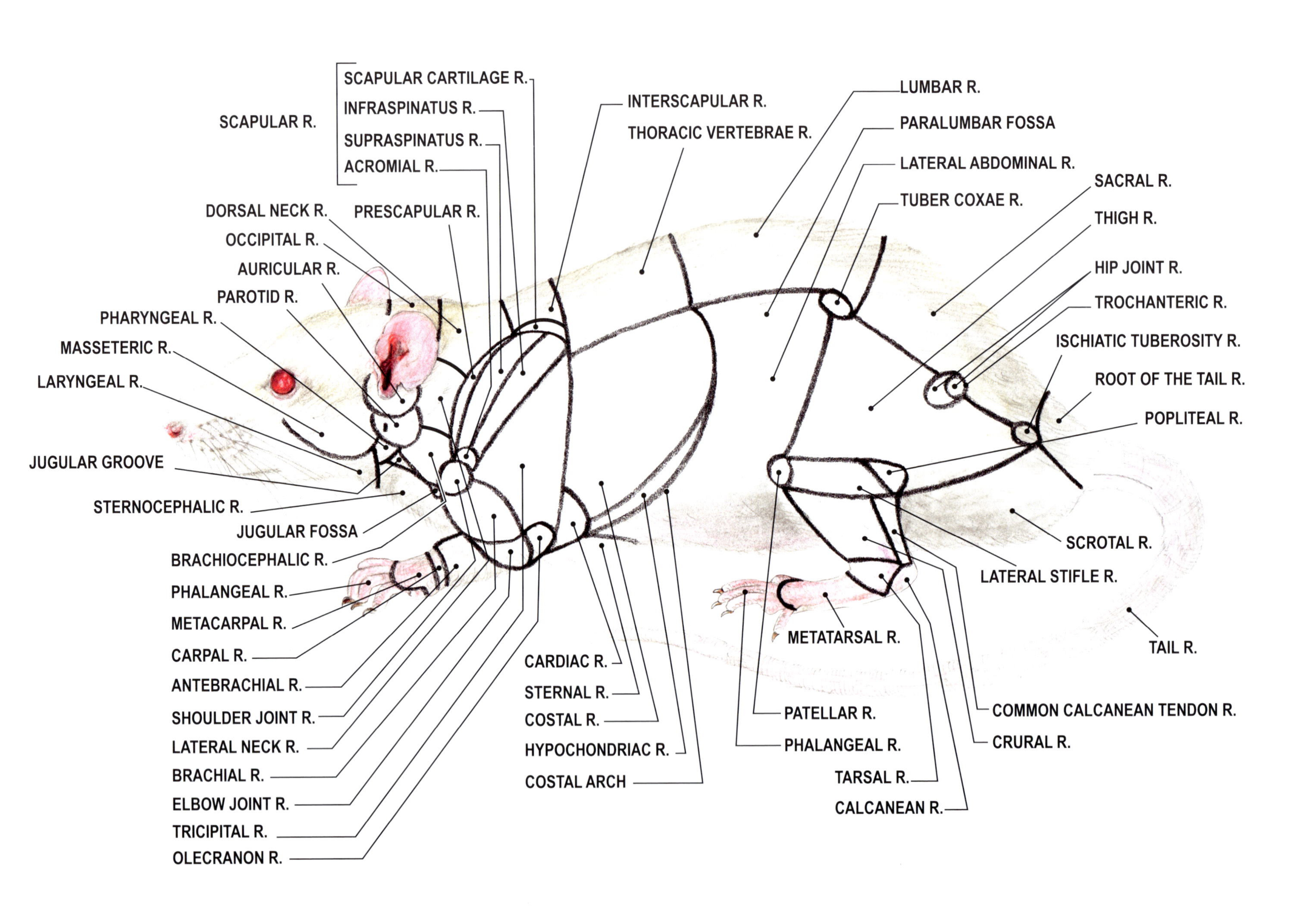

Figure 2. Body regions (ventral aspect), shown in the male rat.

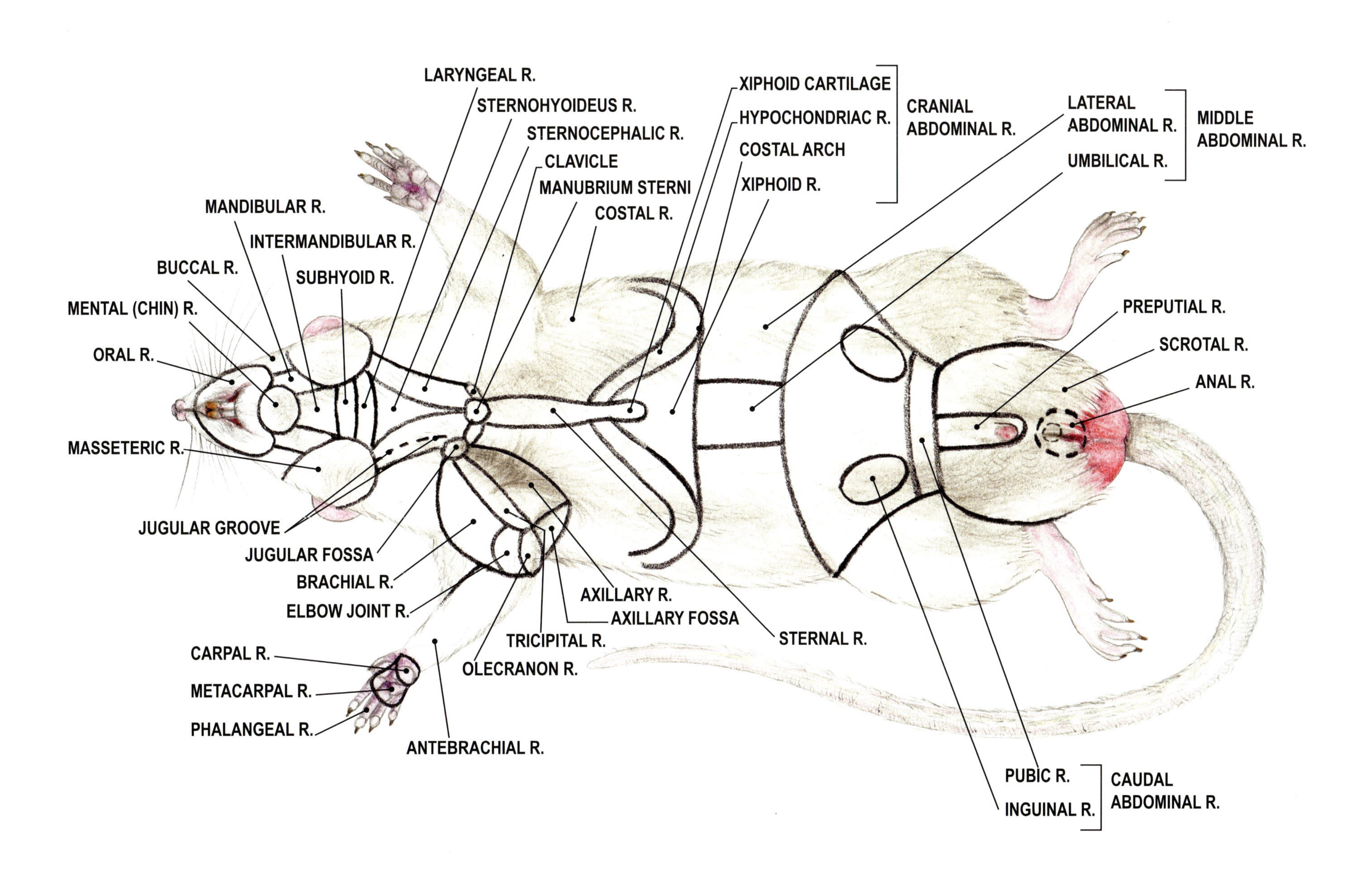

Figure 3. Female reproductive landmarks (ventral aspect), shown in the rat.*

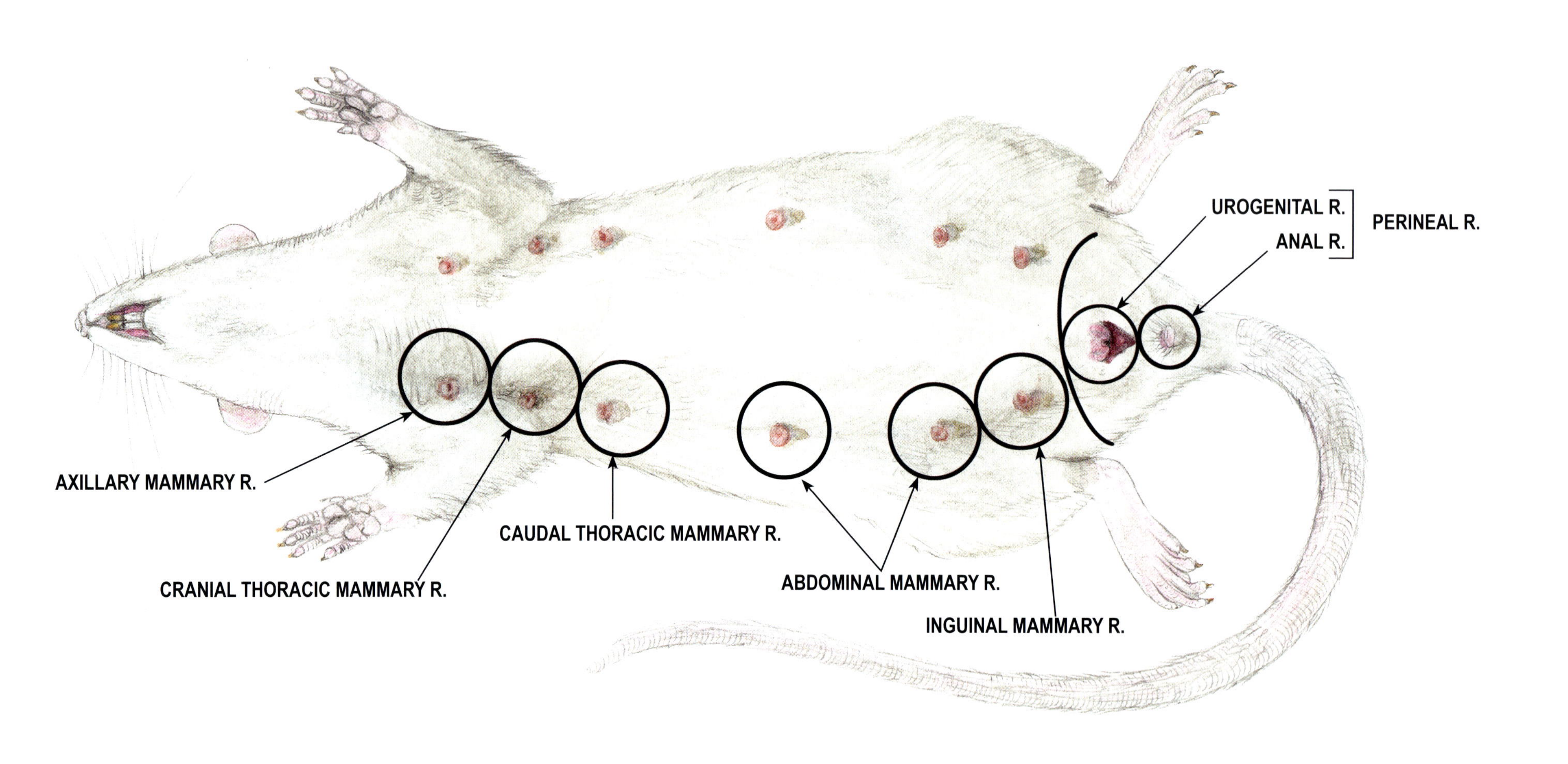

*Due to the variation between species on the extent of the mammary glands over the trunk, the circles in the figure are a schematic for referring to the mammary regions in the rat and the mouse. See Figure 15A and 15B for details on the extent of the mammary glands in both species.

Figure 4. Facial regions of the head (lateral aspect), shown in the rat.

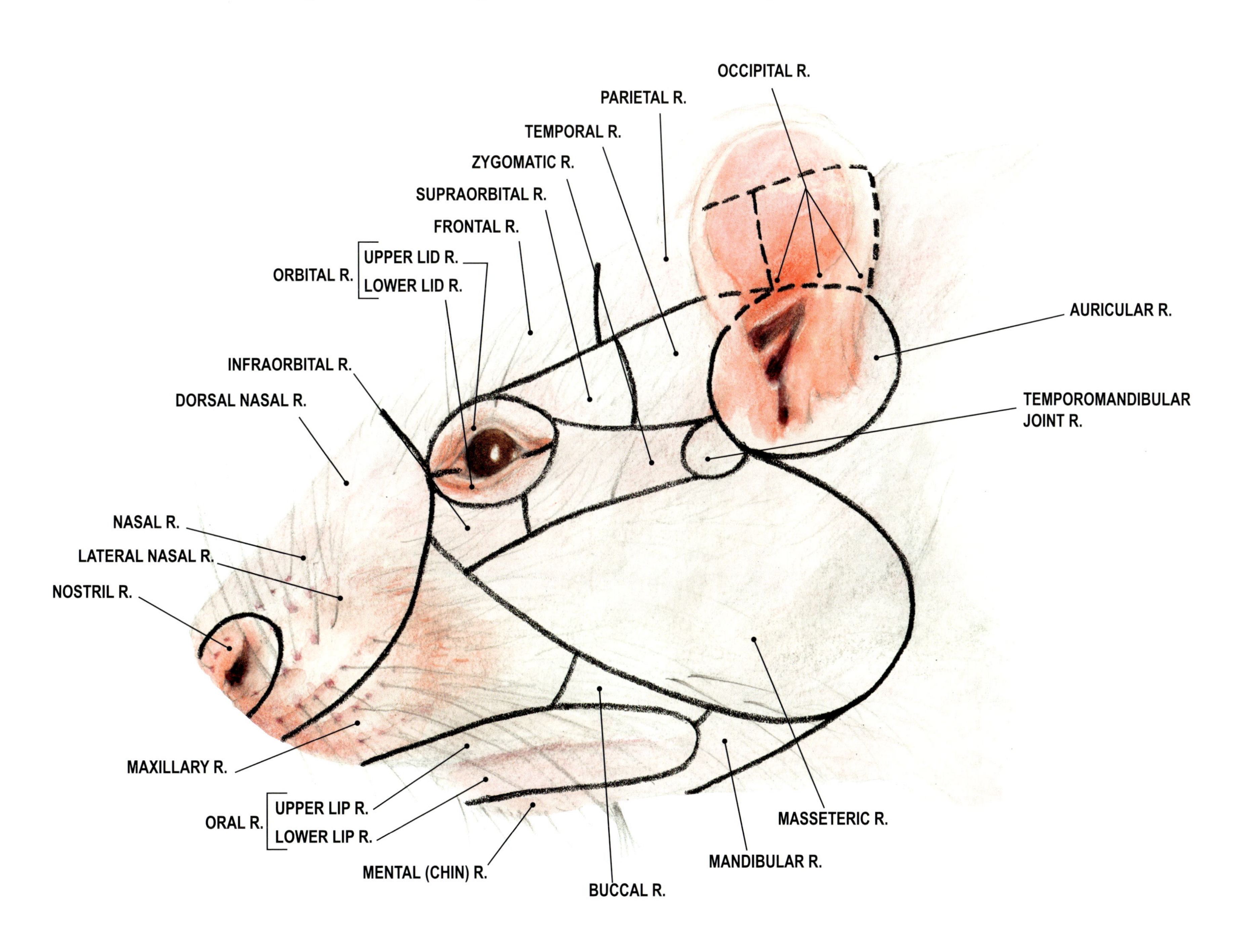

Figure 5. Landmarks for access to cervicothoracic vessels (ventral aspect), shown in the rat.

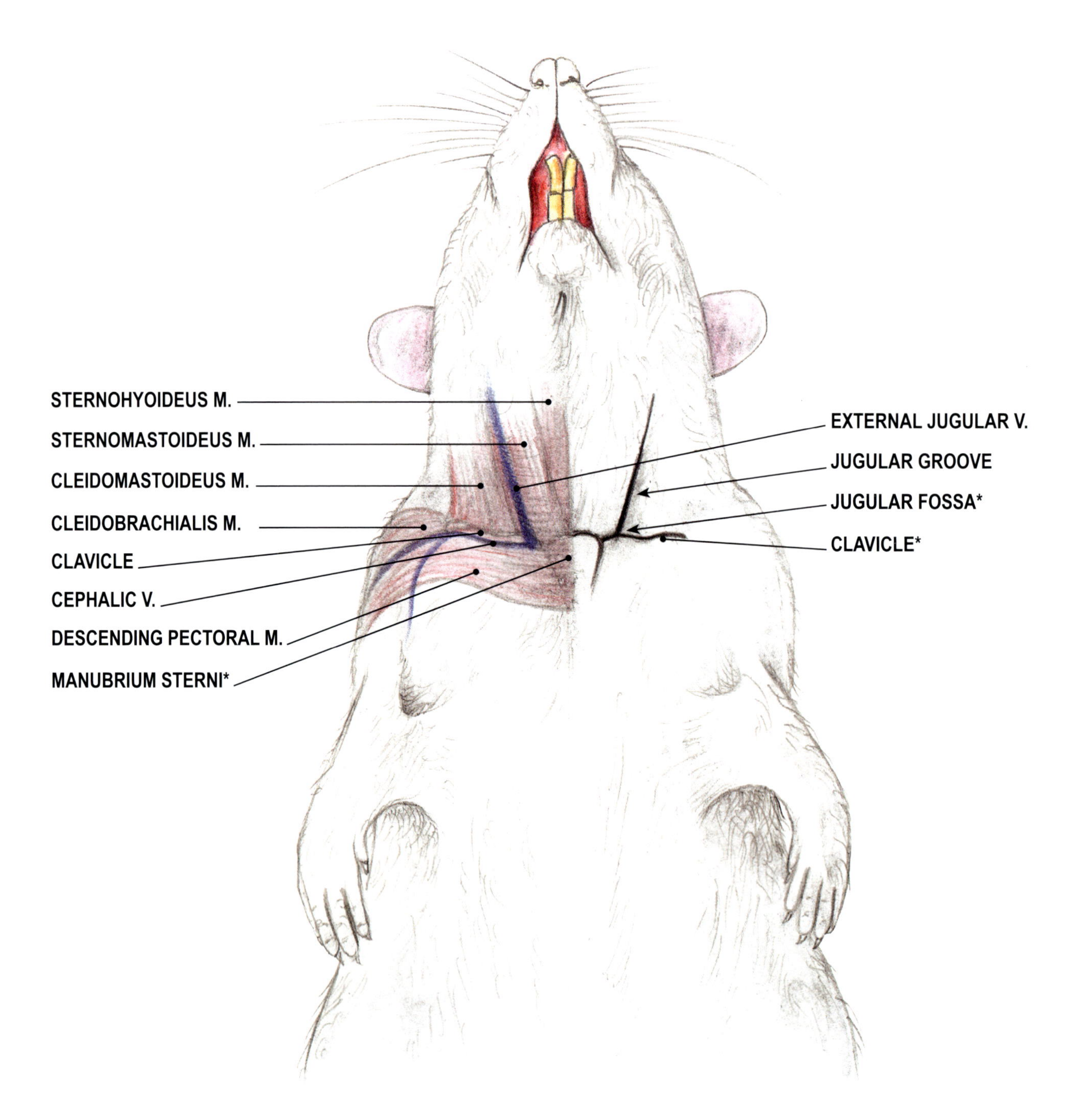

*Landmarks for blood collection from the external jugular V.

Figure 5. Landmarks for access to cervicothoracic vessels (ventral aspect), shown in the rat.

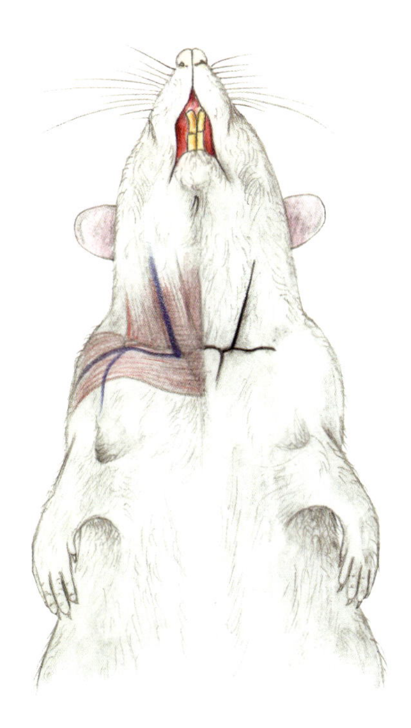

Rat

The point of the shoulder (see the shoulder joint region Figure 1) is a landmark for the jugular groove (furrow) in both rats and mice. The jugular groove on each side of the neck is a landmark for localizing the external jugular vein, the common carotid artery, and the vagosympathetic trunk. The jugular groove is bounded by the cleidomastoideus muscle dorsally, the sternomastoideus muscle ventrally, the trachea on the midline (located deep and not palpable), and the clavicle caudally. The jugular groove is more pronounced caudally, where it ends as the jugular fossa. The jugular fossa is bounded by the clavicle and the manubrium sterni; it is the location for taking blood samples.

To locate the jugular groove in the dorsally recumbent animal, the point of the shoulder is best visualized with the thoracic limbs extended caudally and palms pronated for shoulder rotation. The clavicle and the manubrium sterni are easily palpable.

Juvenile Features and Sex Differentiation

Figure 6A. Mouse. Neonate, albino (lateral aspect), age less than 24 hours, after feeding.

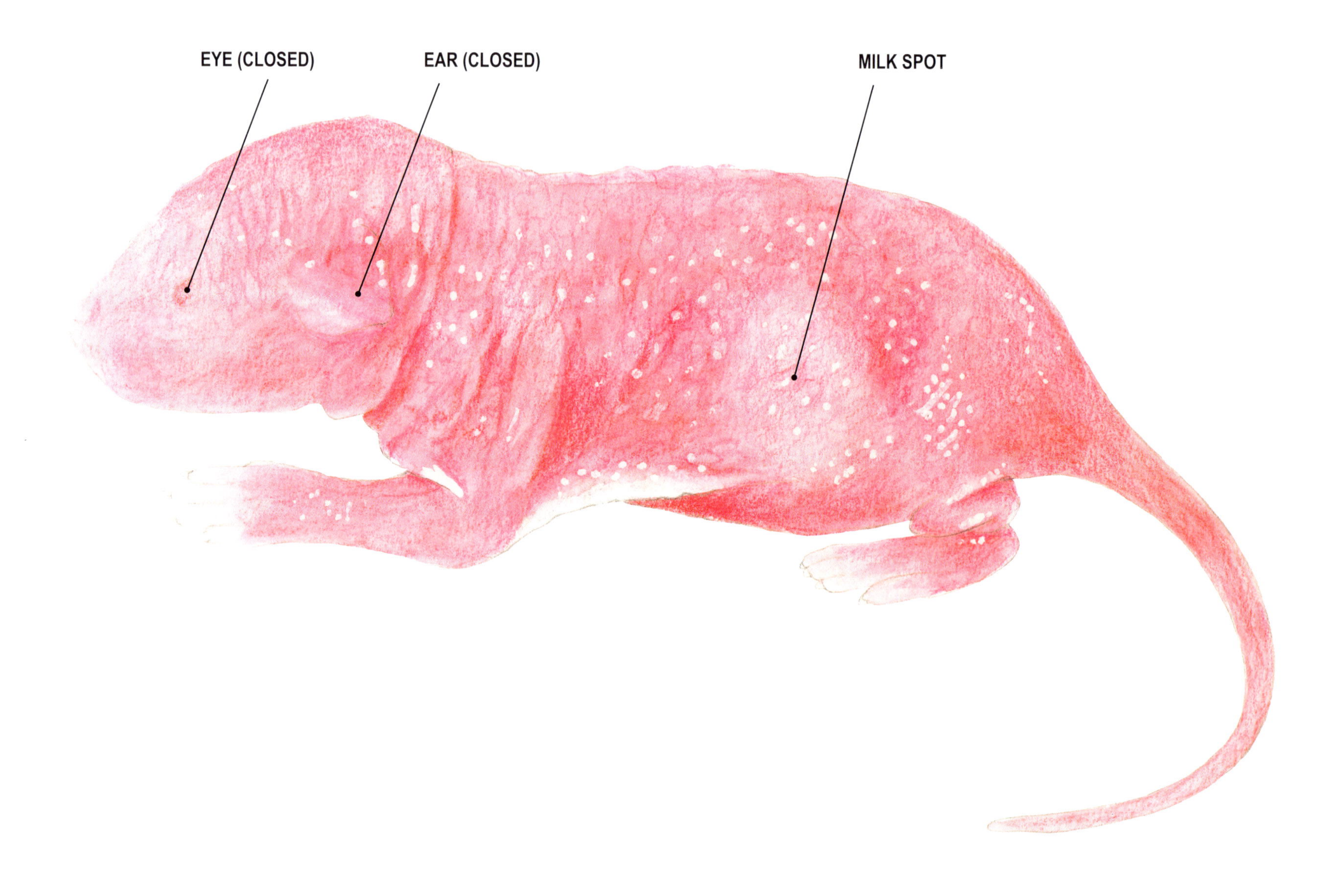

Figure 6B. Rat. Neonate, albino (lateral aspect), age less than 24 hours, after feeding.

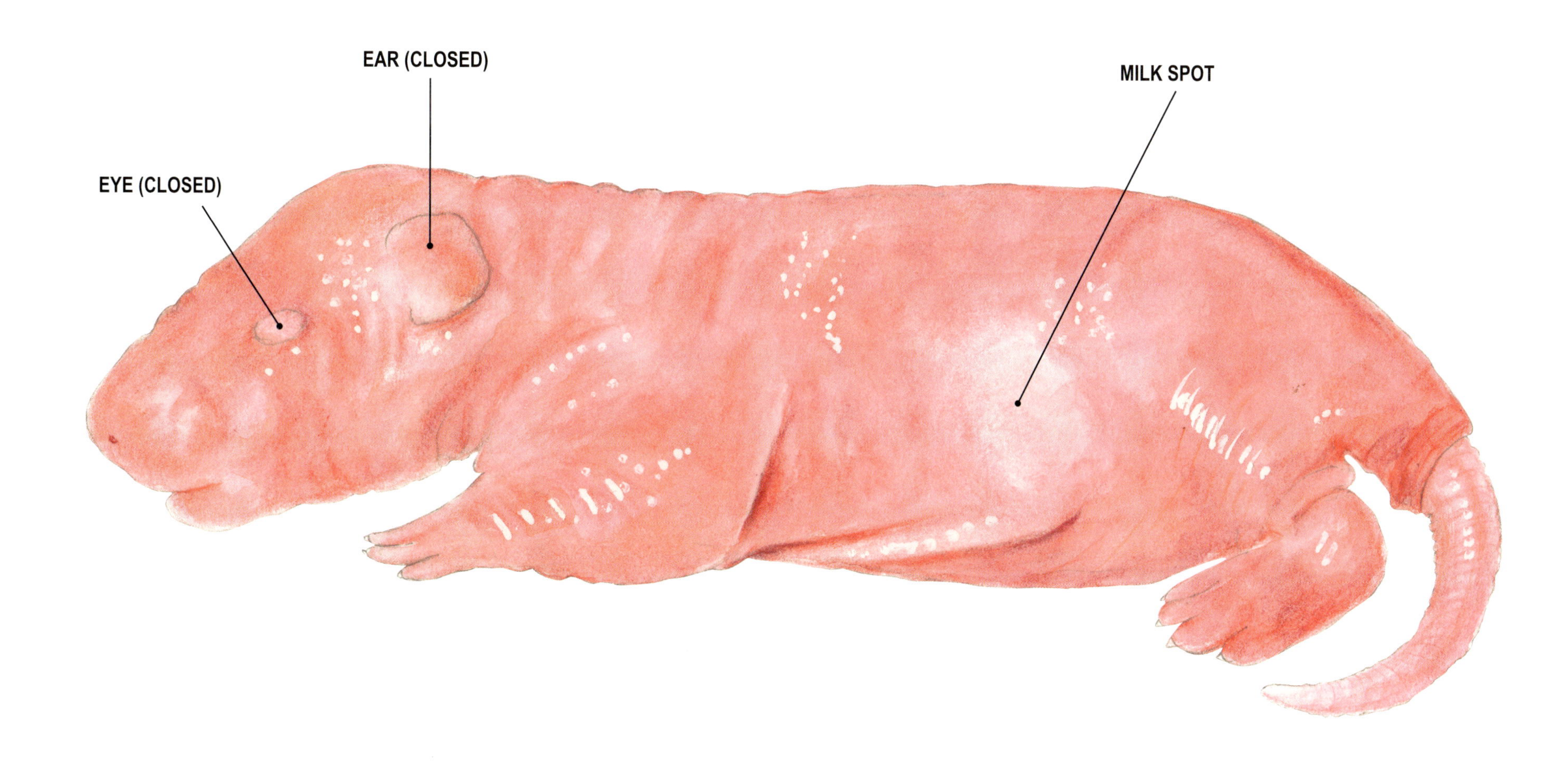

Figure 6A. Mouse. Neonate, albino (lateral aspect), age less than 24 hours, after feeding.

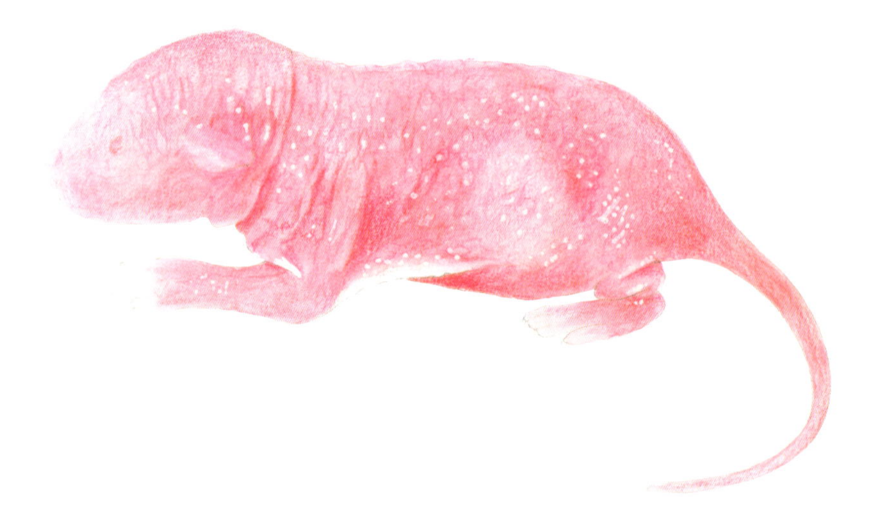

Figure 6B. Rat. Neonate, albino (lateral aspect), age less than 24 hours, after feeding.

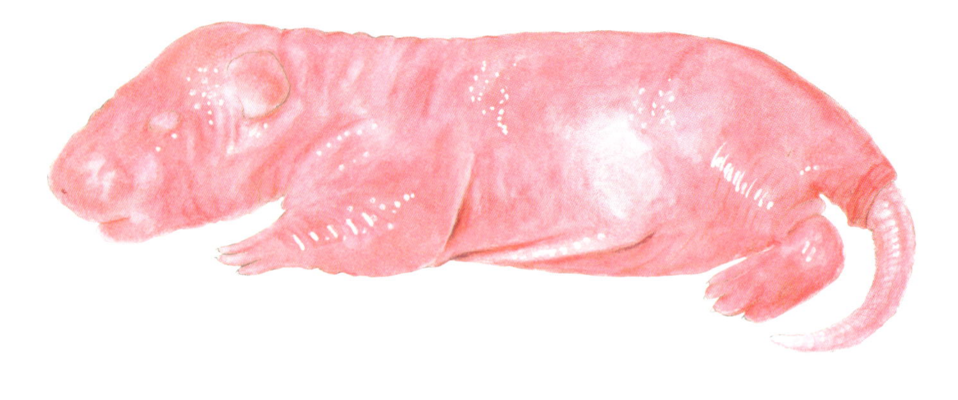

Figure 6. Neonate, albino (lateral aspect), age less than 24 hours, after feeding.

Mouse

The neonates are bright red when born, and gradually fade to pink by 1 day of age.

The head is relatively large compared to the body.

The eyes, not opened, are located in a rostral position.

The milk spot is round.

There are no transverse marks on the tail at this age.

Rat

The neonates are bright red when born, and gradually fade to pink by 1 day of age.

The head and limbs (thoracic and pelvic) are stouter than those of the mouse.

The autopodia (portions of the leg including and distal to the carpus or tarsus) are wide and short.

The eyes, not opened, are located at a position caudal to those of the mouse.

The milk spot is elliptical, with the long axis obliquely oriented ventrocaudally.

The tail has visible transverse marks; the skin of the sacral area folds over the base of the tail and stops abruptly, looking like a sheath.

Figure 7A. Mouse. Sexing juveniles, albino (perineal aspect), 5th day of age.

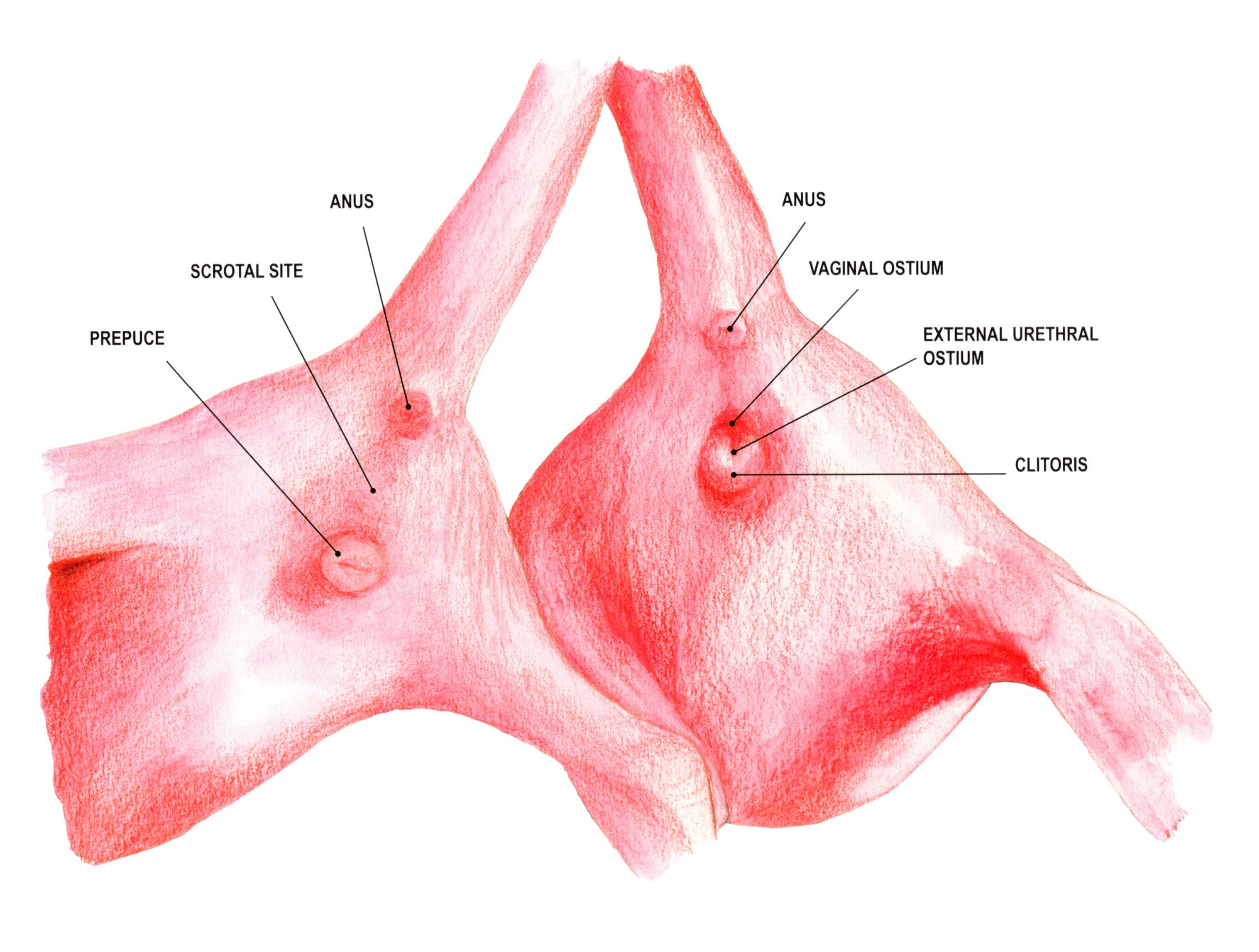

Figure 7B. Rat. Sexing juveniles, albino (perineal aspect), 5th day of age.

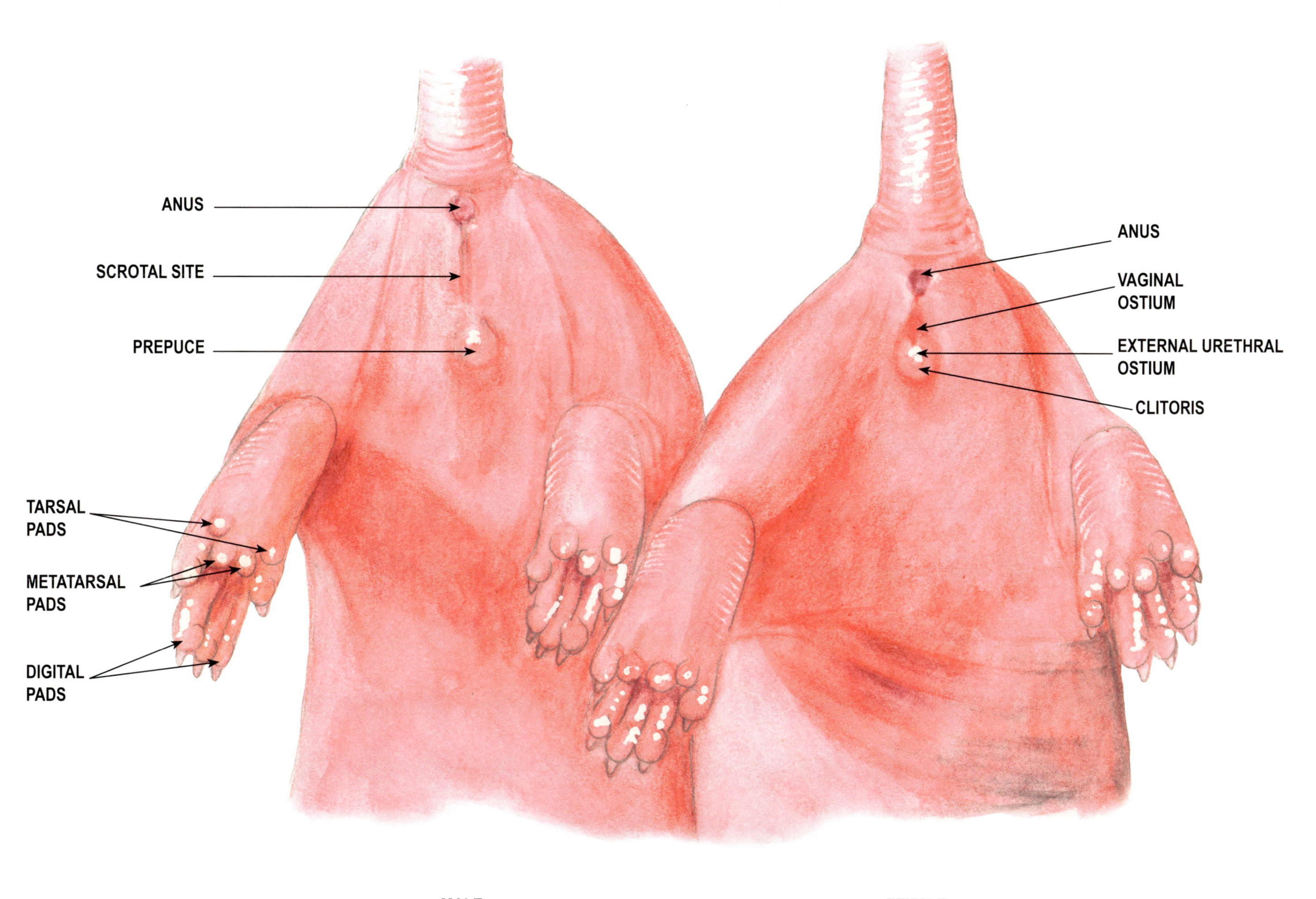

Figure 7A. Mouse. Sexing juveniles, albino (perineal aspect), 5th day of age.

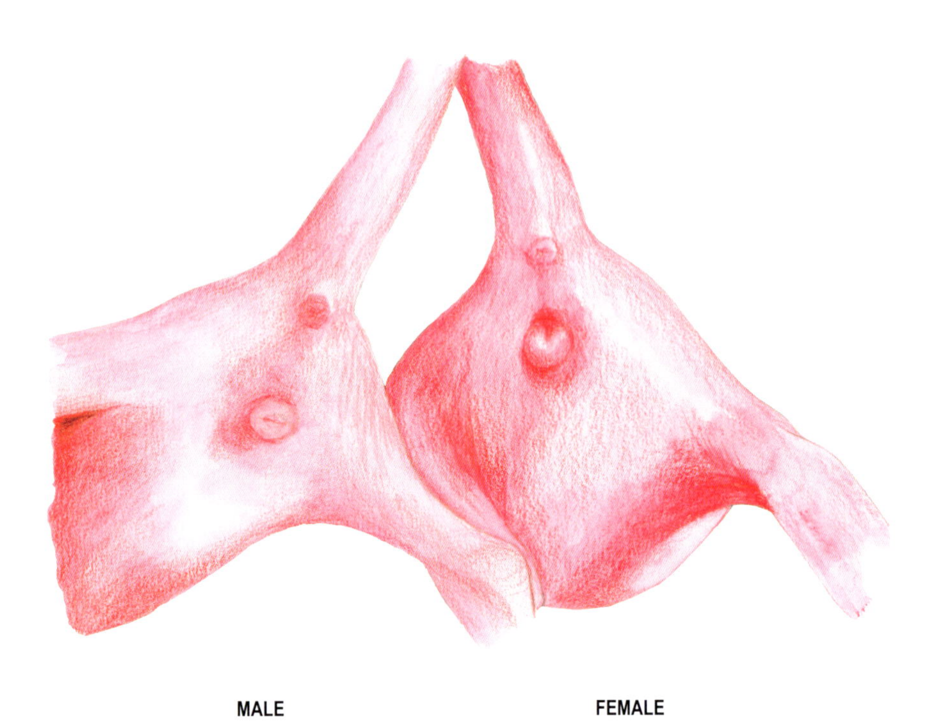

Figure 7B. Rat. Sexing juveniles, albino (perineal aspect), 5th day of age.

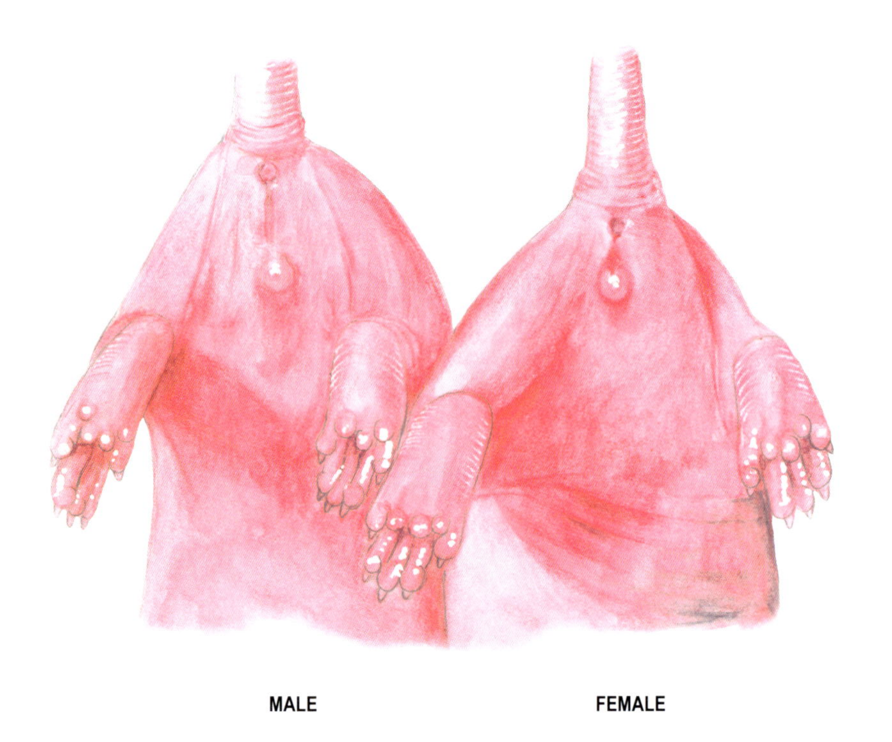

Figure 7. Sexing juveniles, albino (perineal aspect), 5th day of age.

At day 3-4, the anogenital distance (AGD) may be variable in females, especially if the females developed in a location adjacent to a male in utero. Around day 5, sex differentiation in both species becomes clearest, when the AGD becomes more pronounced.

At day 5, the vaginal ostium, the clitoris, and the external urethral ostium are distinguishable in females of both species. In the mouse and rat, the clitoris is associated with the urethra, not the vagina.

The number of ostia seen differentiates the sexes:

- Three ostia can be distinguished in the female caudocranially: anus, vagina (covered by the vaginal plate), and urethral ostium.
- Two ostia are seen in the male: anus caudally and preputial ostium cranially.

In pigmented male mice, pigmentation becomes visible by day 5 when a black spot (not shown) is seen between the anus and the genital papilla. This tissue is the developing bulbourethral gland, which is compact with actively replicating cells and differentiated cells.[3]

The skin is noticeably thicker in both mice and rats at this age, so the milk spot is no longer visible in either species.

On the tail, there are visible transverse marks in the rat, but not in the mouse. The fold of skin over the base of the rat's tail is still present.

Figure 8A. Mouse. Sexing juveniles, agouti (perineal aspect), 11th day of age.

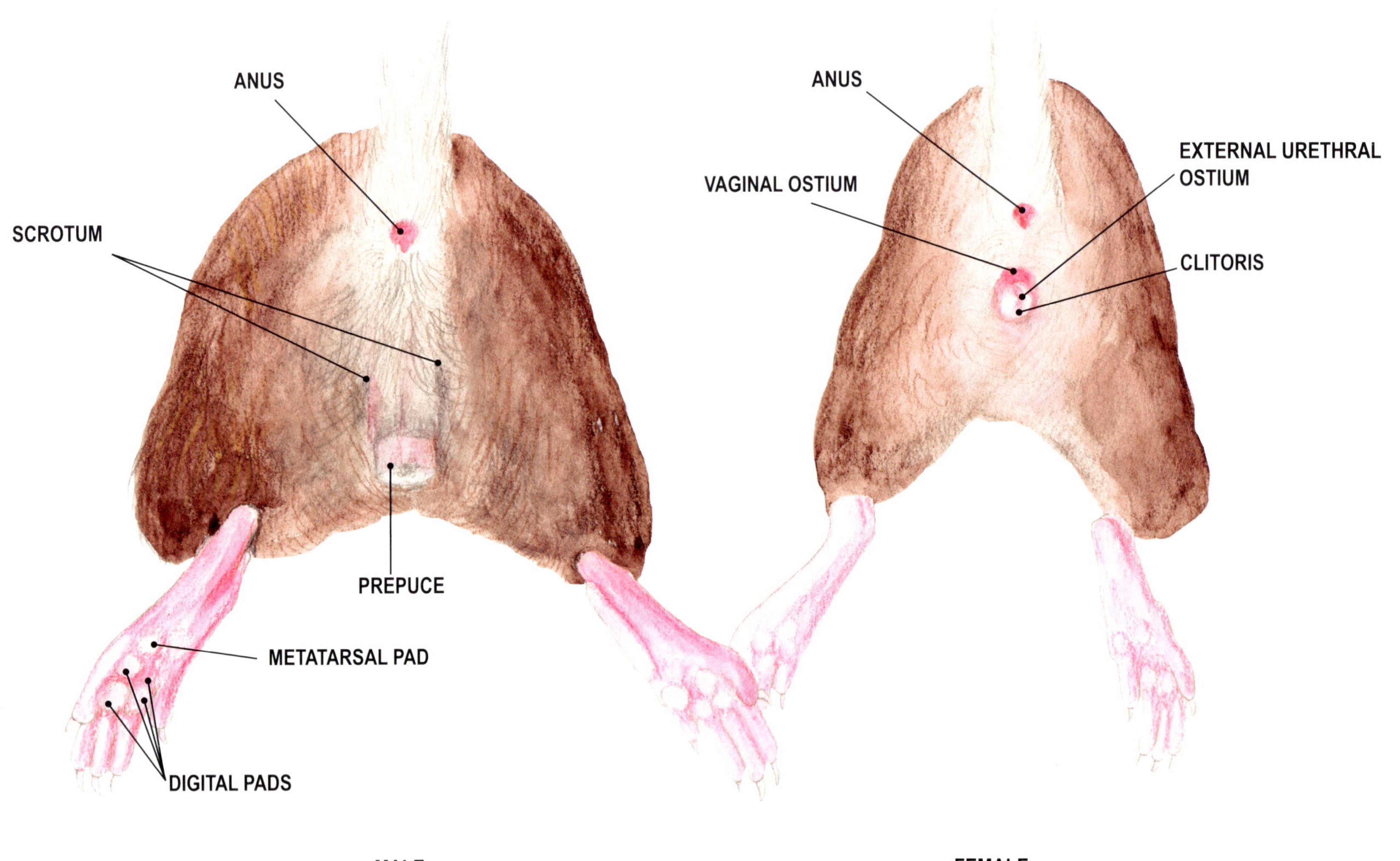

Figure 8B. Rat. Sexing juveniles, albino (positioned with head down, perineal aspect), 11th day of age.

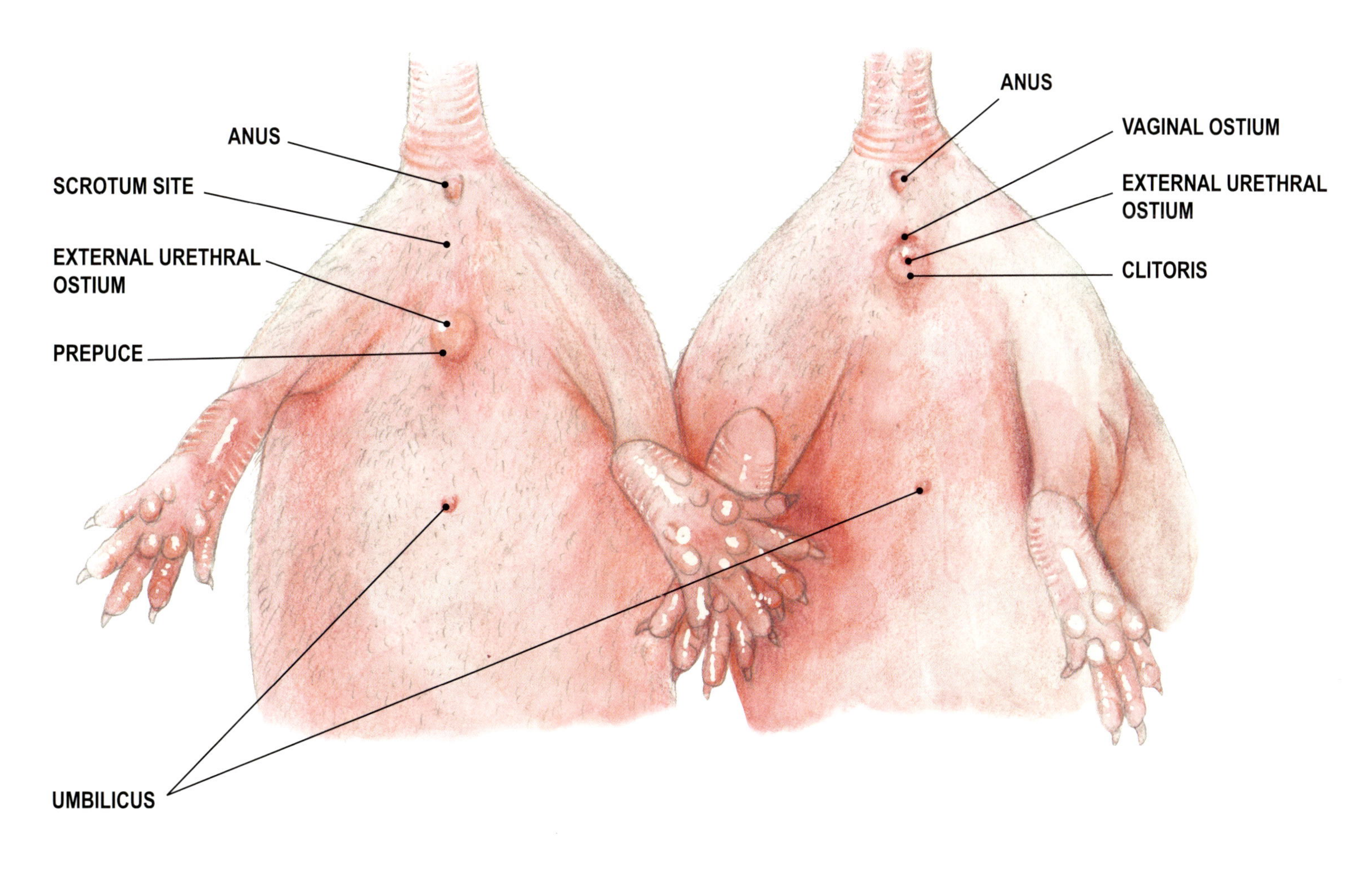

Figure 8C. Rat. Sexing juveniles, albino (positioned with head up, perineal aspect), 11th day of age.

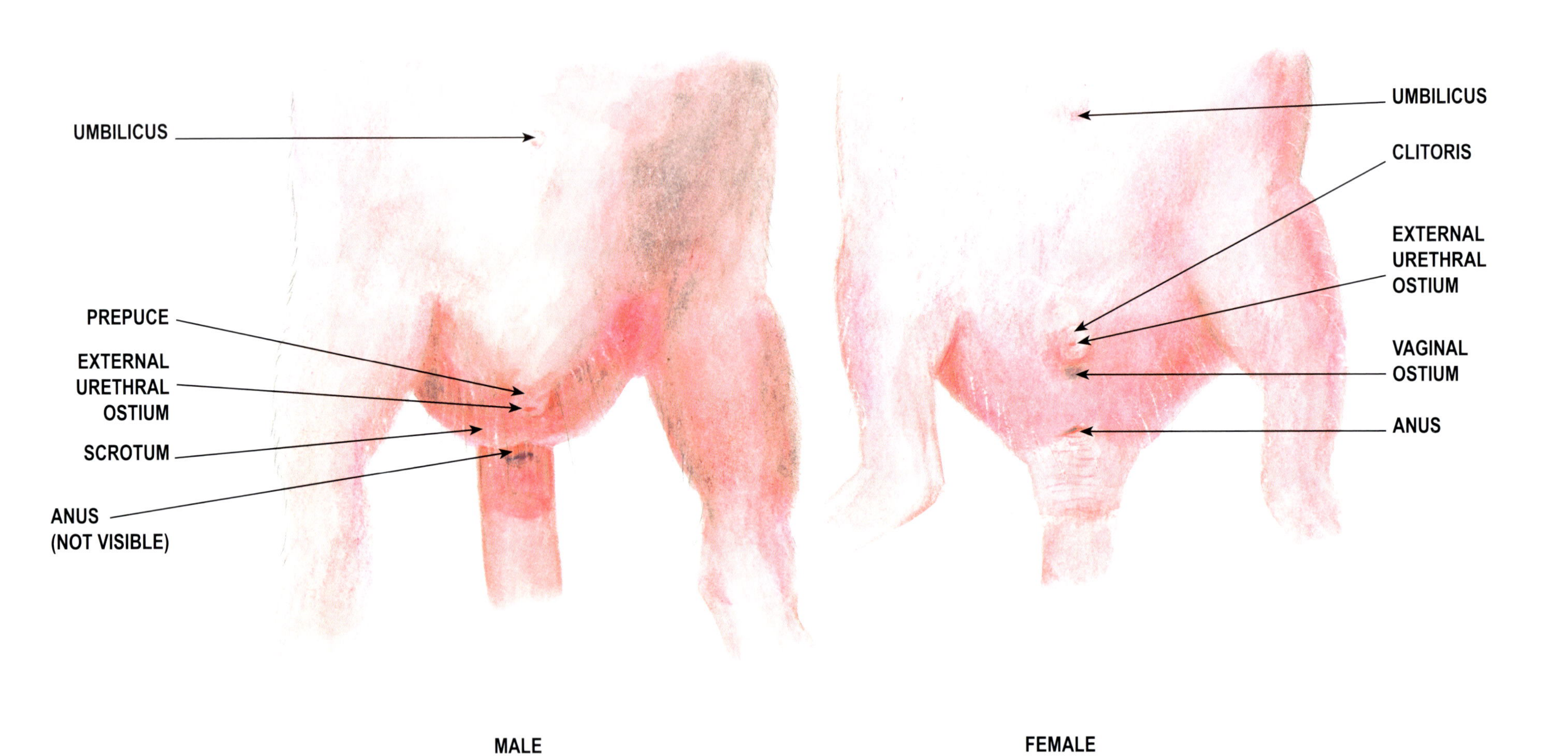

Figure 8. Sexing juveniles, 11th day of age.

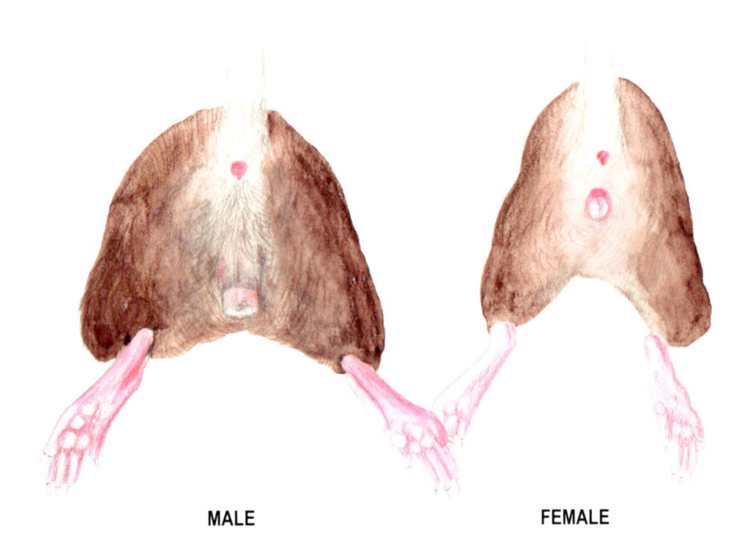

Figure 8A. Mouse, agouti (perineal aspect).

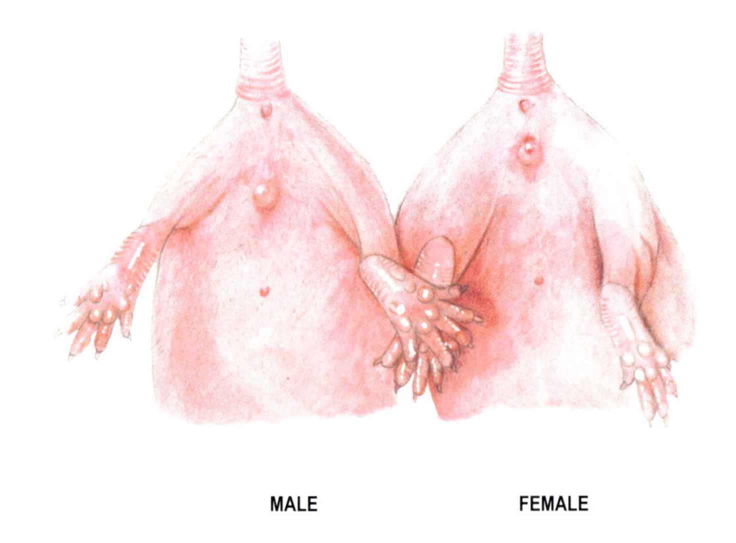

Figure 8B. Rat, albino (positioned with head down, perineal aspect).

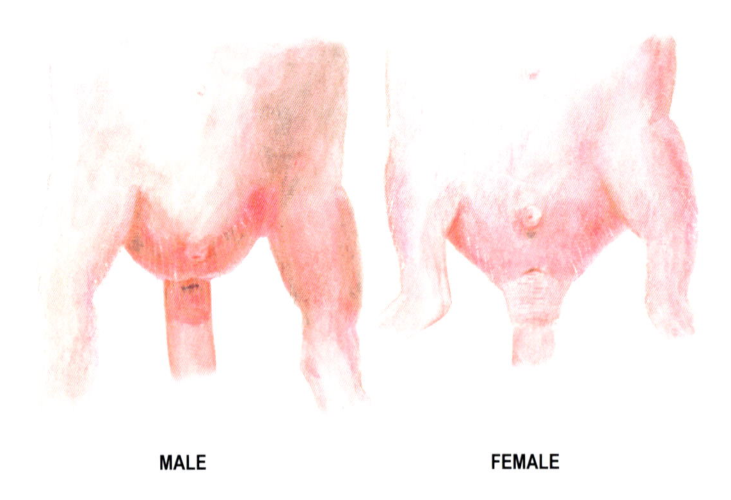

Figure 8C. Rat, albino (positioned with head up, perineal aspect).

Figure 8. Sexing juveniles (perineal aspect), 11th day of age.

The genital papilla in the male is the prepuce, which sheathes the penis. The ostium in the male's genital papilla is the preputial orifice (ostium). The genital papilla in the female is the clitoris cranially, the external urethral ostium, and the vaginal ostium caudally.

Mouse, agouti

The animals shown are from the same litter, but differ in size.

The teeth erupt and the eyes start to open. The hair coat is full but fine.

Sexual dimorphism is clearly visible.

No transverse marks appear on the tail at this age.

Female mice have visible nipples at day 9 (not shown).

Rat, albino

The body is covered with a sparse hair coat.

Sexual dimorphism is clearly visible.

Nipples are not present on the female rat at this age; they appear on day 13, becoming more prominent on day 14.

The umbilicus is visible and located at a shorter distance from the male's genital papilla than the female's. The distance between the vaginal ostium and the anus is much shorter than the distance between the male's prepuce and anus. These differences are helpful for sexing pups.

The skin fold at the base of the tail and striations on the tail, seen in the neonate, remain visible.

Rat pups are easily sexed when held in a position with the head up. In this position, the testicles are pushed caudoventrally into the scrotum, and the prepuce is pointed caudally. This position is recommended for recognizing the scrotum in the rat and differentiating the sexes.

Figure 9A. Mouse. Sexing juveniles (perineal aspect), 21st day of age.

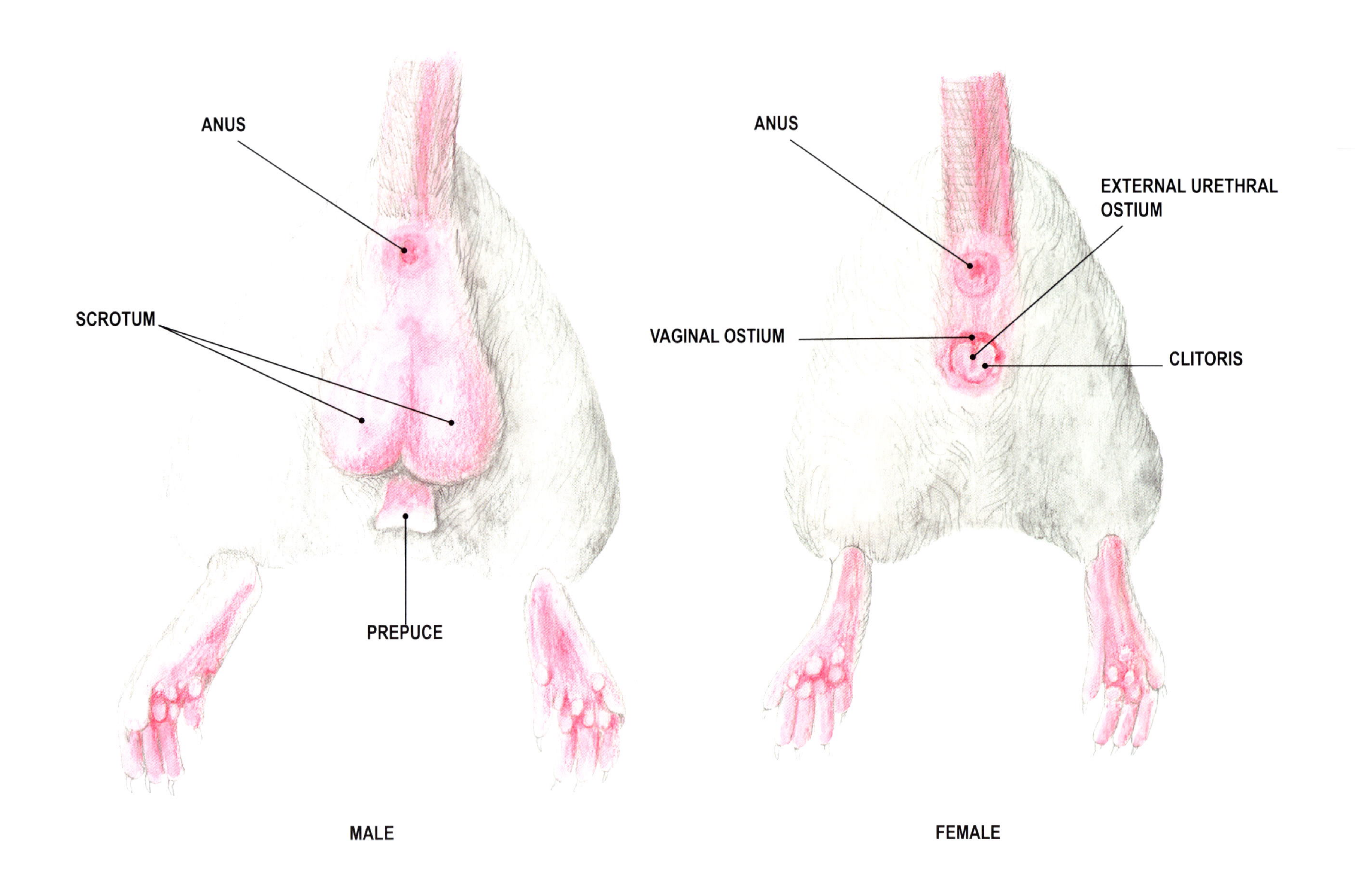

Figure 9B. Rat. Sexing juveniles (perineal aspect), 21st day of age.

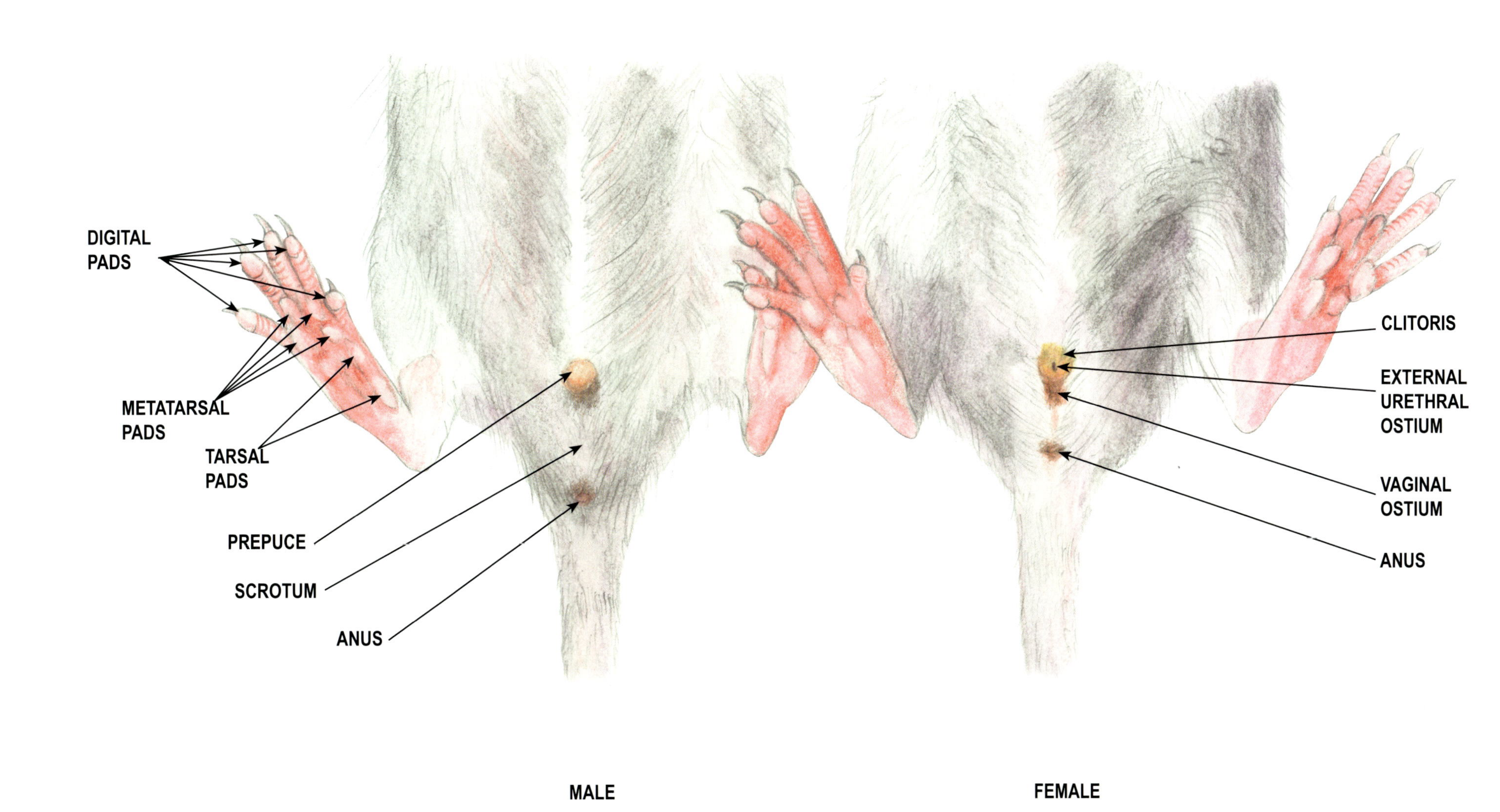

Figure 9A. Mouse. Sexing juveniles (perineal aspect), 21st day of age.

Figure 9B. Rat. Sexing juveniles (perineal aspect), 21st day of age.

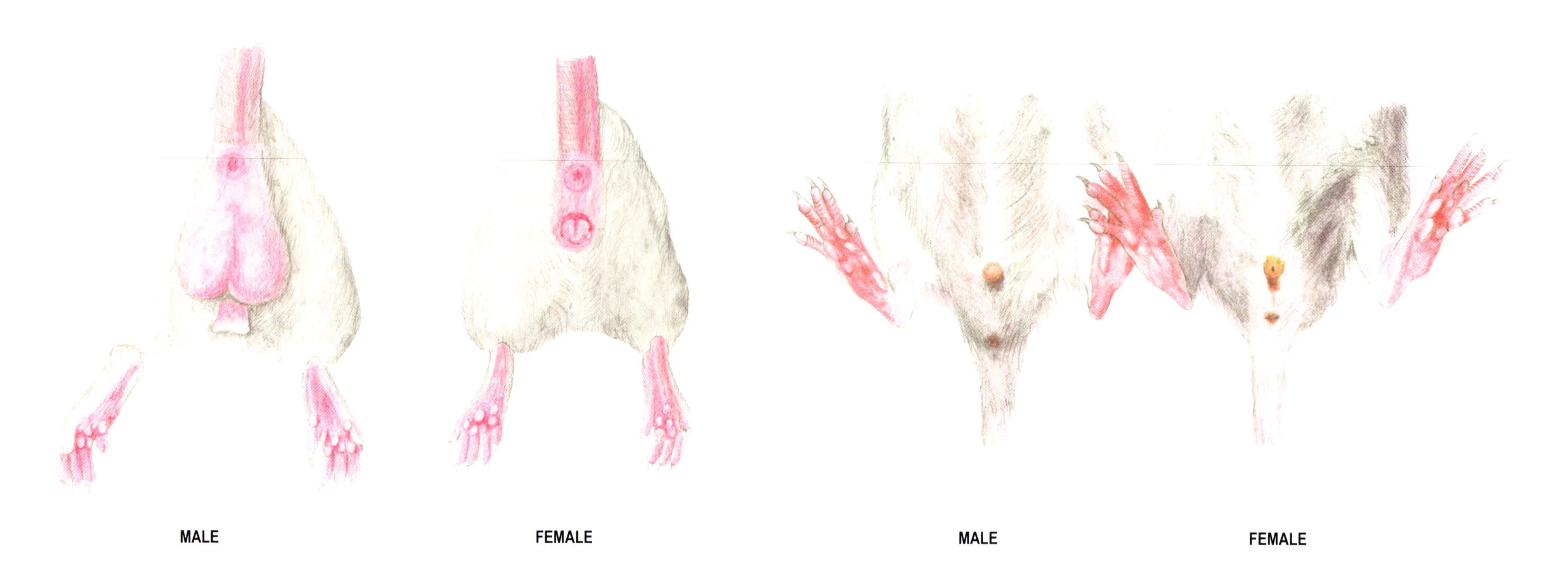

Figure 9. Sexing juveniles (perineal aspect), 21st day of age.

Sexual dimorphism is clearly visible.

The females of both species have a thin layer of epithelium (not shown) overlying the vaginal ostium, known as the vaginal plate. This plate degenerates in puberty, opening the vagina. Magnification is needed to visualize the vaginal plate. The urethra opens separately from the vagina in both the mouse and rat; therefore, no vaginal vestibule is present. (The vaginal plate does not obstruct the urethra.)

Mouse

The scrotum and prepuce in the male and the external genitalia in the female are large and well differentiated.

Transverse marks on the tail become distinguishable.

Female mice at this age have nipples, which developed at age 9 days (not shown).

Rat

Rats are held with the head up, as in Figure 8C, to accentuate the scrotum for greater clarity in sexing.

The hair covering the body is now long, covering the umbilicus.

Figure 10A. Mouse. Sexing adults (perineal aspect).

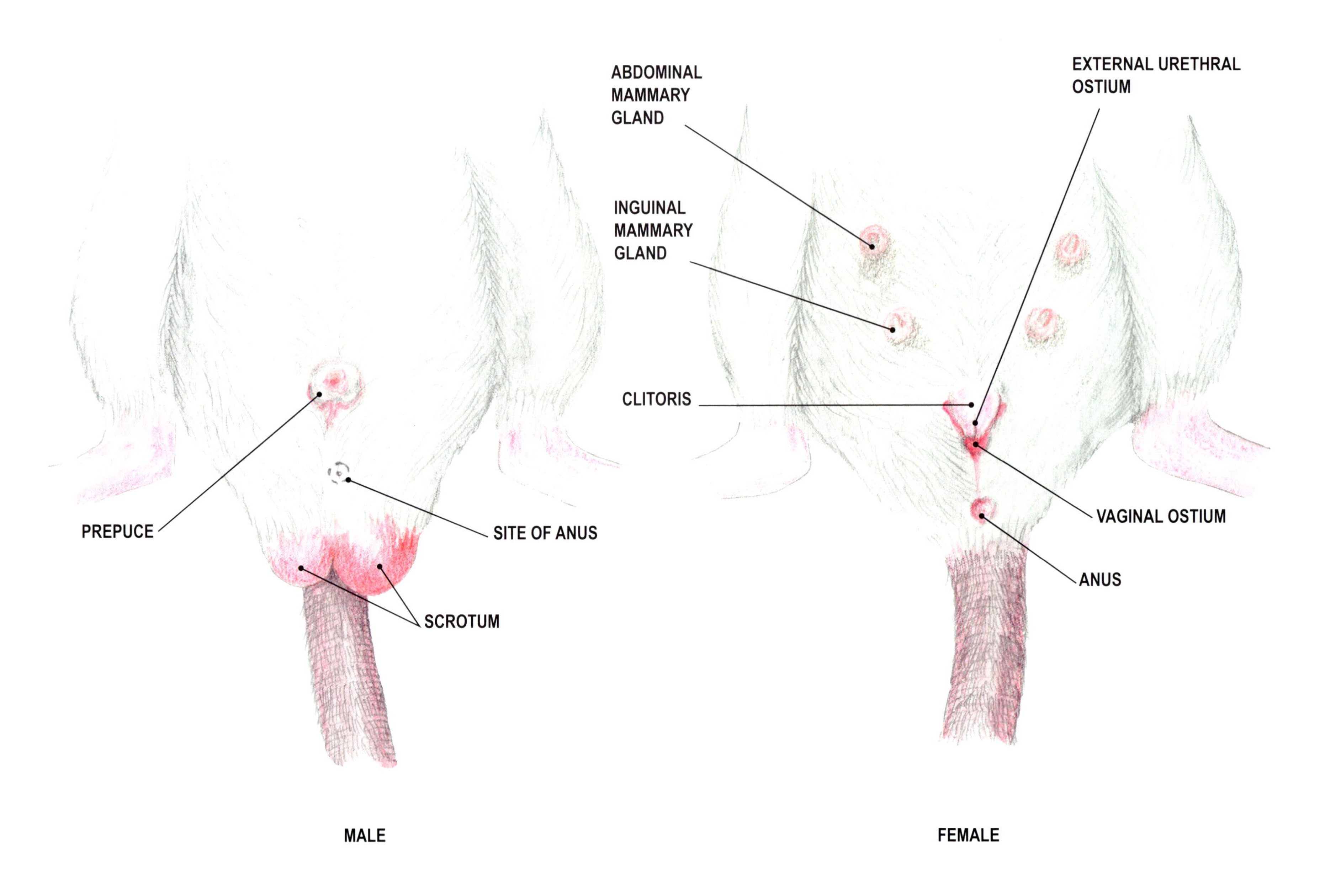

Figure 10B. Rat. Sexing adults (perineal aspect).

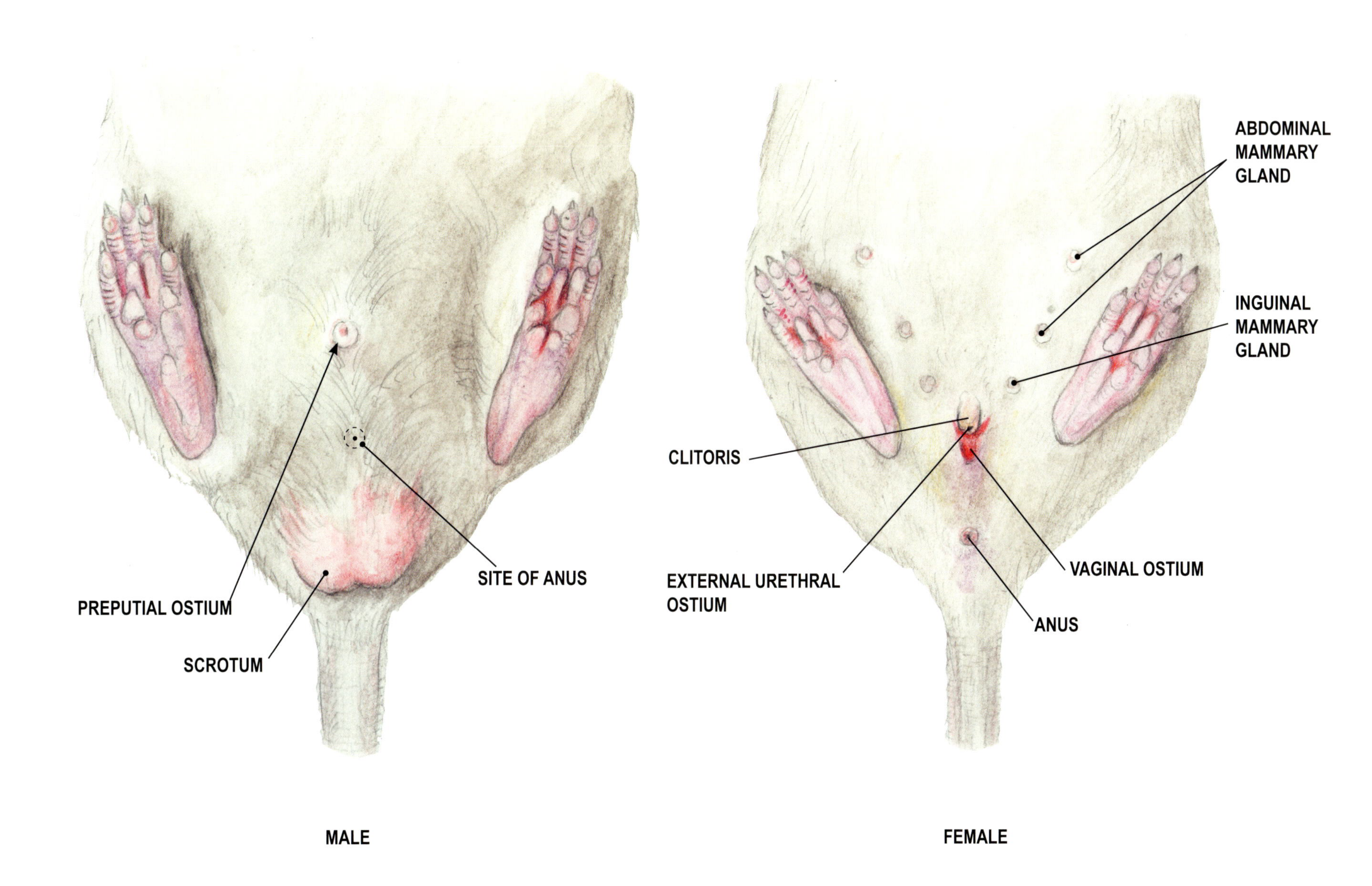

Figure 10A. Mouse. Sexing adults (perineal aspect).

Figure 10B. Rat. Sexing adults (perineal aspect).

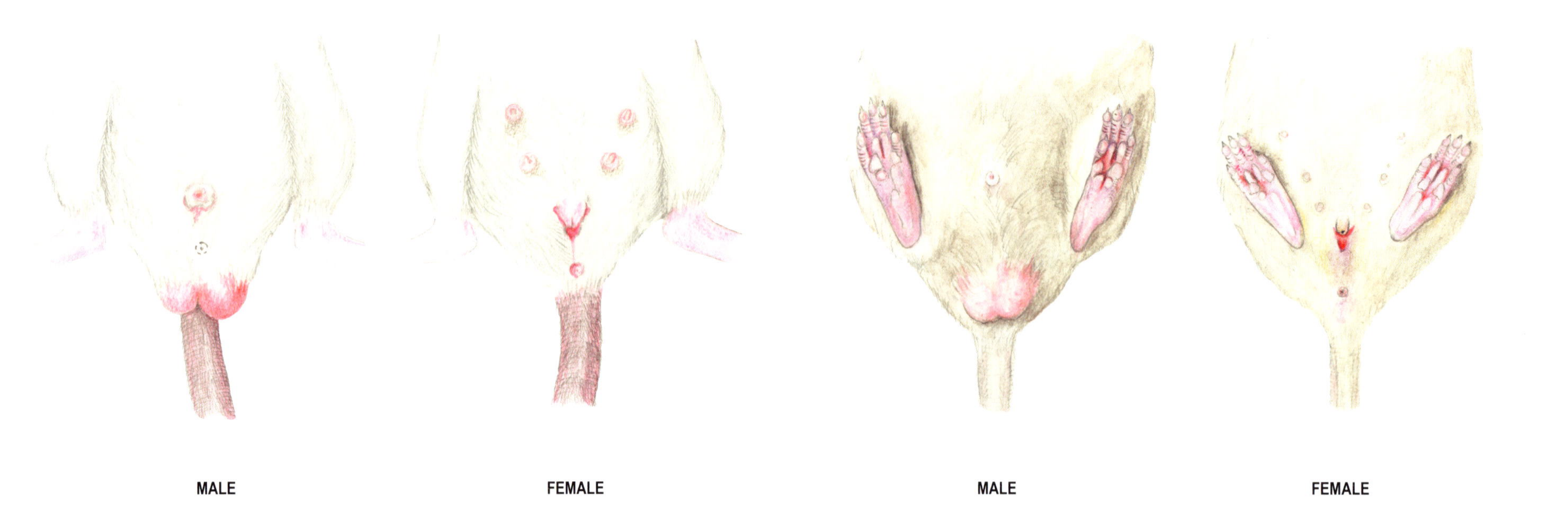

Figure 10. Sexing adults (perineal aspect).

The sexual dimorphism is clearly visible; external genitalia are large and well differentiated. In the male figures, the site of the anus is outlined, being covered by the testicles in both species.

Mouse

Transverse marks on the tail are distinguishable (not shown in the illustration).

The male has a large scrotum, which hangs over the base of the tail. Note that the testicles are not evenly sized in the specimens dissected, which is normal in all species. The prepuce is apparent.

The female has large abdominal and inguinal mammary glands (see Figure 15A).

Rat

The male rat has a proportionately smaller scrotum and prepuce than the mouse.

The female's nipples and genitals (see Figure 15B) are proportionately smaller than in the mouse. The abdominal and inguinal mammary glands are visible.

Some strains of mice (including C57Bl/6 and BALB/c) and stocks of rats (Sprague Dawley and Wistar) have been reported to have an anatomical defect (not shown), of a vaginal septum that obstructs the vagina by bisecting it longitudinally, transversely, or obliquely.[4,5,12,13] This defect reduces fertility in affected females. When present, the septum can be seen as a band of tissue distally in the vagina upon dilatation.

External Features

Figure 11A. Mouse. Male adult (dorsal aspect).

Figure 11B. Rat. Male adult (dorsal aspect).

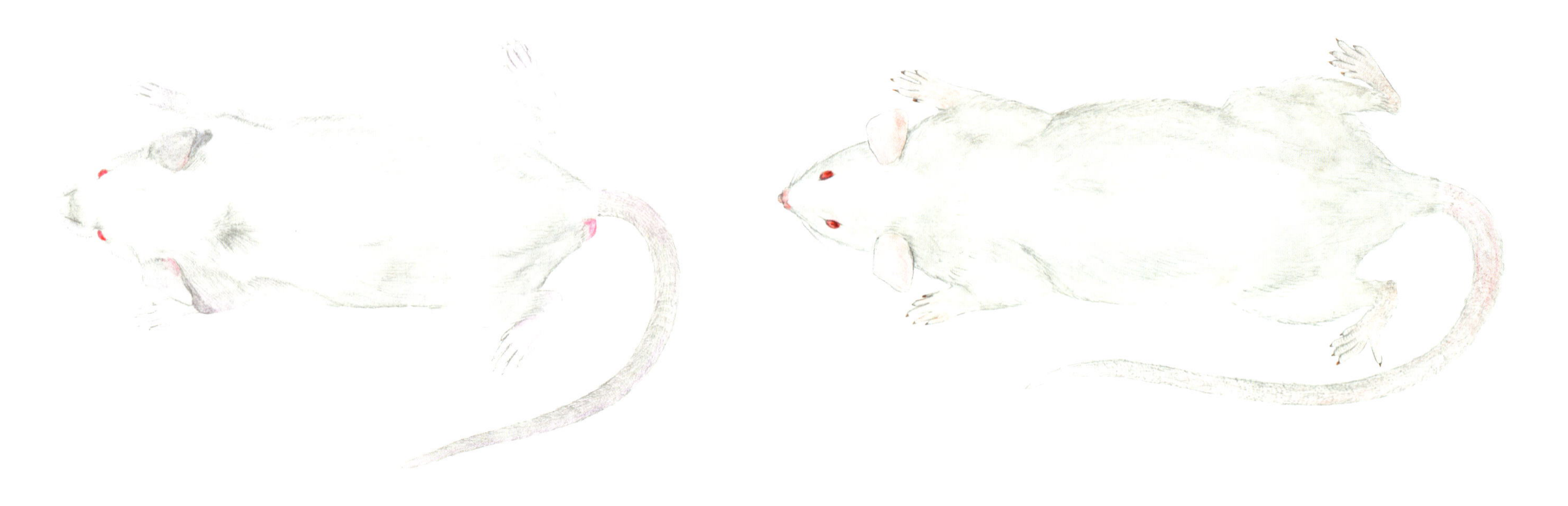

Figure 11A. Mouse. Male adult (dorsal aspect).

Figure 11B. Rat. Male adult (dorsal aspect).

Figure 11. Male adult (dorsal aspect).

The head is more elongate and triangular in shape in the rat than in the mouse.

The snout is more pointed in the rat, more rounded in the mouse.

The ears are proportionally larger in the mouse than in the rat.

The testicles in the scrotum are more visible in the mouse than in the rat.

In both species, males are easily differentiated from females by looking at the dorsum and noting the scrotum protruding at the tail base.

Figure 12A. Mouse. Male adult (ventral aspect).

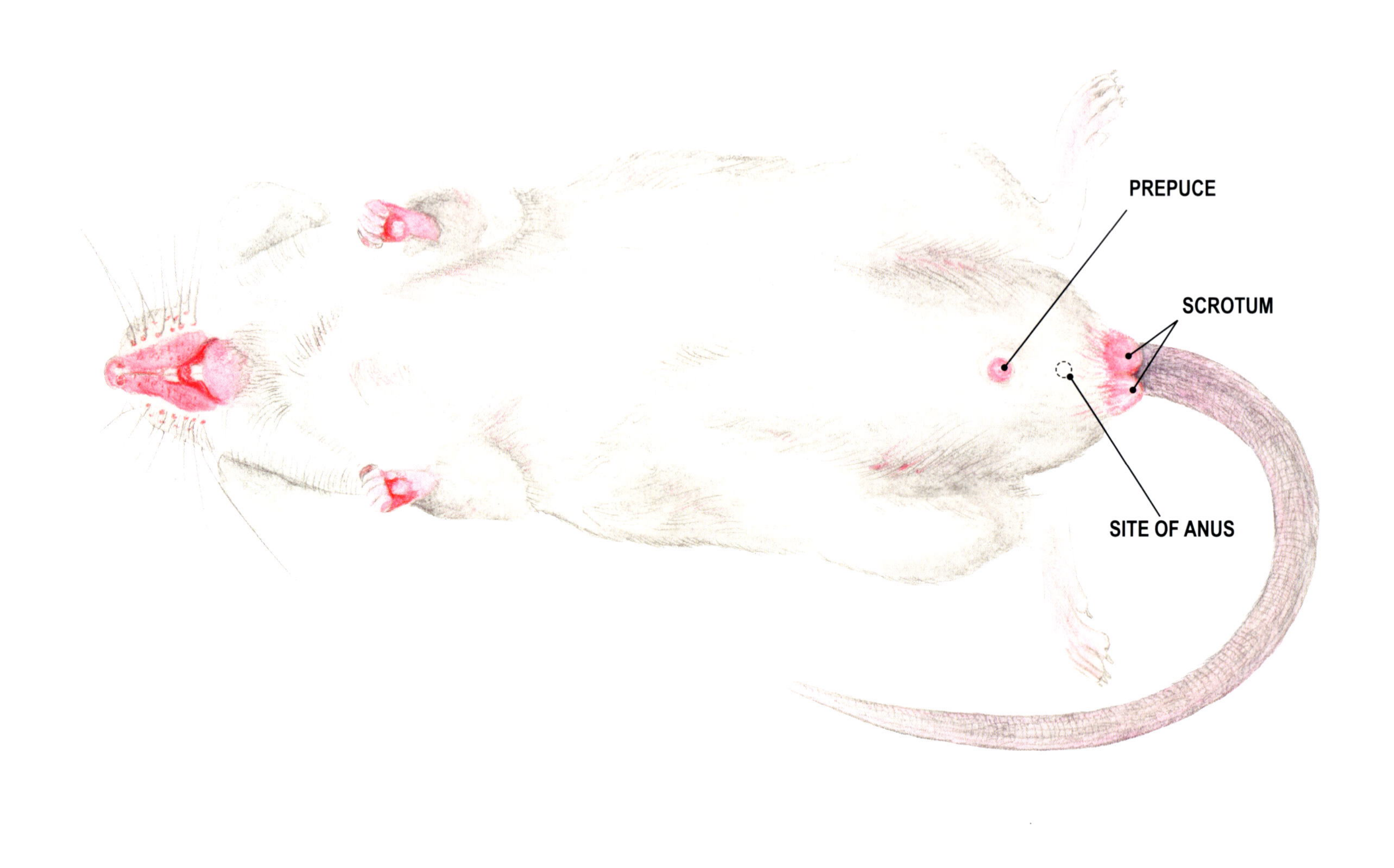

Figure 12B. Rat. Male adult (ventral aspect).

Figure 12A. Mouse. Male adult (ventral aspect).
SITE OF ANUS
Figure 12B. Rat. Male adult (ventral aspect).
SITE OF ANUS

Figure 12. Male adult (ventral aspect).

The head is more elongate and triangular in shape in the rat than in the mouse.

The snout is more pointed in the rat, more rounded in the mouse.

The ears are proportionally larger in the mouse than in the rat.

The caudal poles of the testicles within the scrotum are pointed in the mouse, rounded in the rat. In both species, the testicles overlie the anus, the site of which is outlined by a circle (see left).

The hairless areas (ears, lips) are more pink in the mouse than in the rat.

In both species, iron deposits in the enamel can turn teeth yellow. In the specimens examined, the teeth were prominently yellow in the rat. Depending on age and strain, some mice can have yellow-tinged incisors.

In both species, the vibrissae (whiskers) are engorged with blood; innervation is controlled by a brain receptor center. There are five categories of vibrissae; see Figure 16 for more information.

Both species have five digits on the front and rear feet. Digit 1 has two phalanges; all other digits have three phalanges. Toenails are present on all digits. See Figure 59.

Figure 13A. Mouse. Female adult (ventral aspect).

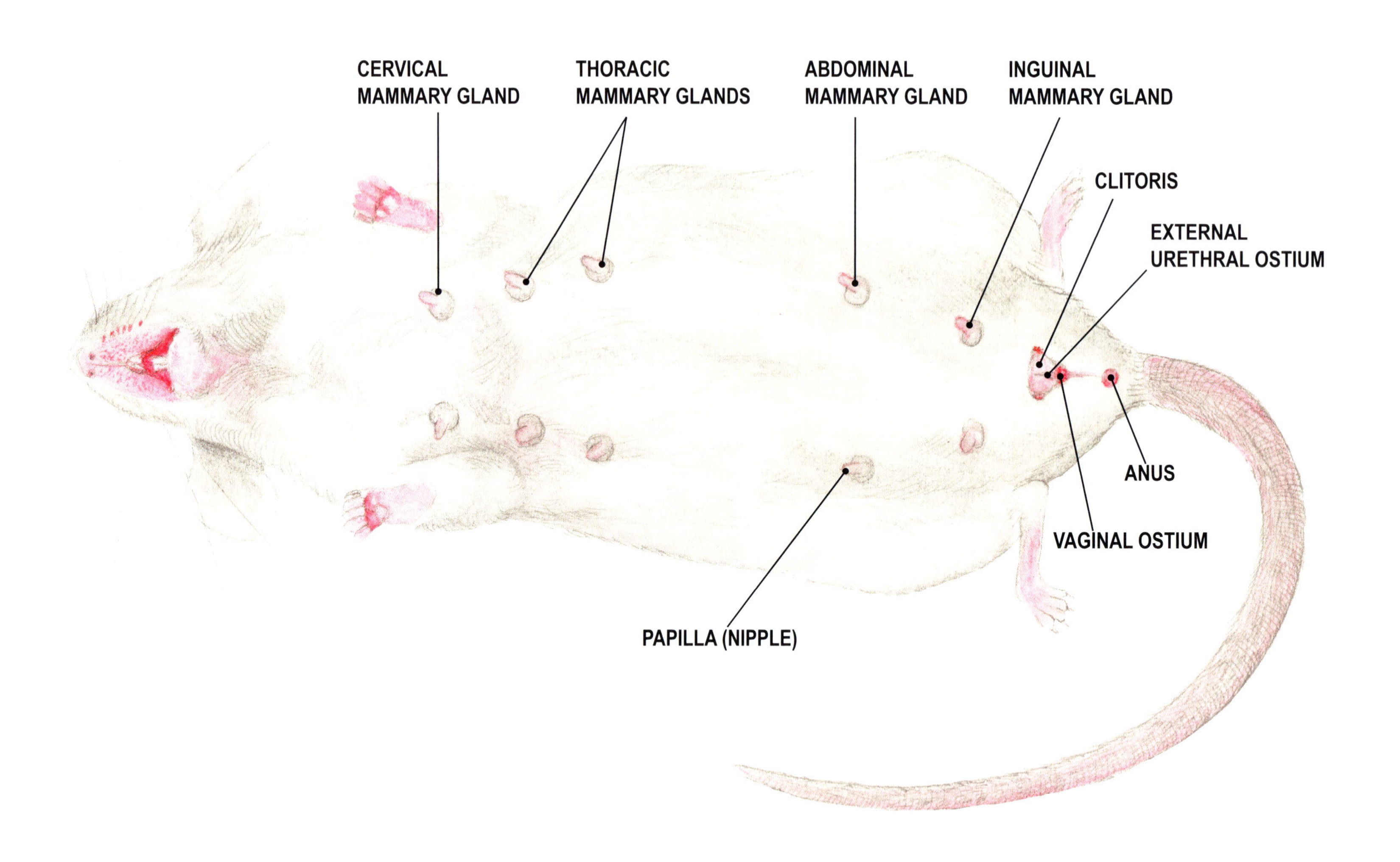

Figure 13B. Rat. Female adult (ventral aspect).

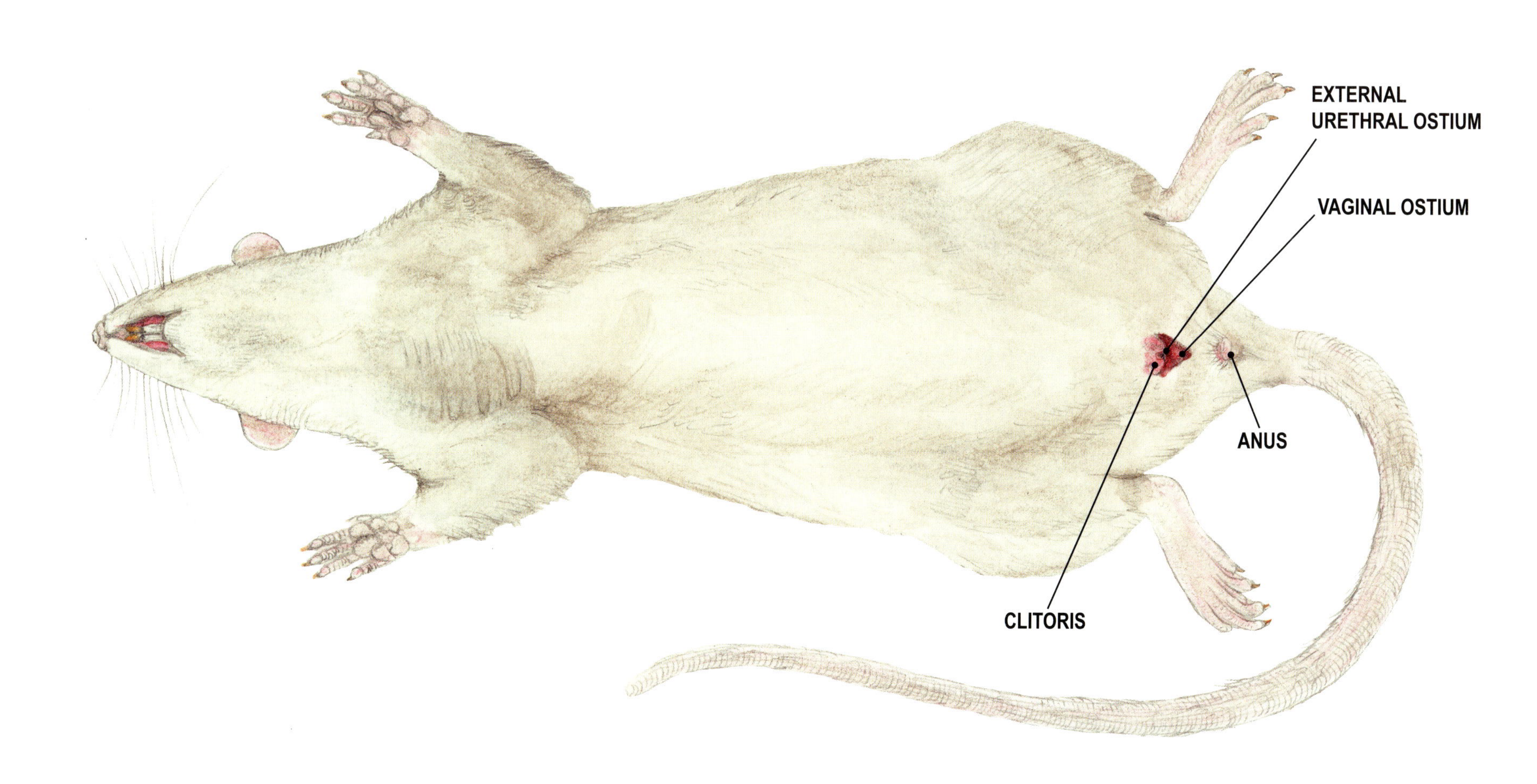

Figure 13A. Mouse. Female adult (ventral aspect).

Figure 13B. Rat. Female adult (ventral aspect).

Figure 13. Female adult (ventral aspect).

The head is more elongate and triangular in shape in the rat than in the mouse.

The snout is more pointed in the rat, more rounded in the mouse.

The ears are proportionally larger in the mouse than in the rat.

In both species, the vibrissae (whiskers) are engorged with blood; innervation is controlled by a brain receptor center. There are five categories of vibrissae; see Figure 16 for more information.

Both species have five digits on the front and rear feet. Digit 1 has two phalanges; all other digits have three phalanges. Toenails are present on all digits. See Figure 59.

The mammary glands are large in the mouse and small in the rat, unless the rat is nursing. (For mammaries in the rat, see Figures 10B and 15B.)

Figure 14A. Mouse. Feet (palmar and plantar aspects).

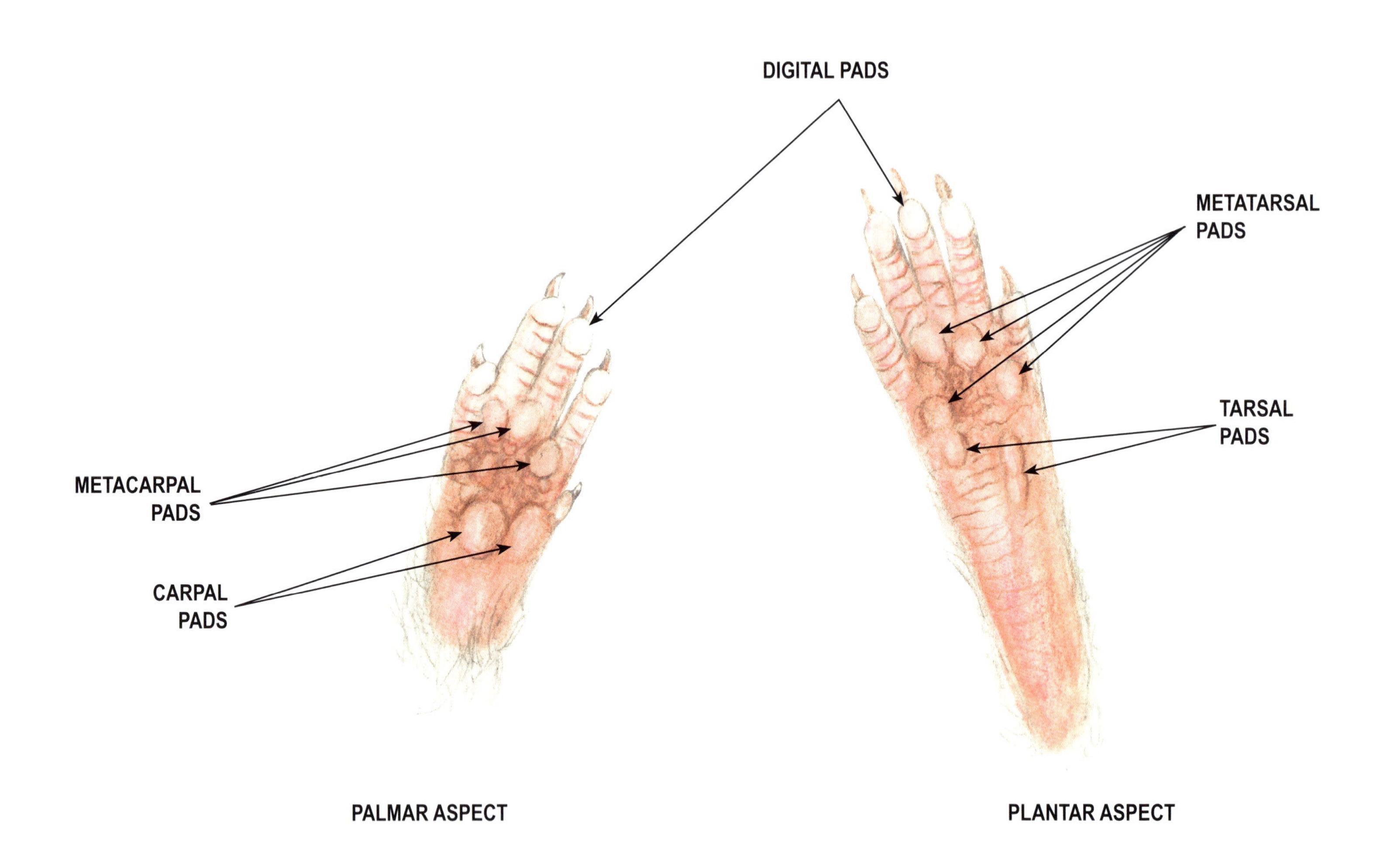

Figure 14B. Rat. Feet (palmar and plantar aspects).

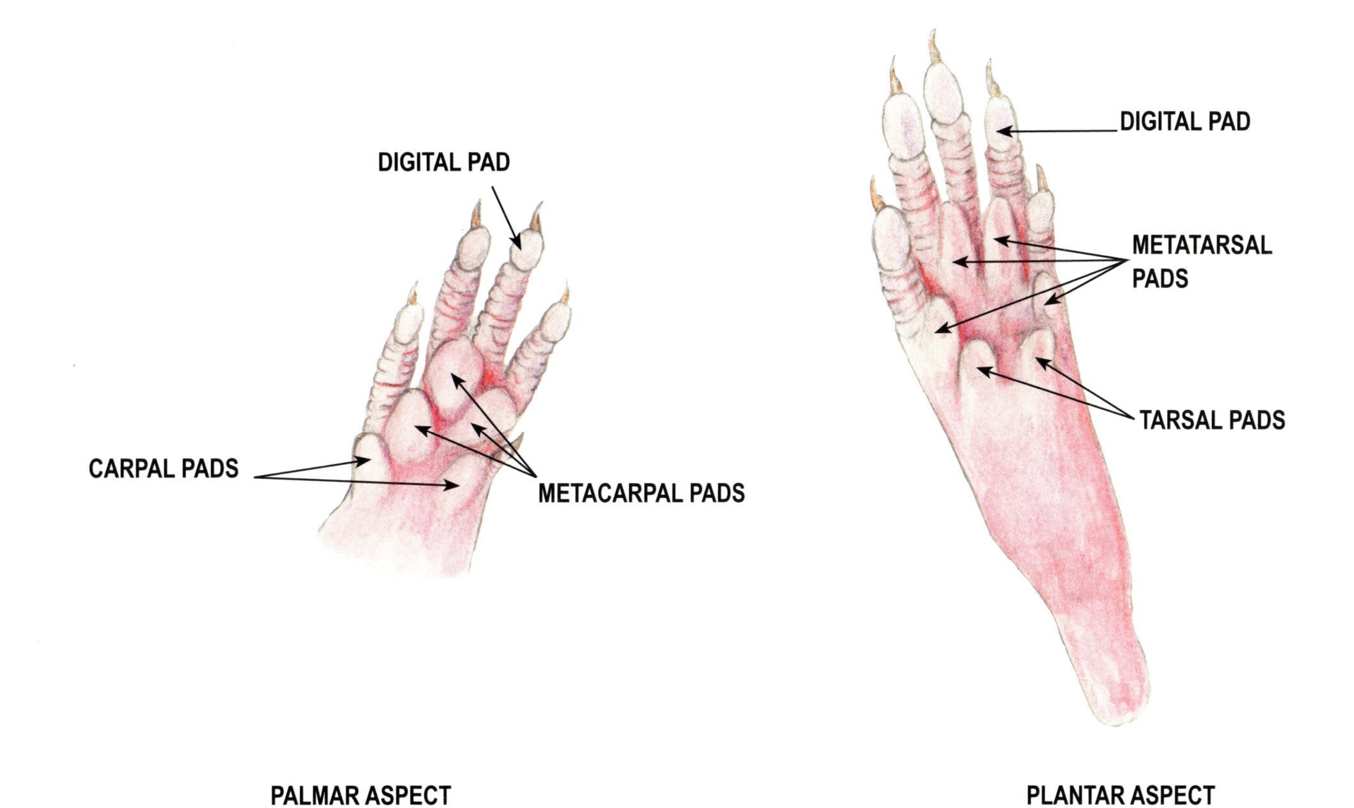

Figure 14A. Mouse. Feet (palmar and plantar aspects). **Figure 14B. Rat. Feet (palmar and plantar aspects).**

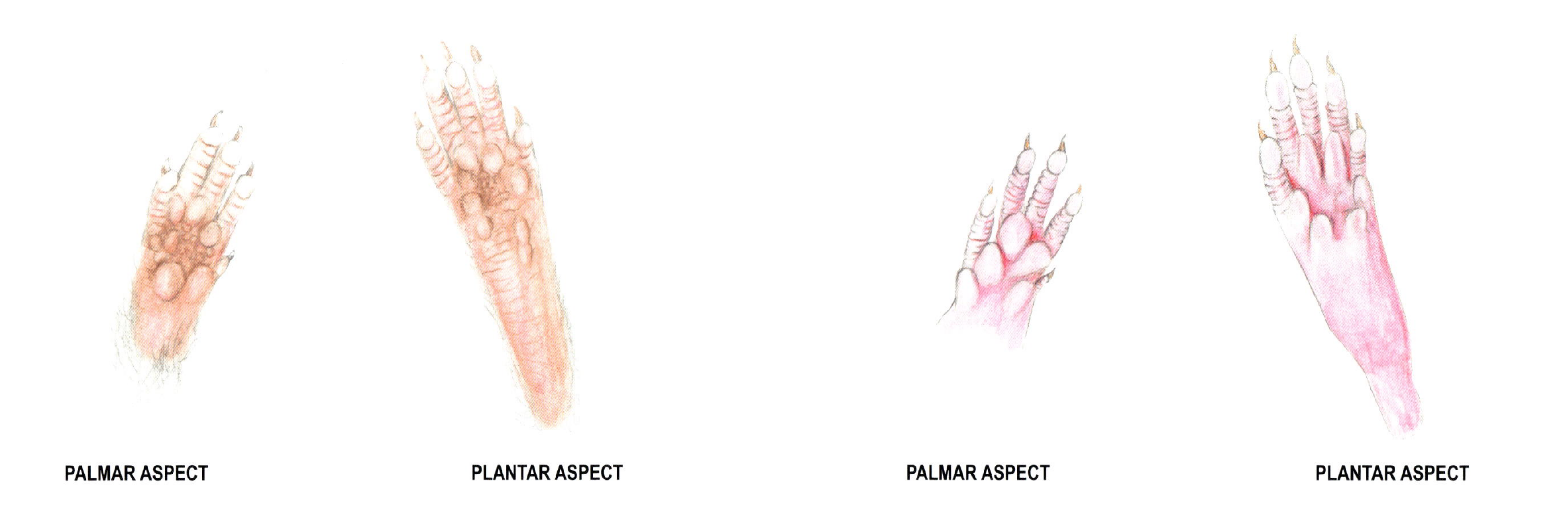

Figure 14. Feet (palmar and plantar aspects).

Mouse

Five digital pads (one for each digit) are present on the thoracic and pelvic limbs.

Three metacarpal and two carpal pads are present on the thoracic limb. The borders of the carpal pads touch.

Four metatarsal and two to three tarsal pads are present on the pelvic limb.

Rat

Five digital pads are present on each limb, similar to those of the mouse.

Four metacarpal pads and one carpal pad, all rounded, are present on the thoracic limb.

Four metatarsal and two tarsal pads, all cylindrical and prominent, are present on the pelvic limb.

Mammary Glands

Figure 15A. Mouse. Mammary glands in lateral, ventral, and dorsal aspects.

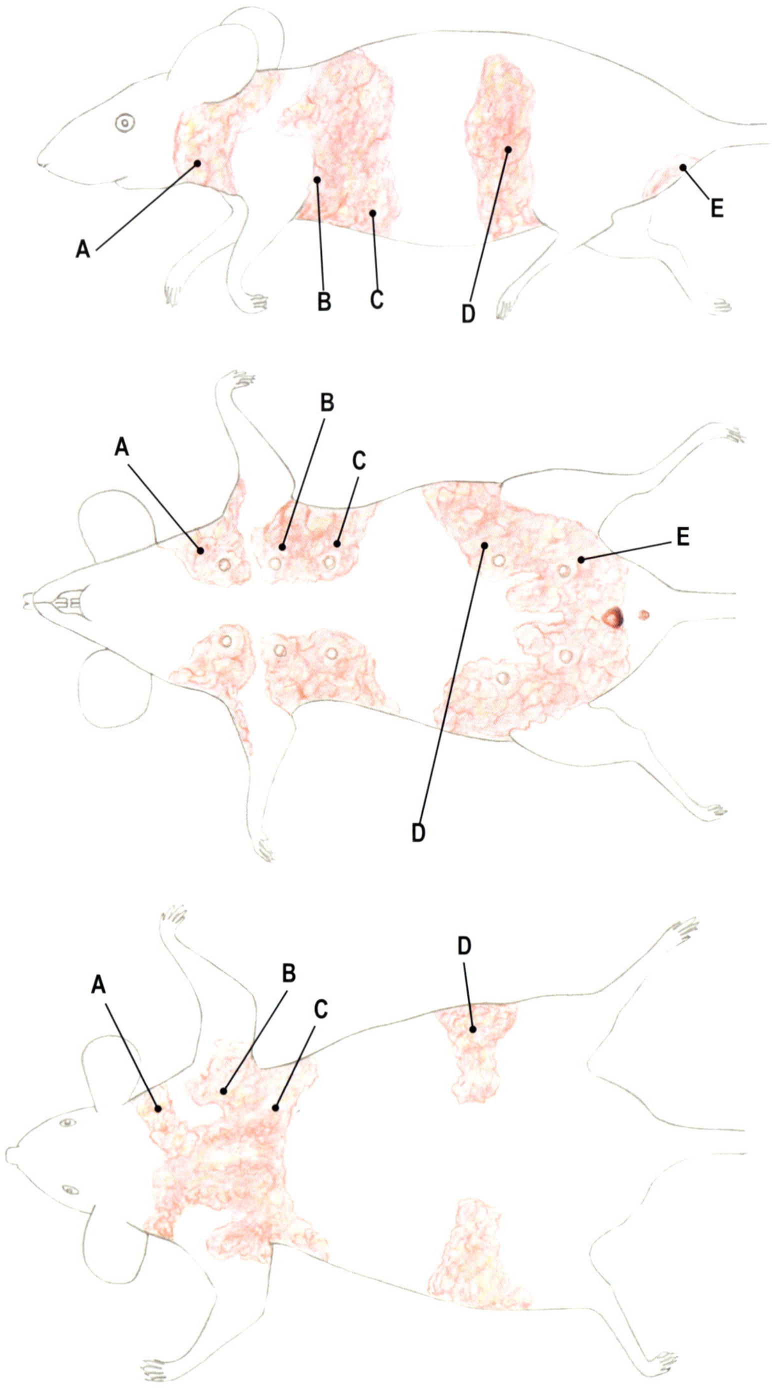

(A) AXILLARY (B) CRANIAL THORACIC (C) CAUDAL THORACIC (D) ABDOMINAL (E) INGUINAL

Figure 15B. Rat. Mammary glands in lateral and ventral aspects.

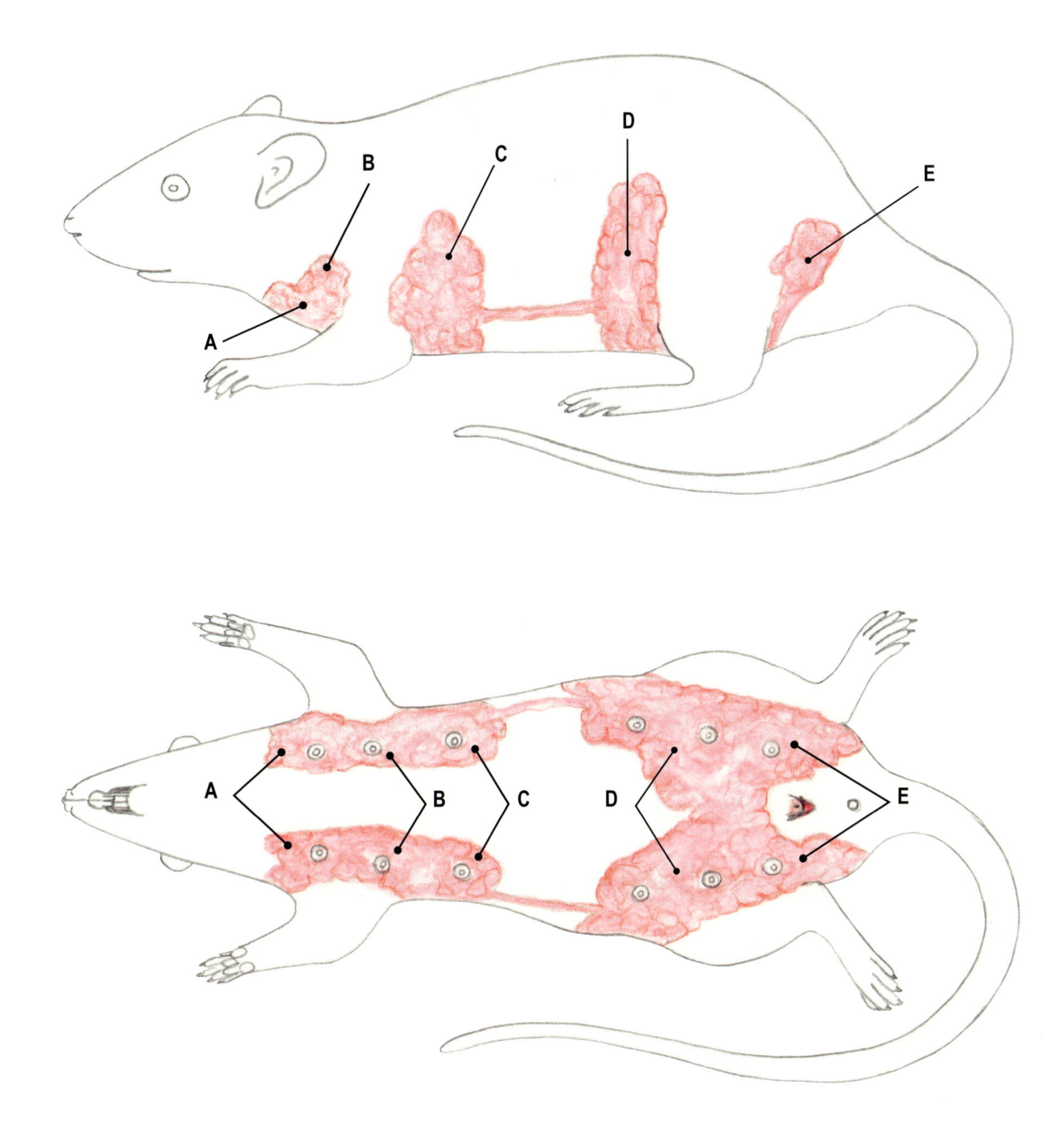

(A) AXILLARY (B) CRANIAL THORACIC (C) CAUDAL THORACIC (D) ABDOMINAL (E) INGUINAL

Figure 15A. Mouse. Mammary glands in lateral, ventral, and dorsal aspects.

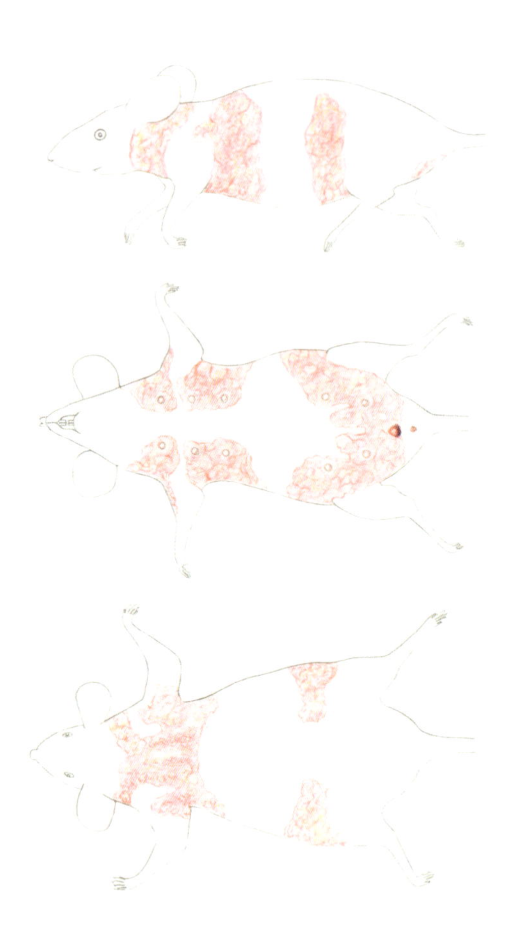

Figure 15B. Rat. Mammary glands in lateral and ventral aspects.

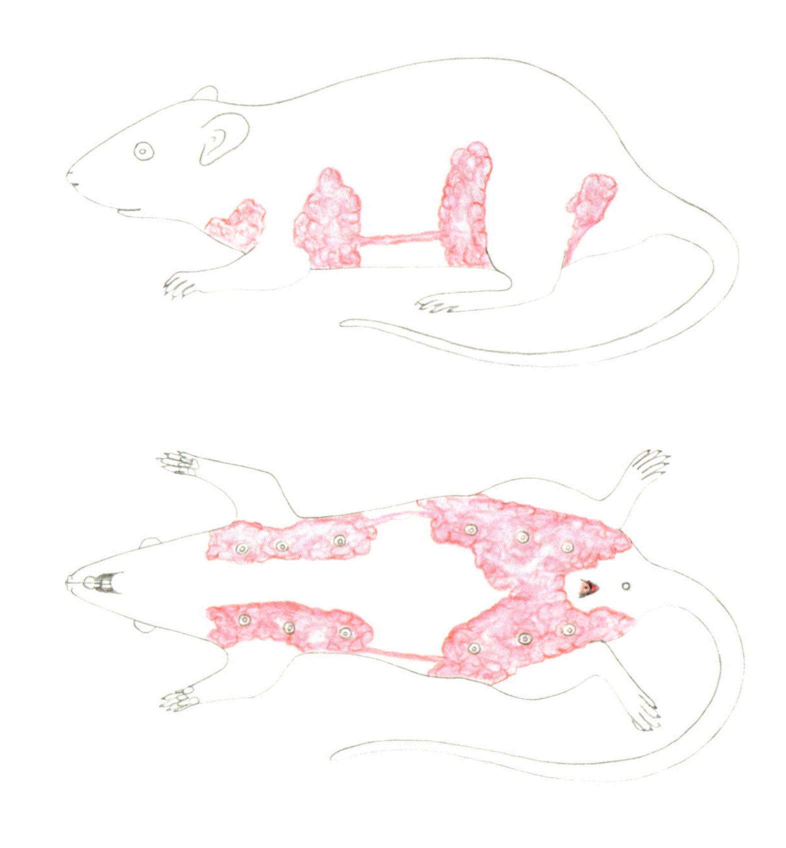

Figure 15. Mammary glands in lateral, ventral, and dorsal (mouse only) aspects.

Also known as mammae, the mammary glands are symmetrical between the left and right aspects of the body. The mammary glands are embedded in fat, often referred to as a fat pad.

Common sites for mammary/fat pad injection are between the nipples of the 2nd and 3rd mammary glands (the 1st and 2nd thoracic mammary glands, respectively).

Mouse

The mammary glands in the mouse are proportionately larger and extend more dorsally than in the rat.

The axillary and thoracic mammary glands (cranial and caudal) are separated from each other on the lateral and ventral aspects of the body. The axillary and thoracic mammary glands are fused dorsally by a median bridge.

The abdominal and inguinal mammary glands are fused ventrally. The abdominal mammary glands extend to the dorsum. The inguinal mammary glands extend laterally from the perineal area and completely encircle the external genitalia ventrally. The inguinal mammary glands do not extend dorsally.

Rat

In non-lactating animals, the mammary glands are more extensive in the mouse than in the rat.

The rat mammary glands are distinguished from those of the mouse by these differences:

- None of the mammary glands reaches the dorsum.
- The axillary and thoracic mammary glands are fused ventrally on each side, whereas these glands maintain a separation on the ventrum in the mouse (but are fused on the dorsum).
- There are two pairs of abdominal mammary glands in the rat, but only one in the mouse.
- A narrow band of tissue, located on the lateral aspect of the body, bridges the thoracic and the inguinal mammary glands.
- The inguinal mammary glands surround but do not reach the external genitalia.

Structures of the Head and Neck

Figure 16A. Mouse. Head, facial (lateral aspect), showing vibrissae.

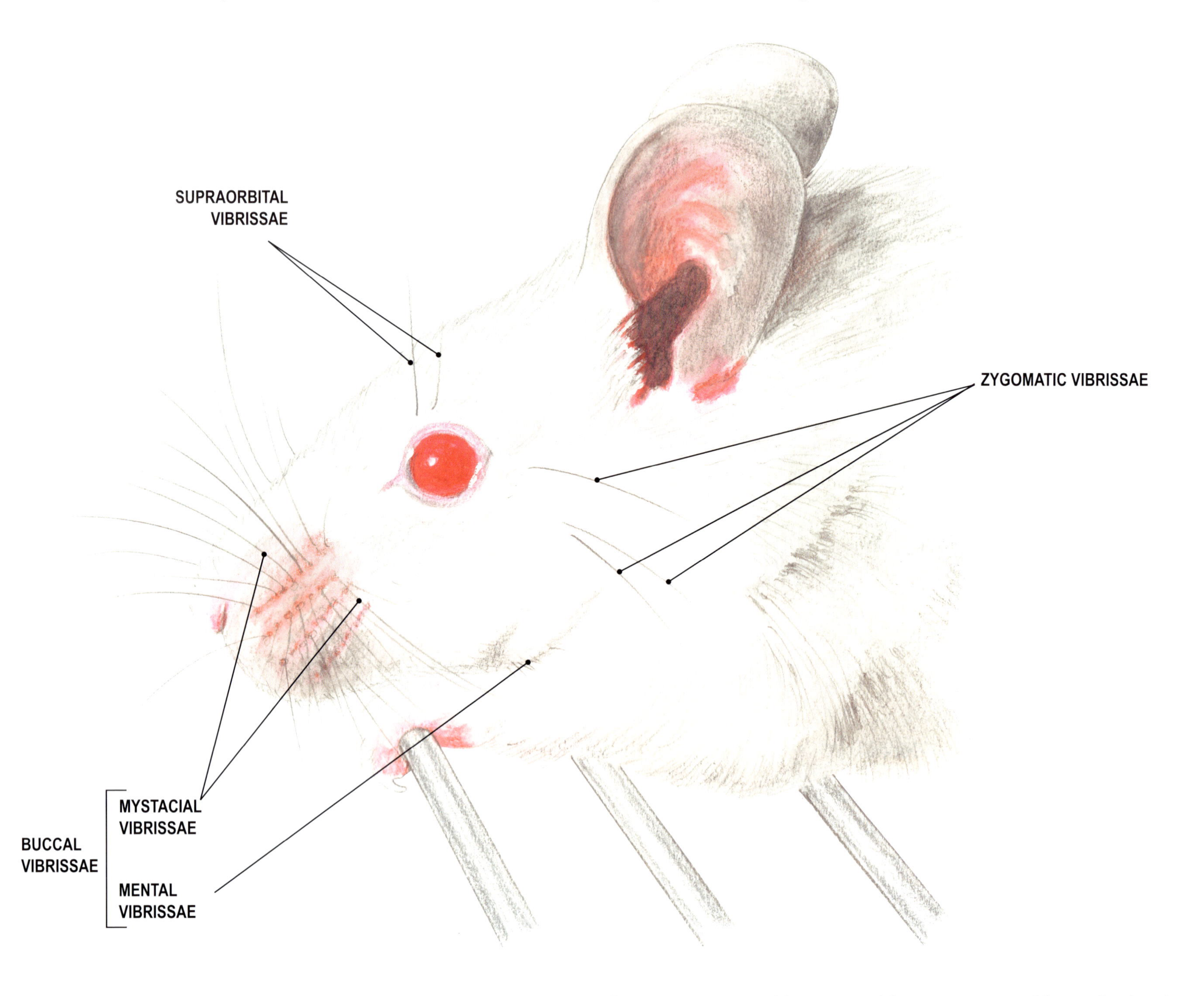

Figure 16B. Rat. Head, facial (lateral aspect), showing vibrissae.

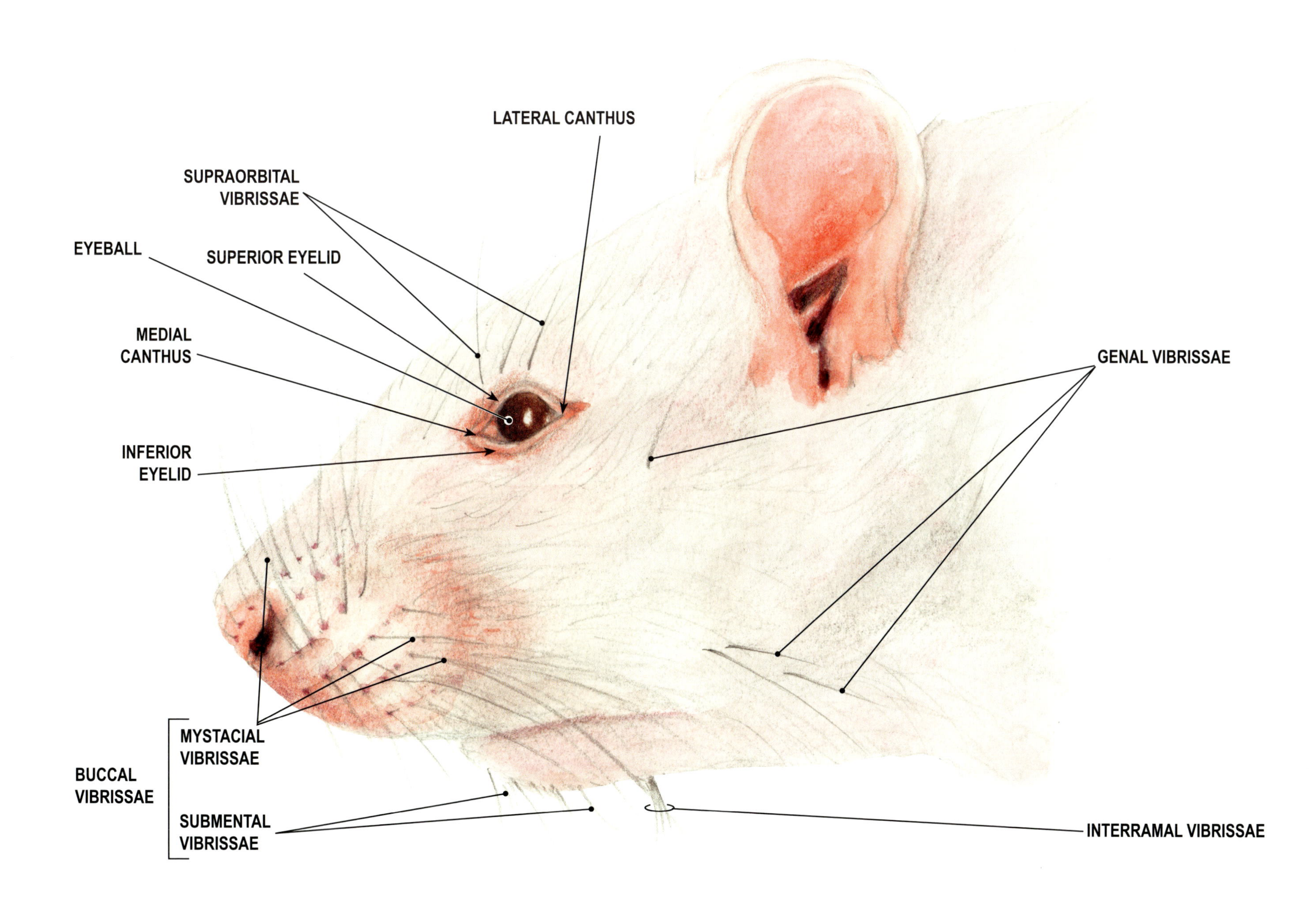

Figure 16A. Mouse. Head, facial (lateral aspect), showing vibrissae.

Figure 16B. Rat. Head, facial (lateral aspect), showing vibrissae.

Figure 16. Head, facial (lateral aspect), showing vibrissae.

The mouse has proportionately larger eyeballs and ears than the rat.

The location and quantity of vibrissae vary between mice and rats. The vibrissae function as specialized tactile hairs that "feel" the environment and convey information to the brain. The females of some strains of mice barber (remove hairs and vibrissae by chewing) their cage mates, often removing the vibrissae preferentially. An animal barbered of its whiskers has a reduced sensory perception of its environment, similar to a blind person, and may behave abnormally as a result. Therefore, animals lacking vibrissae may not be suitable candidates for behavioral studies.

Note on terms: In the Nomina Anatomica Veterinaria (NAV), the term "vibrissa" refers to hair on the nose, and the term "tactile hair" refers to other specialized hairs on the head.[7] In other references, including this publication, the term "vibrissa" refers more generally to specialized tactile hair on the head, including the nose.[2] The NAV terms, based on body regions (see Figure 4) are provided in parentheses below:

Supraorbital vibrissae (supraorbital tactile hairs). These are found dorsocaudal to the eye in the supraorbital region.

Genal vibrissae (zygomatic tactile hairs). These are found ventrocaudal to the eye in the zygomatic region.

Buccal vibrissae include both the mystacial and submental vibrissae (below).

Mystacial vibrissae (superior labial tactile hairs). They are associated with the upper lip in the maxillary region and with the lateral nasal region and the nasal region.

Submental vibrissae (inferior labial tactile hairs and tactile hairs). These are located on the lower lip and in the mental region (chin) in both species.

Interramal vibrissae (mental vibrissae). These are found in the intermandibular region between the bodies of the mandibles on the ventral midline of the head.

Mouse

Supraorbital vibrissae: Fewer in the mouse than in the rat.

Genal vibrissae (labeled zygomatic vibrissae in figure): Closer together in the mouse than in the rat.

Mystacial vibrissae: Up to four rows in the mouse; the rat can have more rows.

Submental vibrissae: Visible in the illustration.

Interramal vibrissae: The view of these vibrissae is obstructed by the position of the thoracic limb.

Rat

Supraorbital vibrissae: More in the rat than in the mouse.

Genal vibrissae: One ventrocaudal to the eye, and three dorsocaudal to the commissure of the lips (distal to those of the mouse).

Mystacial vibrissae: Up to six rows in the rat; the mouse has fewer rows.

Submental vibrissae: Visible in the illustration.

Interramal vibrissae: Visible in the illustration.

Figure 17A. Mouse. Head, lacrimal and salivary glands (lateral aspect).

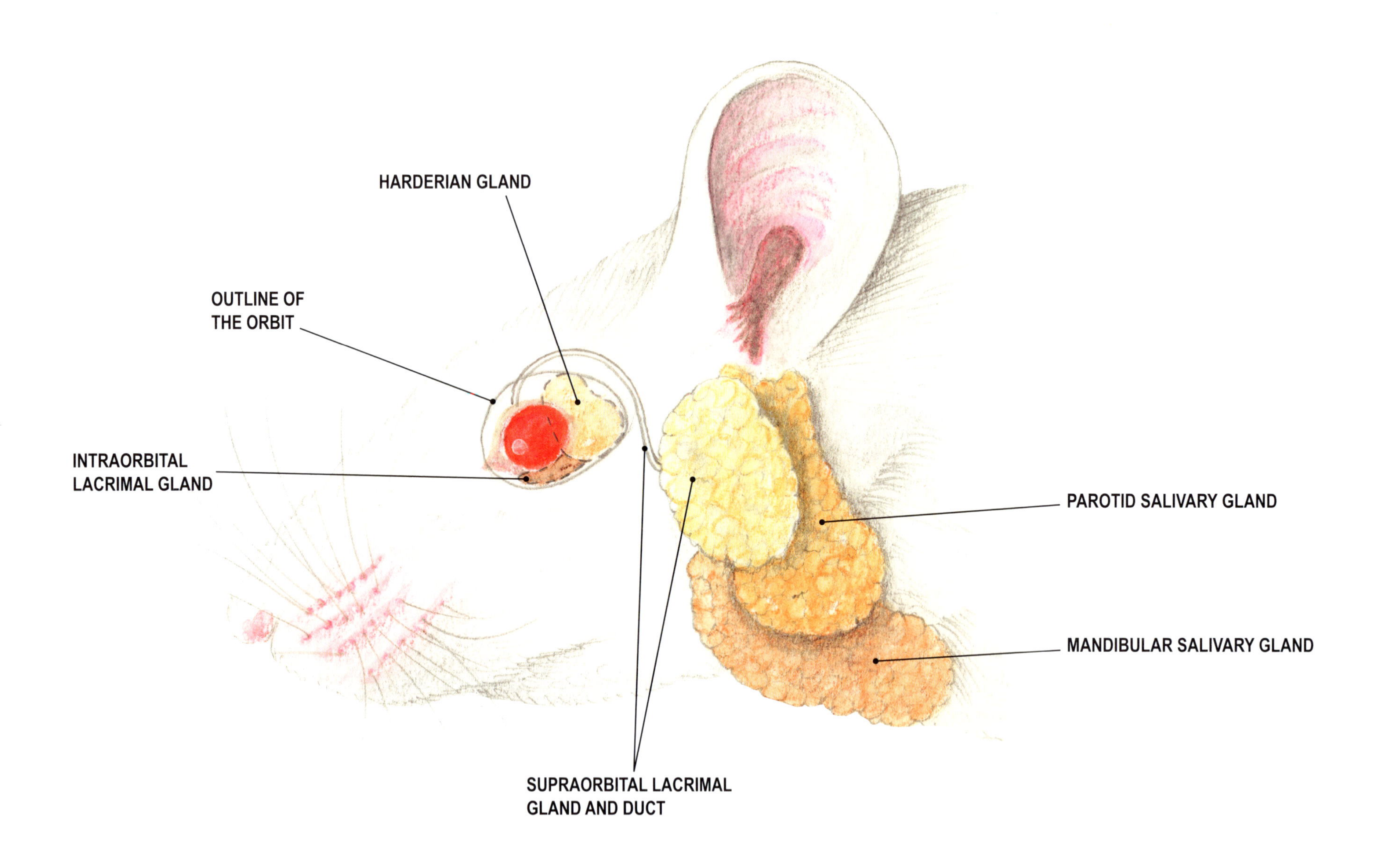

Figure 17B. Rat. Head, lacrimal and salivary glands (lateral aspect).

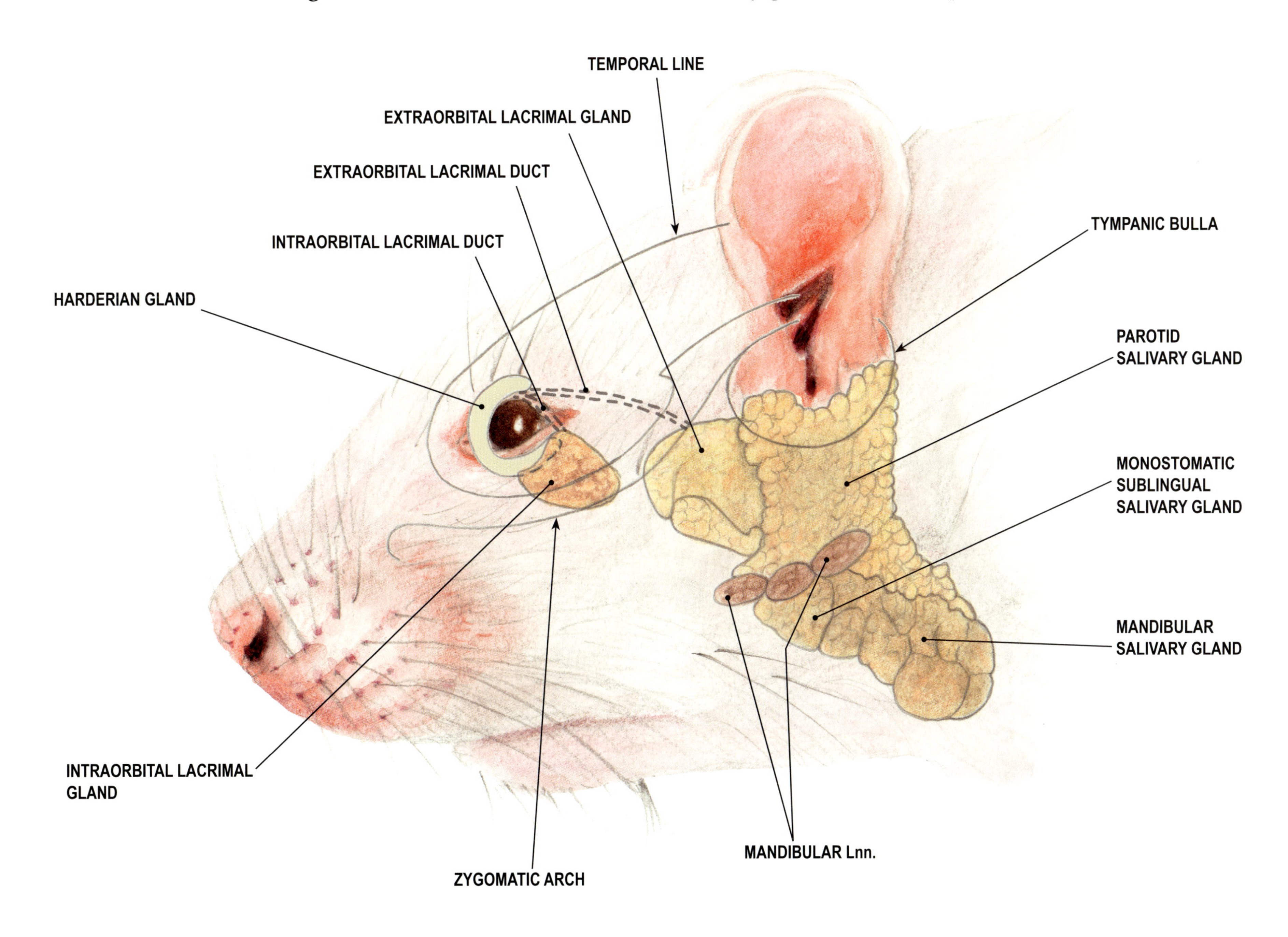

Figure 17A. Mouse. Head, lacrimal and salivary glands (lateral aspect).

Figure 17B. Rat. Head, lacrimal and salivary glands (lateral aspect).

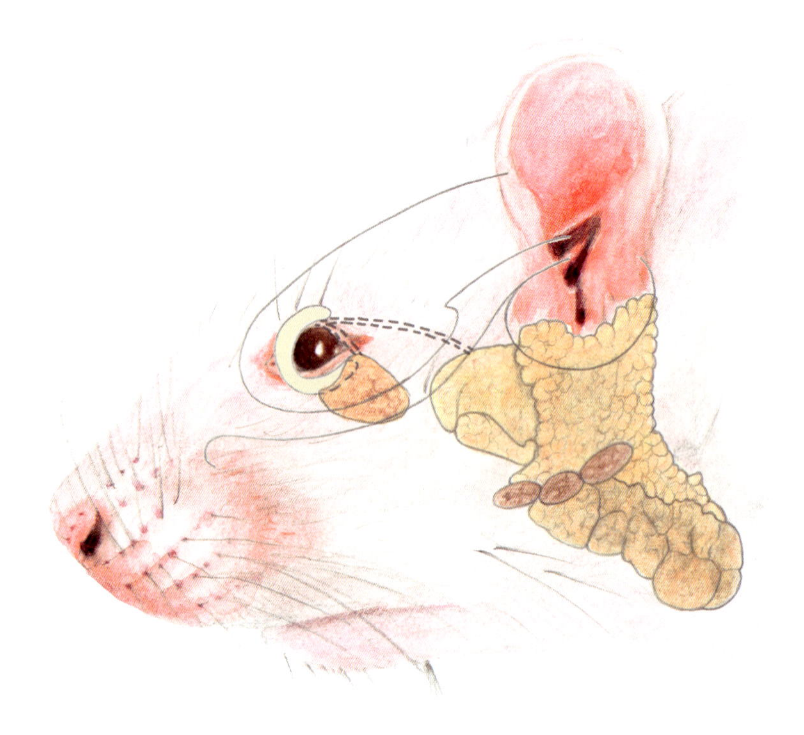

Figure 17. Head, lacrimal and salivary glands (lateral aspect).

The Harderian gland is the deep lacrimal gland of the third eyelid.

Mouse

The mouse has a large Harderian gland, retrobulbar, extended dorsally and ventrally. The best access is through the lateral canthus; in the rat, access is best through the medial canthus.

The intraorbital lacrimal gland is ventral to the eyeball, on the floor of the orbit.

The extraorbital lacrimal gland covers the parotid salivary gland rostrodorsally. The excretory duct enters the orbit in a dorsorostral direction.

The parotid salivary gland is elliptical, with the long axis vertically oriented. The parotid gland is large in males, as compared to females.

The mandibular salivary gland is obliquely oriented. It is partially covered by the ventral extent of the parotid salivary gland and by the ventral end of the extraorbital lacrimal gland. In mice, the mandibular salivary gland in males is almost twice the size as in females.

The monostomatic sublingual salivary gland lies deep to the mandibular salivary gland.

Rat

Outlines of the temporal line, the tympanic bulla, and the zygomatic arch are drawn to provide bony landmarks to the orbital structures illustrated in the figure.

The Harderian gland has a semilunar shape and surrounds the eyeball dorsorostroventrally. Its ventral end, caudally oriented, is identified by a dashed line where it is covered by the intraorbital lacrimal gland. (For best access, see Mouse.)

The intraorbital lacrimal gland is located ventrocaudal to the eyeball, on the orbit floor in an oblique position. It covers the ventrocaudal end of the Harderian gland. The extraorbital lacrimal gland touches the rostral border of the parotid salivary gland.

The ducts of the intraorbital and extraorbital lacrimal glands are shown by dashed lines. The duct of the intraorbital gland is oriented dorsally and joins the duct of the extraorbital gland in a common duct which drains onto the surface of the eye.

The mandibular salivary gland lies ventral to the caudal half of the ventral border of the parotid salivary gland.

The monostomatic sublingual salivary gland is located ventral to the parotid gland, and rostral to the mandibular gland. At the border between the parotid and the monostomatic salivary glands, a group of mandibular lymph nodes can be palpated under the skin.

Figure 18A. Mouse. Orbital veins and venous plexus (lateral aspect).

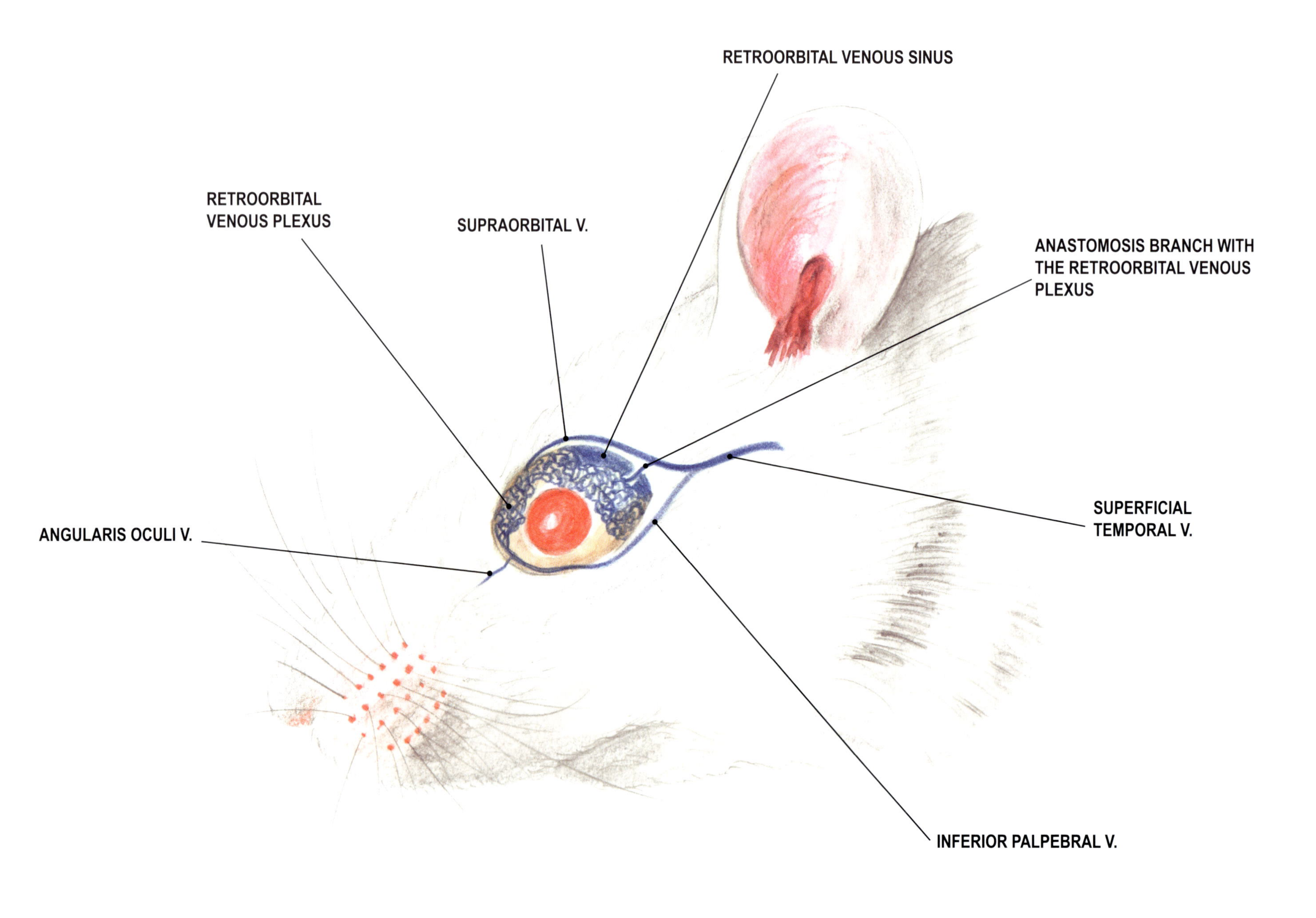

Figure 18B. Rat. Orbital veins and venous plexus (lateral aspect).

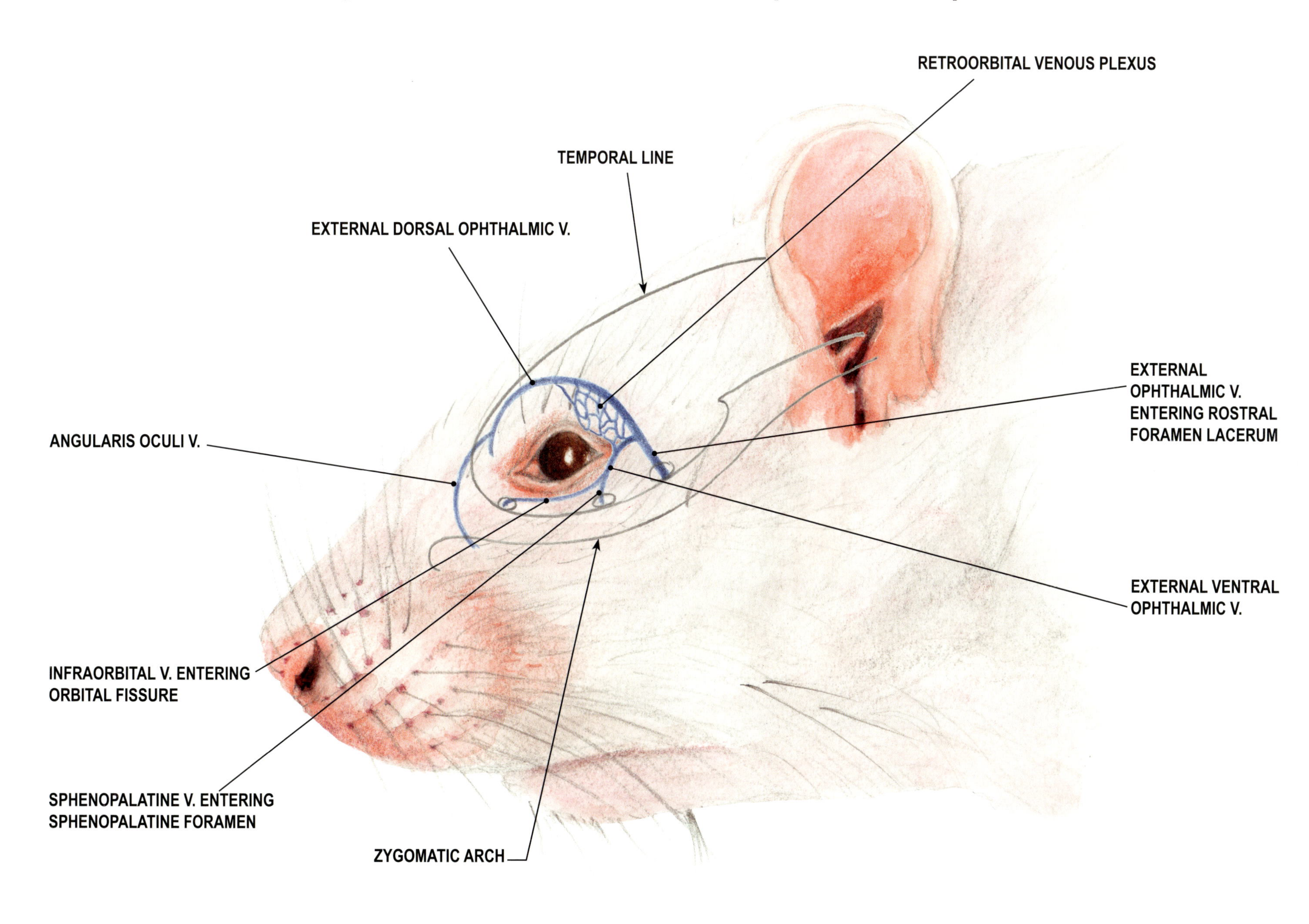

Figure 18A. Mouse. Orbital veins and venous plexus (lateral aspect).

Figure 18B. Rat. Orbital veins and venous plexus (lateral aspect).

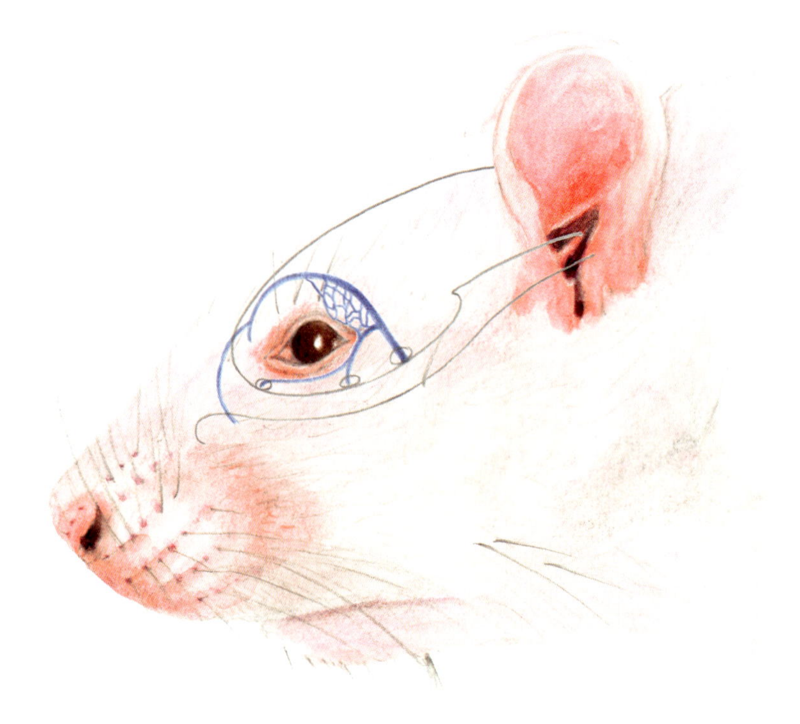

Figure 18. Orbital veins and venous plexus (lateral aspect).

Mouse

A rich retroorbital venous plexus, moon-shaped, surrounds the eyeball rostrodorsocaudally, becoming larger from the medial to the lateral angles of the eye.

There is a retroorbital venous sinus, associated with the retroorbital venous plexus, on the dorsocaudal border of the plexus.

The drainage of the retroorbital venous plexus is provided by three large veins:

- The supraorbital vein originates from the rostrodorsal aspect of the plexus.
- An anastomosis branch with the retroorbital venous plexus originates from the plexus and the sinus.
- The inferior palpebral vein originates from the rostral end of the plexus.

The angularis oculi vein joins the origin of the inferior palpebral vein.

The retroorbital venous plexus drains into the supraorbital and inferior palpebral veins, which finally converge to the superficial temporal vein.

The retroorbital venous plexus and sinus cover most of the Harderian gland, which is located in a different location in the mouse than in the rat (caudal to the eyeball; see figure). The venous plexus extends caudal to the dorsocaudal end of the Harderian gland. The dorsal ophthalmic vein parallels the external curvature of the Harderian gland.

Rat

Outlines of the temporal line and the zygomatic arch are drawn to provide bony landmarks to the orbital structures illustrated in the figure.

The retroorbital venous plexus is reduced in the rat compared with the mouse. This plexus is located at the dorsolateral angle of the eye.

The drainage of the retroorbital plexus is provided by the following veins:

- The dorsal ophthalmic vein surrounds the Harderian gland dorsocaudally, collects blood from the retroorbital venous plexus, and joins the ventral ophthalmic vein to form the ophthalmic vein.
- The angularis oculi vein joins the dorsal ophthalmic vein close to its origin.
- The infraorbital vein is the most rostral vein ventral to the eyeball; it enters the skull through the infraorbital foramen.
- The sphenopalatine vein joins the preceding vein and enters the sphenopalatine foramen.
- The point where the infraorbital and sphenopalatine veins merge is the origin of the ventral ophthalmic vein.
- The dorsal and the ventral ophthalmic veins are the roots of the ophthalmic vein, which enters the rostral foramen lacerum.

Most of the infraorbital vein in the orbit, the entire sphenopalatine vein, and the origin of the ventral ophthalmic vein are in close contact with the Harderian gland.

Figure 19A. Mouse. Brain with arterial and venous vasculature (dorsal aspect).

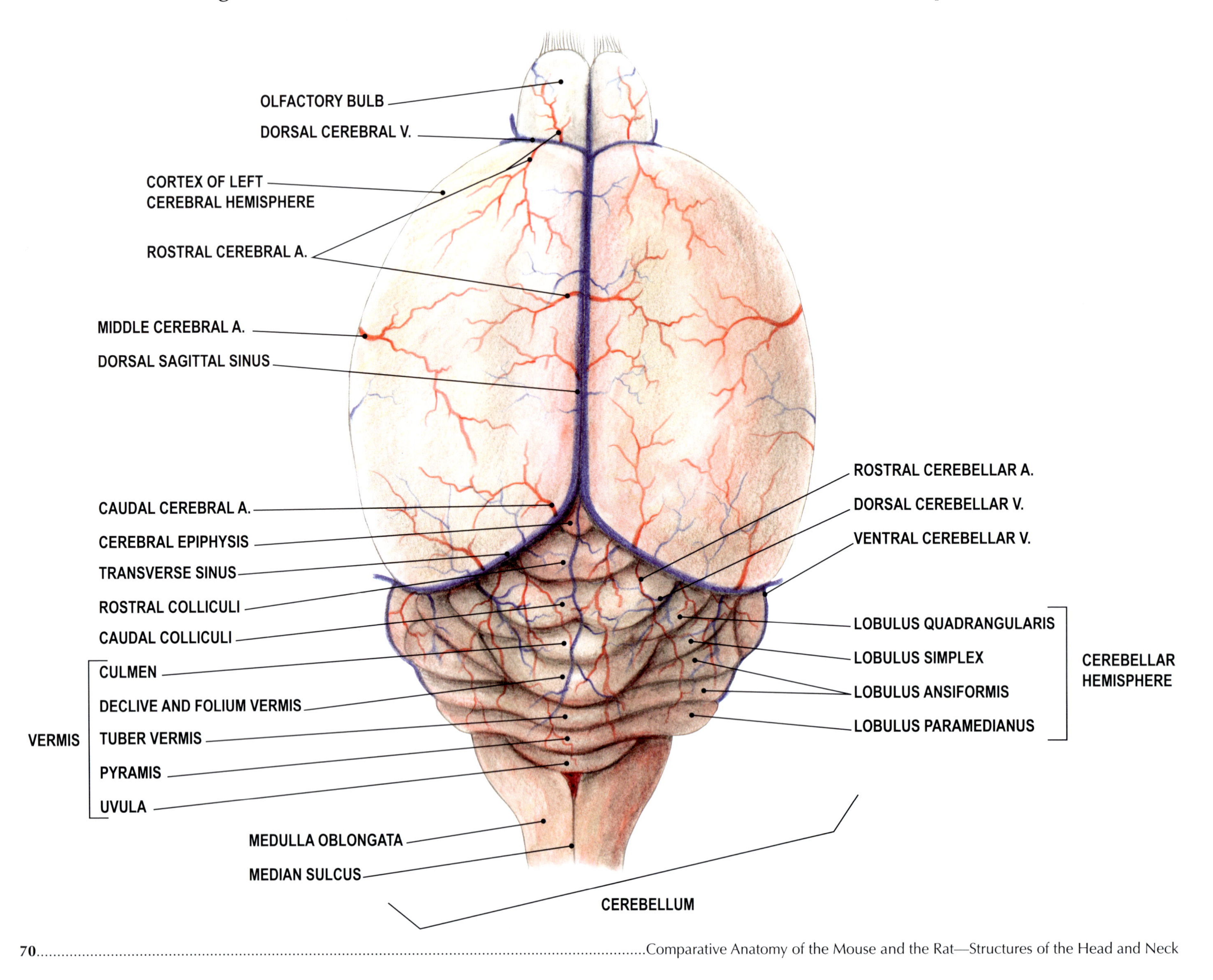

Figure 19B. Rat. Brain with arterial and venous vasculature (dorsal aspect).

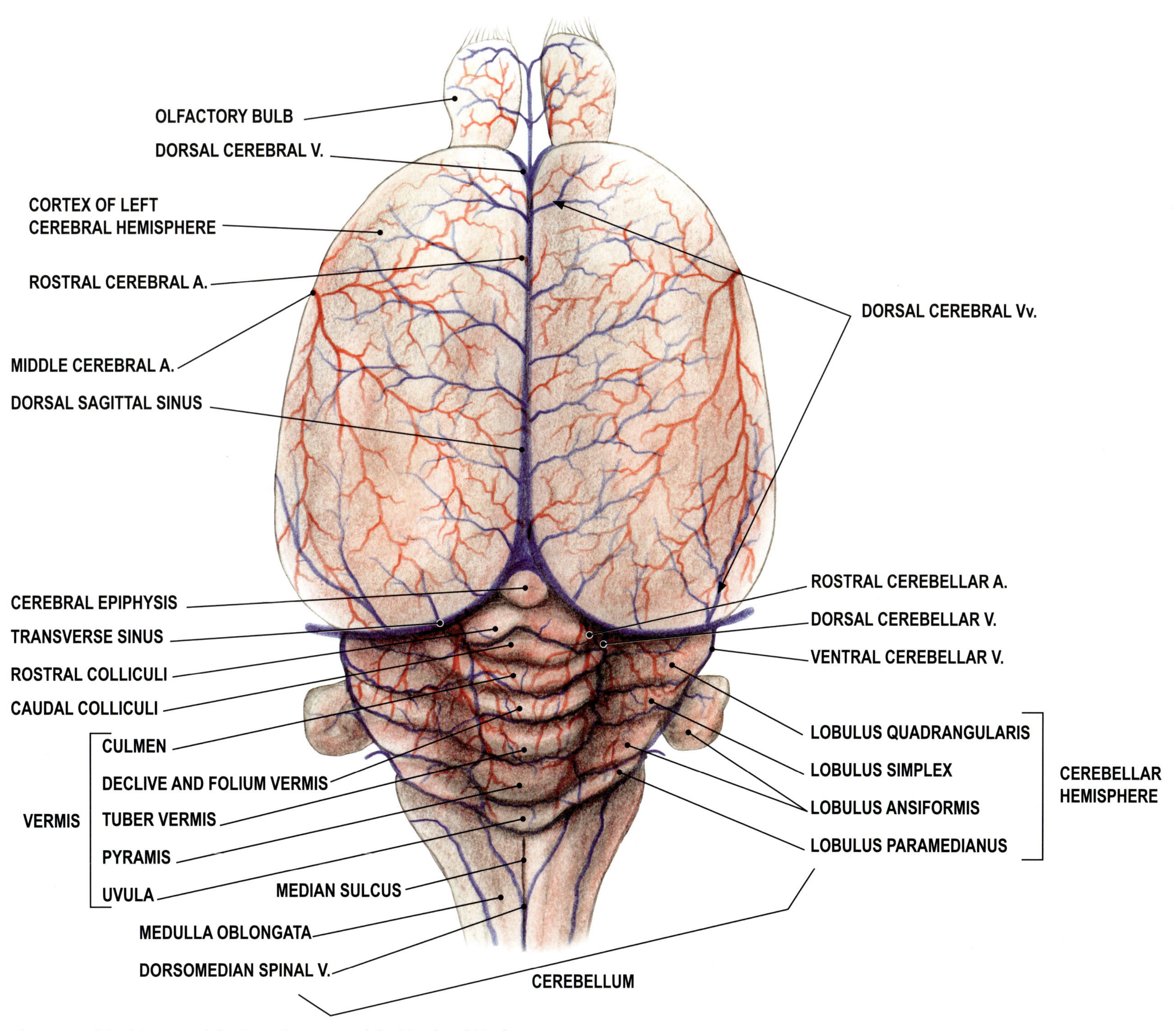

Figure 19A. Mouse. Brain with arterial and venous vasculature (dorsal aspect).

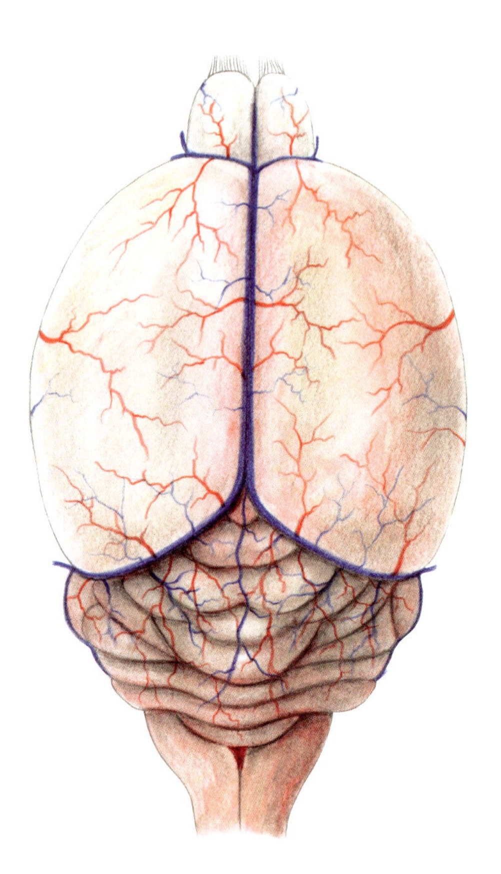

Figure 19B. Rat. Brain with arterial and venous vasculature (dorsal aspect).

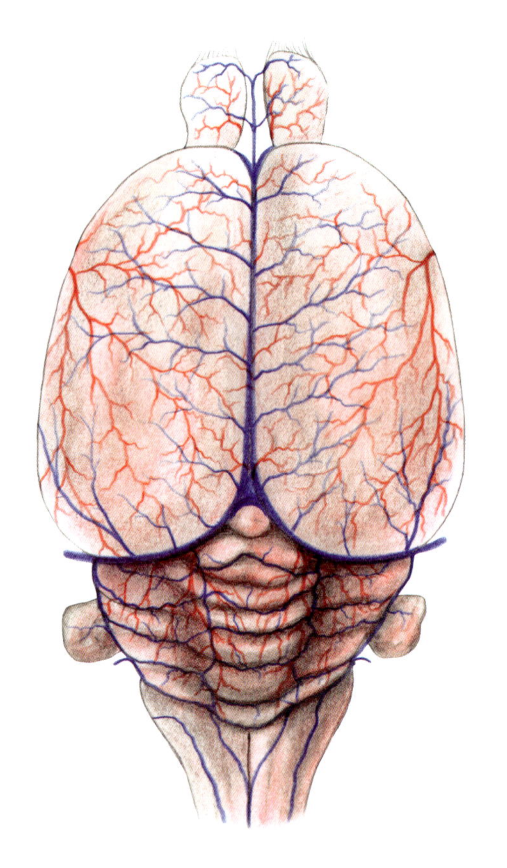

Figure 19. Brain with arterial and venous vasculature (dorsal aspect).

Microtechniques involving the central nervous system of rodents have underscored the need to understand their neuroanatomy. Although some of these procedures require a brain atlas and stereotaxic techniques, techniques such as dissecting large areas of tissue as slices or for organ culture, collecting cerebrospinal fluid from the ventricle or cerebellomedullar cistern, canulating the lateral ventricles, or exposing pathways for tracer injection all require knowledge of anatomic brain landmarks. In some instances, smaller regions, such as dentate hippocampal neurons, must be collected to verify findings of lymphocytic choriomeningitis, which can be identified on the basis of neuroanatomic landmarks. Surgically, vascular occlusion and separation of the corpus callosum are commonly done in both mice and rats. Sometimes the cortex of specific lobes of the brain must be exposed surgically for deliberately damaging the pial surface or for electroporation. All of these types of procedures require anatomical landmarks, which are presented in Figures 19-21.*

Figures 19B, 20B, and 21B are modified and illustrated in color from Paxinos and Watson.[10] The greater abundance of the arterial supply in the rat compared with the mouse specimens in these figures is due to an intraarterial injection of the Batson No. 17 anatomical corrosion compound.[11]

The differences in shape and design of the dorsal surface of the brain between the two species are minor, yet important for research.

*These comments thanks to Gayle Johnson, DVM, PhD, Associate Professor of Veterinary Pathology at the College of Veterinary Medicine, University of Missouri.

Figure 20A. Mouse. Brain with arterial vasculature (ventral aspect).

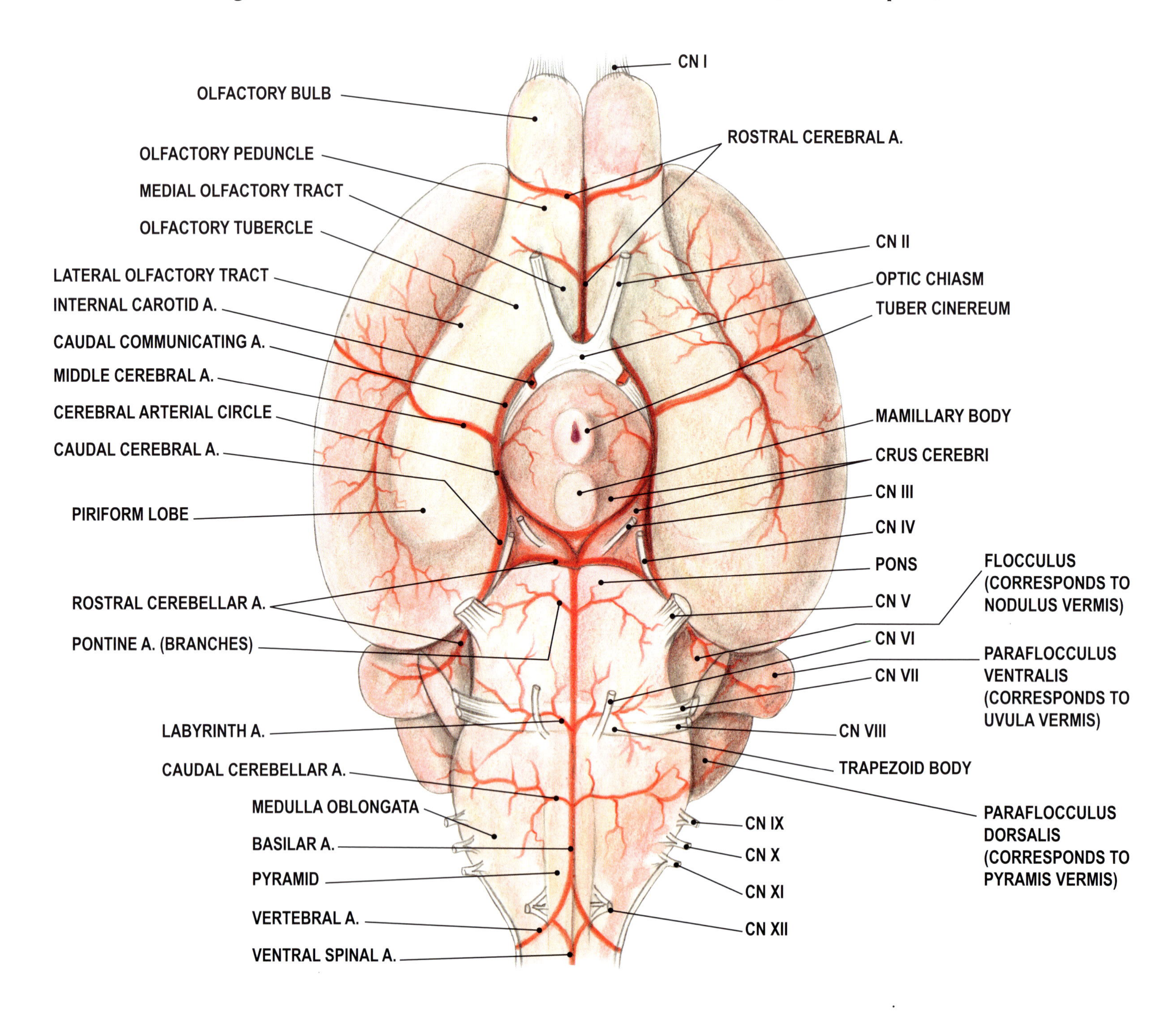

Figure 20B. Rat. Brain with arterial vasculature (ventral aspect).

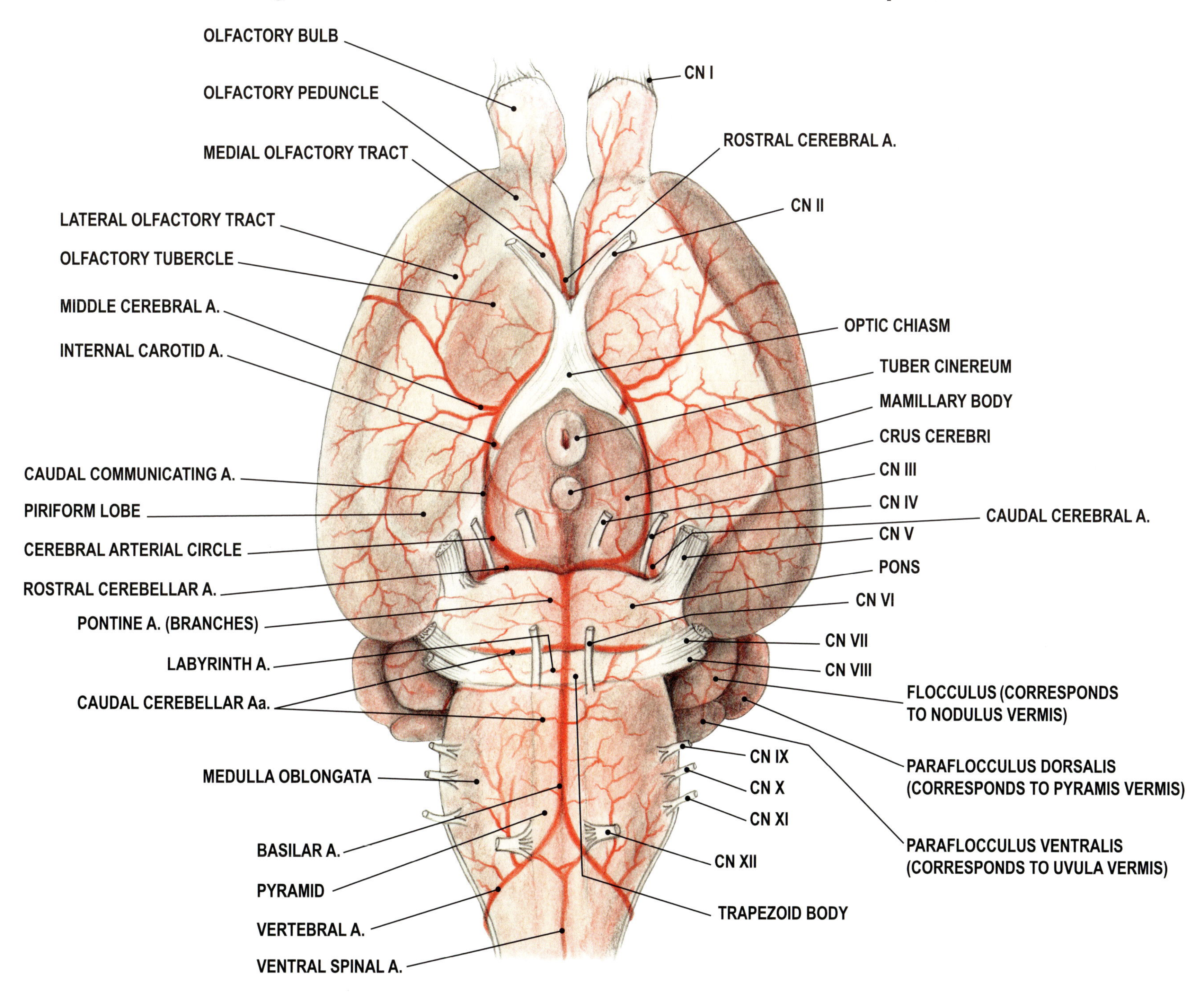

Figure 20A. Mouse. Brain with arterial vasculature (ventral aspect).

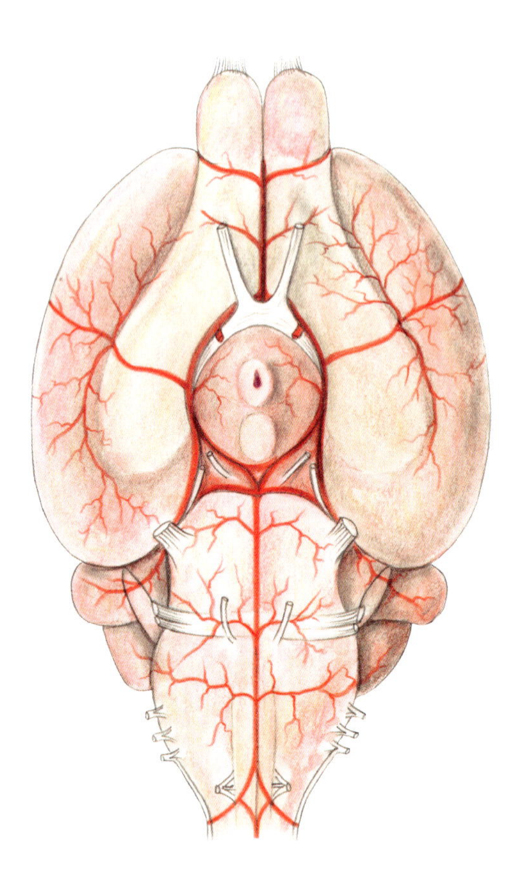

Figure 20B. Rat. Brain with arterial vasculature (ventral aspect).

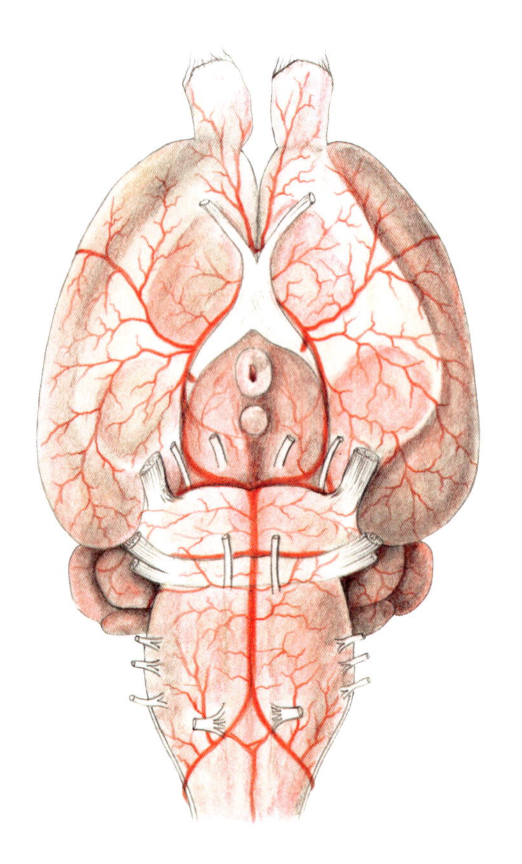

Figure 20. Brain with arterial vasculature (ventral aspect).

The ventral aspect shows distinct differences between the two species as far as the extent of the arterial supply and some components of the telencephalon, mesencephalon, and rhombencephalon.

In the mouse and the rat, the rhinencephalon, or olfactory brain, exposed on the ventral aspect of the brain, consists of the olfactory bulb, olfactory peduncle, olfactory tracts, olfactory tubercle and the piriform lobe. The rhinencephalon is very well developed in these species, and serves their keen sense of smell. The medial olfactory tract is larger in the rat than in the mouse. The large cerebellum accommodates the fine motor coordination, which is essential for these animals' daily activities. See Figures 19 and 22 for the dorsal aspect and the median section of the brain.

The hypophysis was removed to expose the tuber cinereum and the mamillary body. The hypophyseal recess is shown in the middle of the tuber cinereum, inside of the hypophyseal infundibulum, which makes the connection between the diencephalon and the hypophysis.

In the cerebral arterial circle, the caudal cerebral artery in the mouse is much longer than in the rat. In the rat it is a very short artery between the CN IV and CN V.

Figure 21A. Mouse. Brain with arterial vasculature (left lateral aspect).

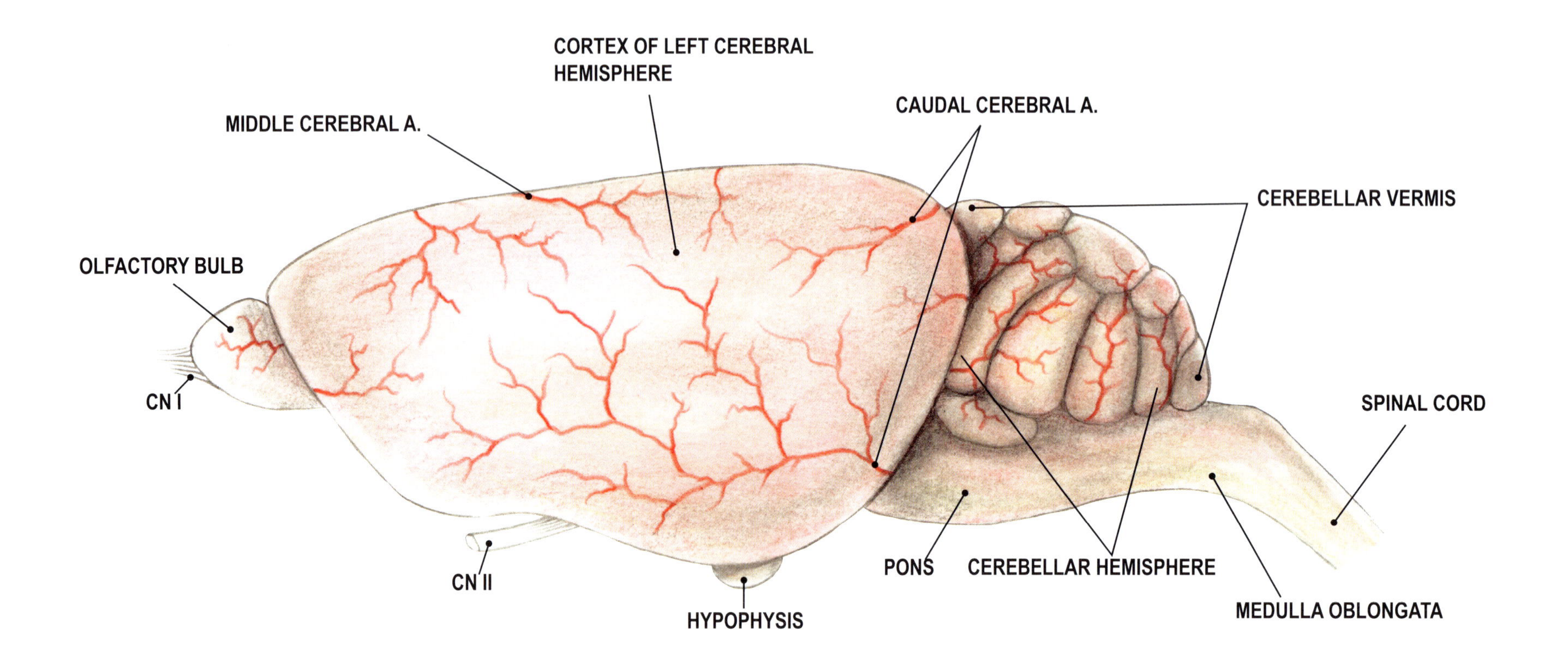

Figure 21B. Rat. Brain with arterial vasculature (left lateral aspect).

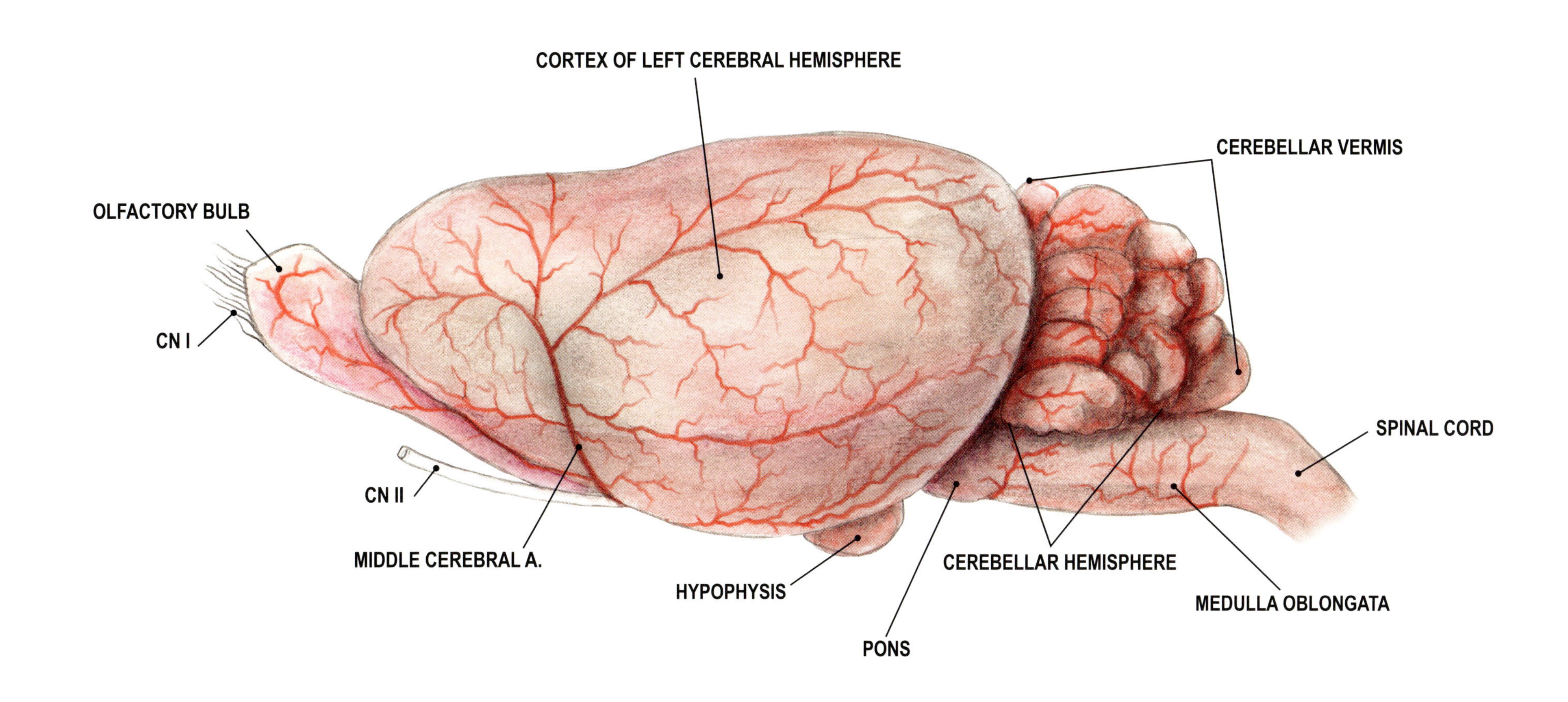

Figure 21A. Mouse. Brain with arterial vasculature (left lateral aspect).

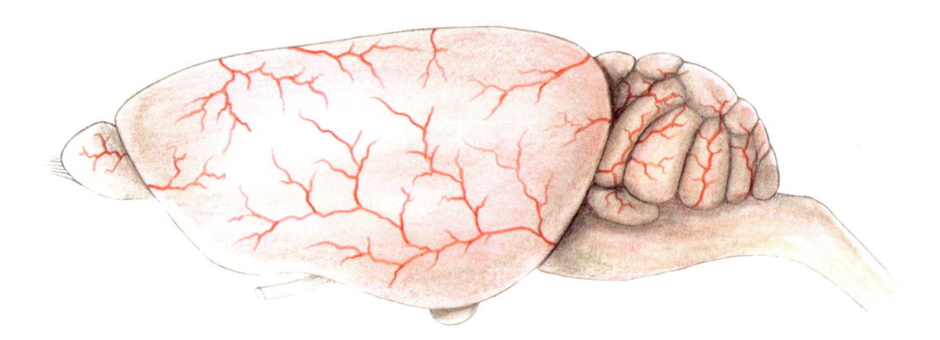

Figure 21B. Rat. Brain with arterial vasculature (left lateral aspect).

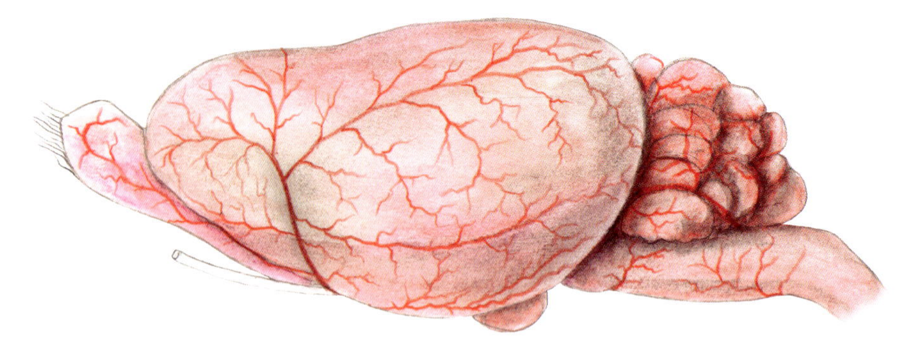

Figure 21. Brain with arterial vasculature (left lateral aspect).

The lateral aspect of the cerebral and cerebellar cortices are similar in both species. The variations in the cerebellar vermis are due to individual variation, not species difference.

The arterial supply is distinctly different between the two species, as shown by the illustrations.

Figure 22A. Mouse. Brain (median section).*

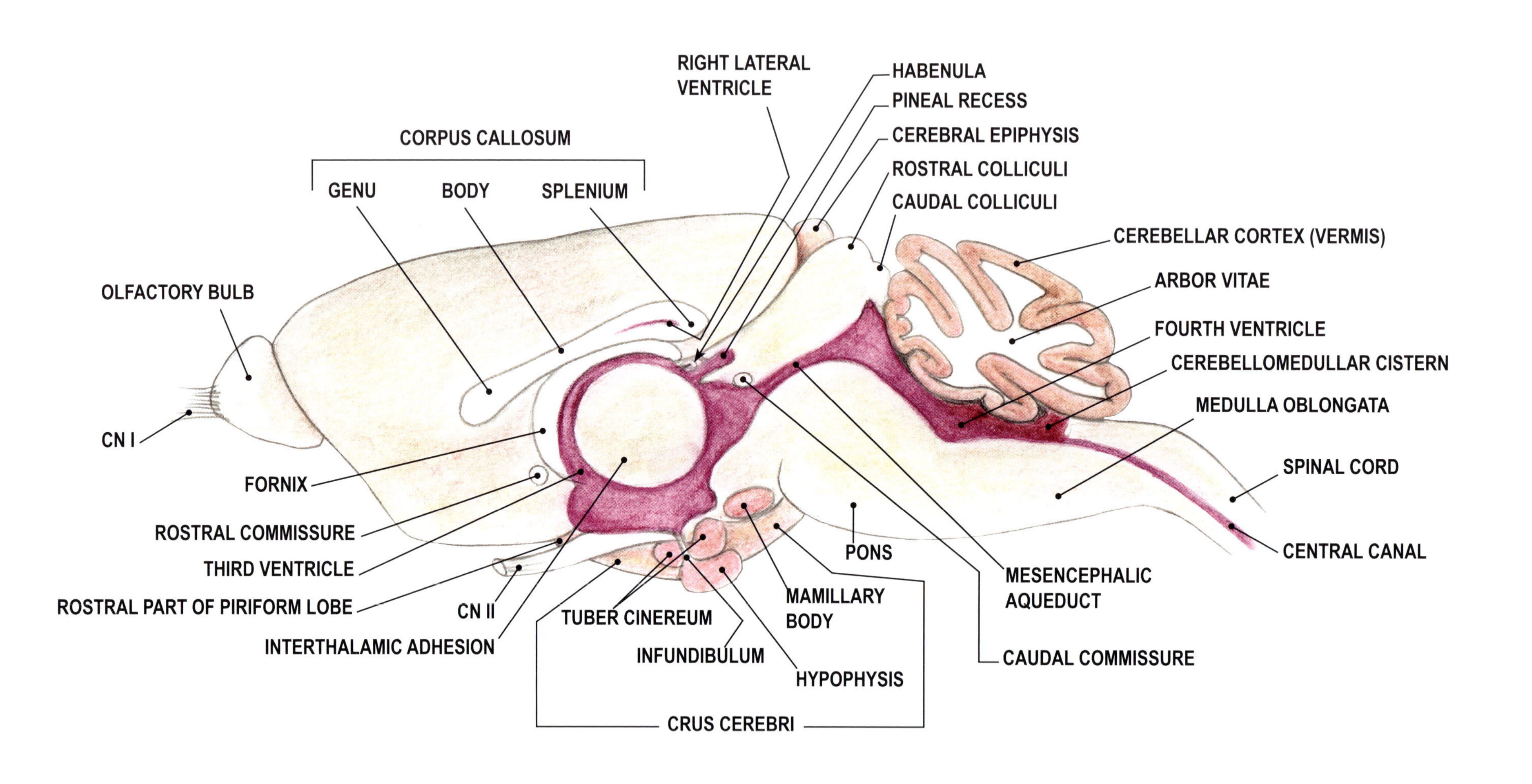

*Coloration added for illustrative purposes to give contrast between structures.

Figure 22B. Rat. Brain (median section).*

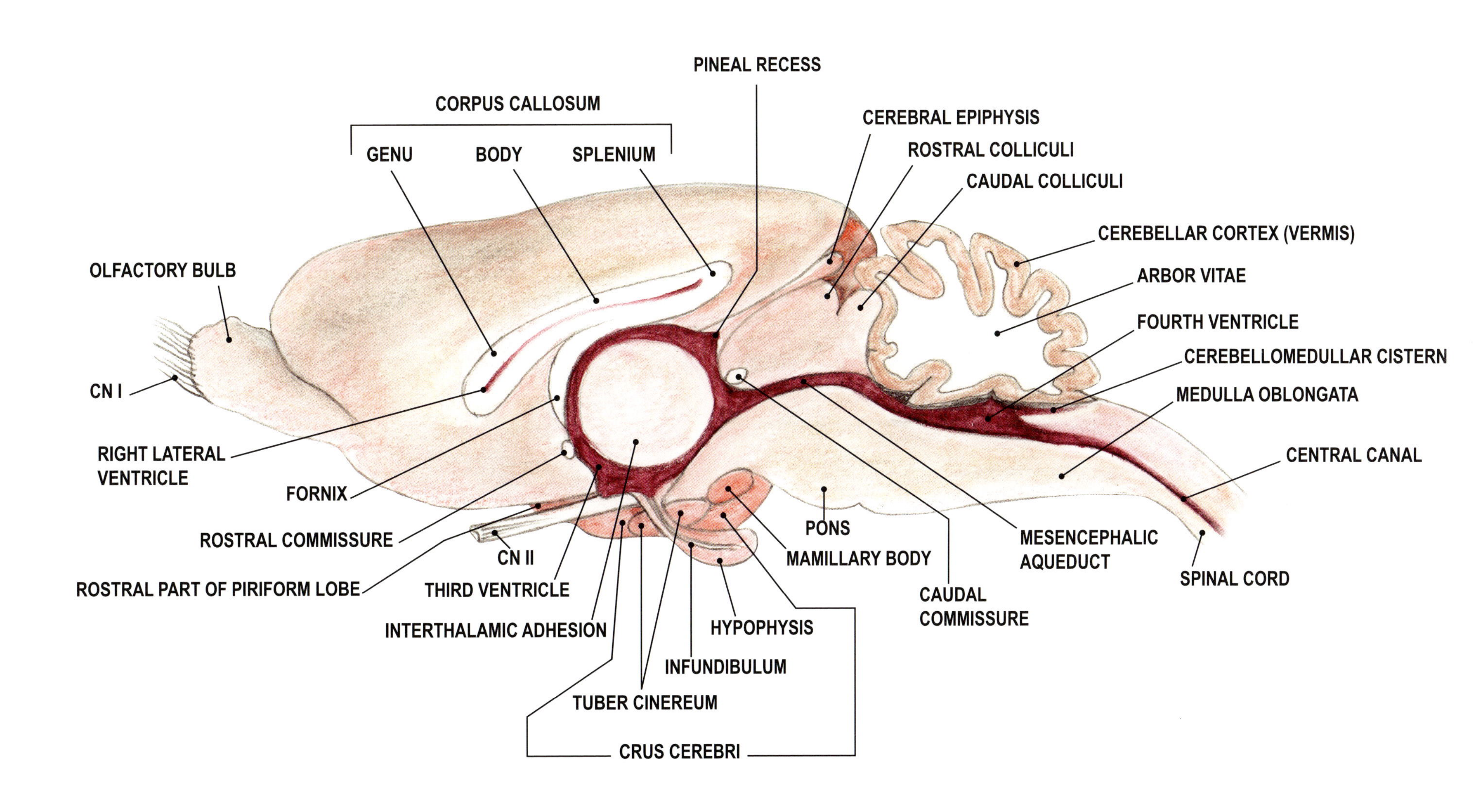

*Coloration added for illustrative purposes to give contrast between structures.

Figure 22A. Mouse. Brain (median section).*

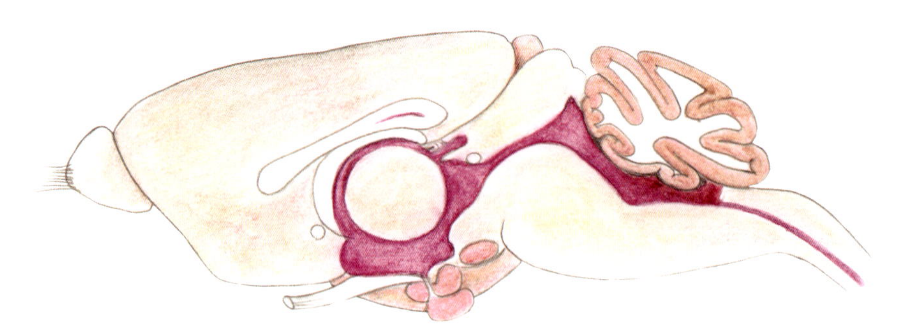

Figure 22B. Rat. Brain (median section).*

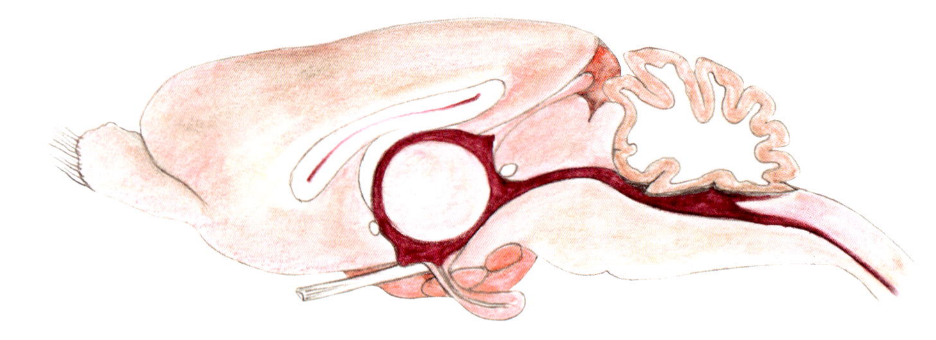

*Coloration added for illustrative purposes to give contrast between structures.

Figure 22. Brain (median section).

Differences noted in the mouse as compared to the rat:

- A larger extent of the ventral part and the caudal part of the third ventricle.
- The presence of a dorsal niche (pineal recess) of the third ventricle.
- A wider fourth ventricle.
- A superficial position of the cerebral epiphysis (pineal body) and mesencephalic tectum (rostral and caudal colliculi).

Figure 23A. Mouse. Head (median section).

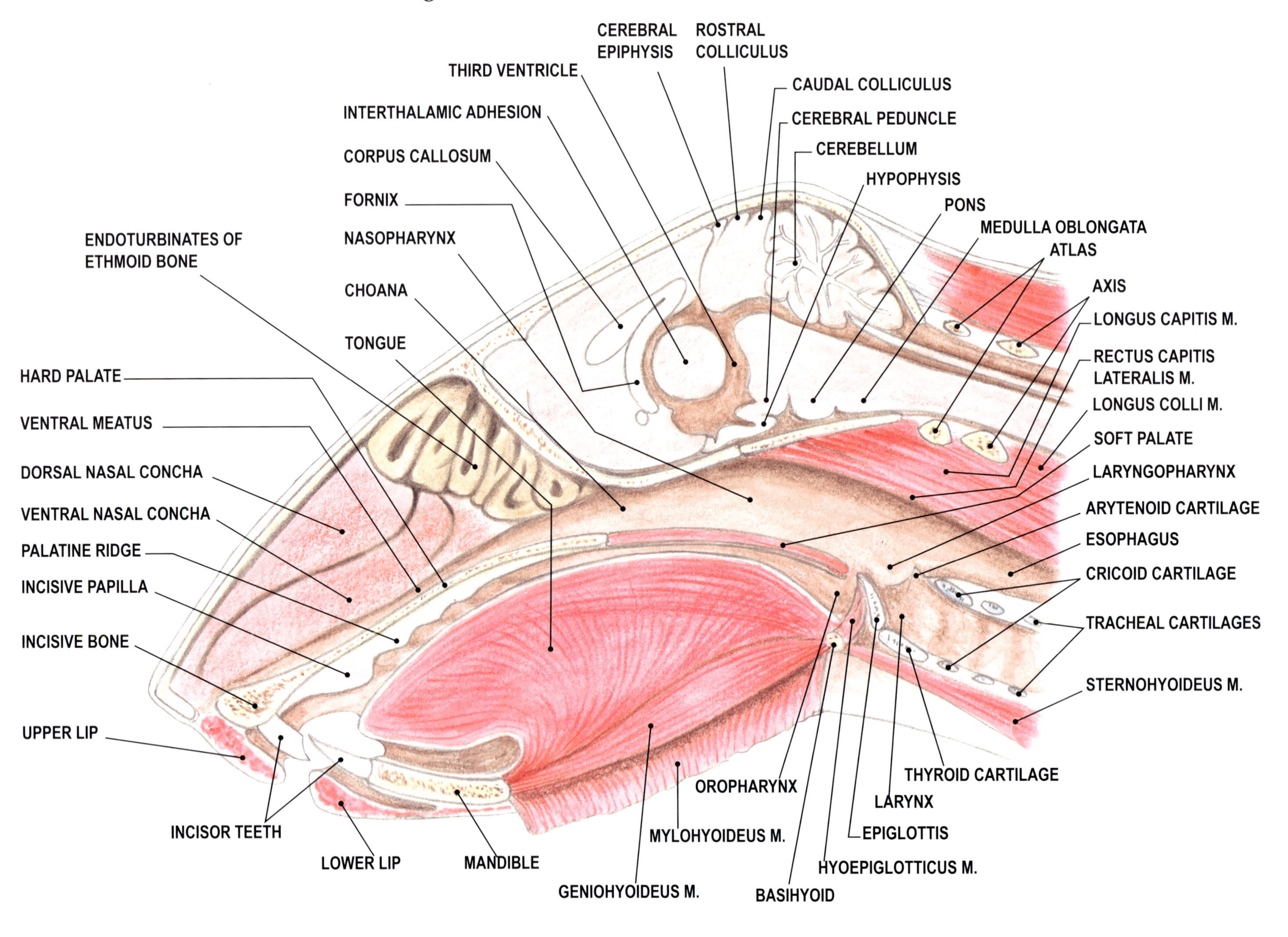

Figure 23B. Rat. Head (median section).

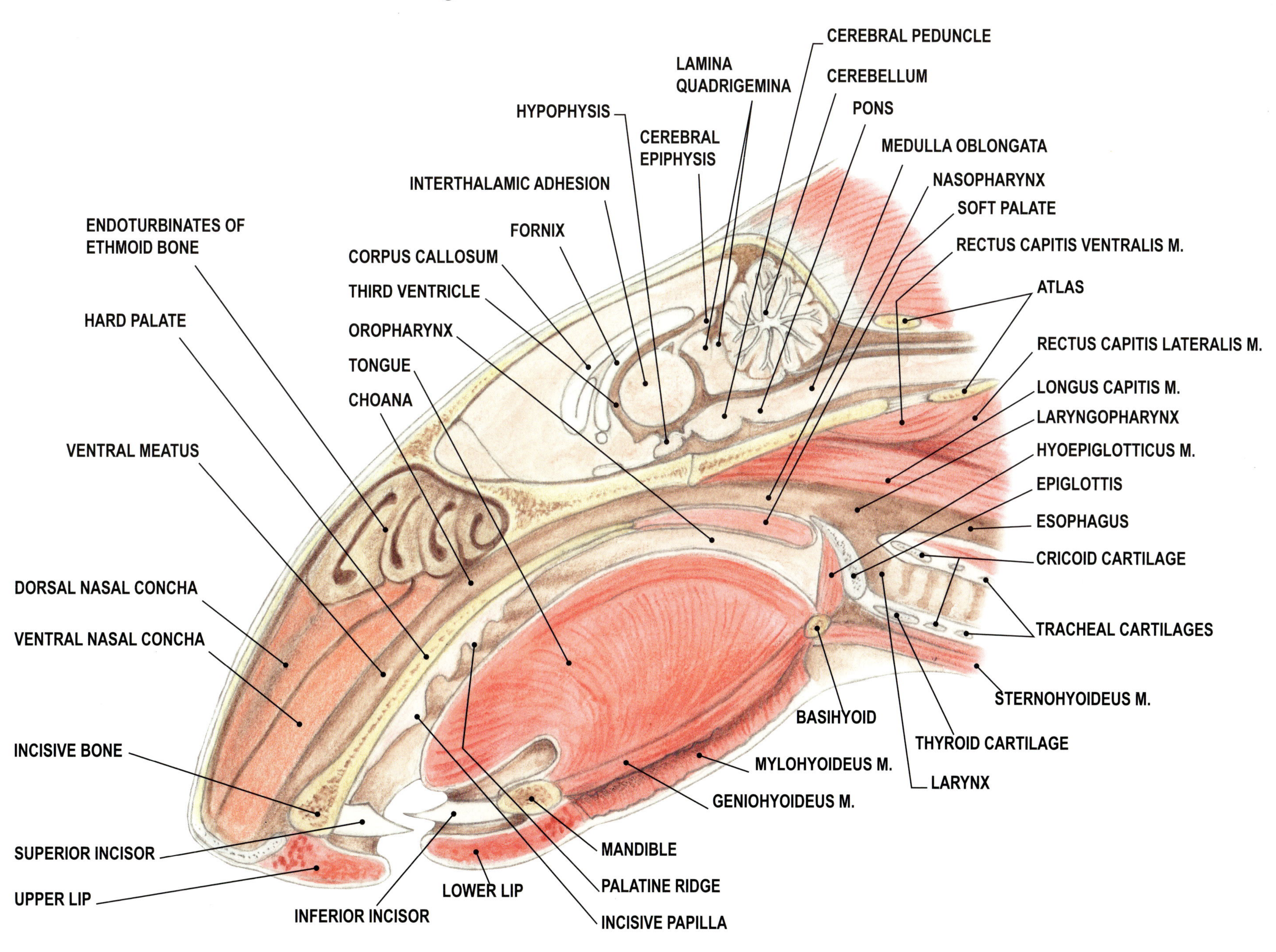

Figure 23A. Mouse. Head (median section).

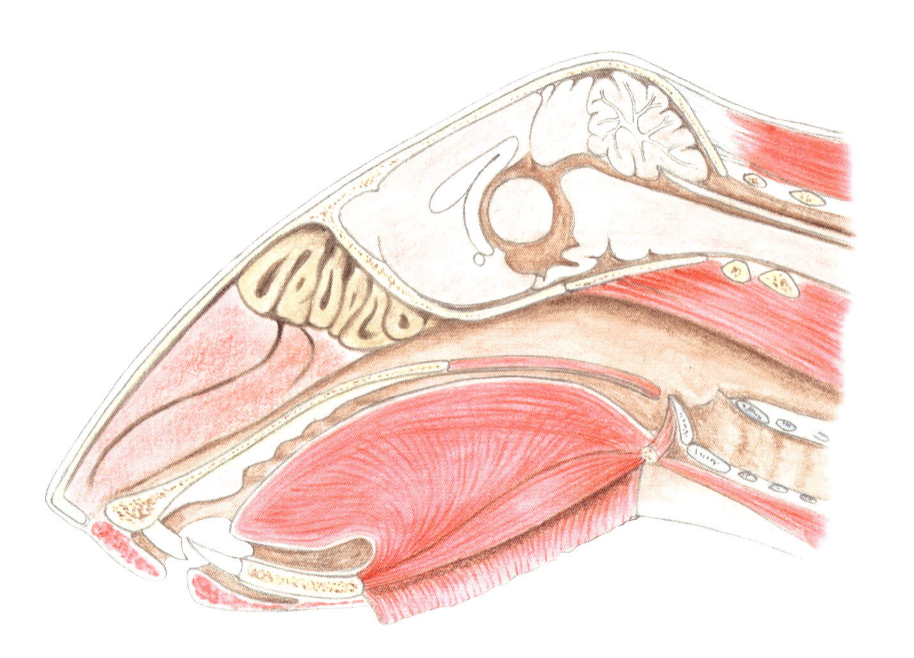

Figure 23B. Rat. Head (median section).

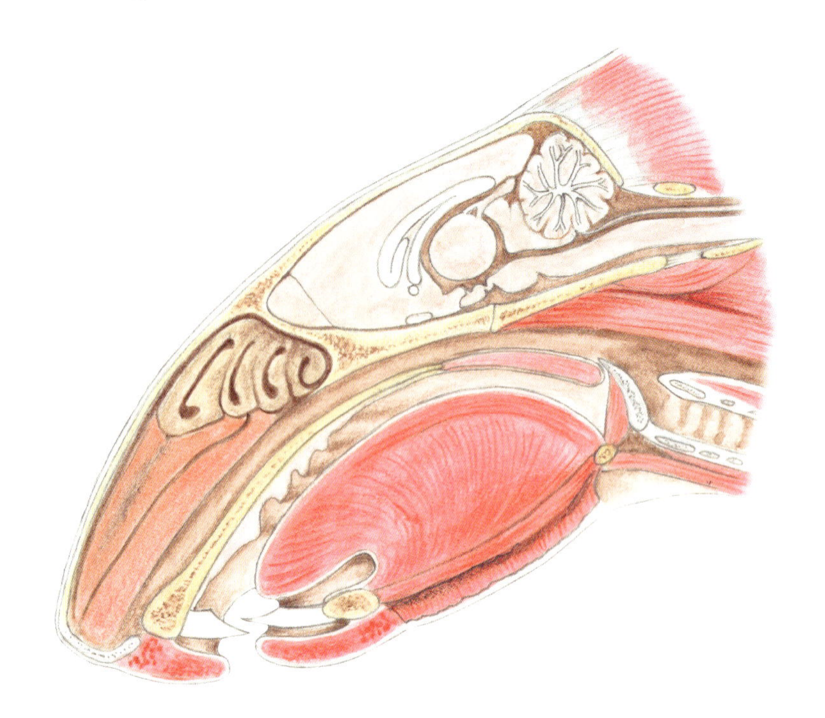

Figure 23. Head (median section).

Mouse

The narrow ventral meatus is the only passage to the nasopharynx, thus communicating with the laryngopharynx, larynx, and esophagus. The dorsal meatus, which separates the dorsal concha from the roof of the nasal cavity, is narrower.

The nasopharynx and the laryngopharynx are wide in comparison to the narrow ventral meatus.

The soft palate is long and thin.

The dorsal and ventral nasal conchae are separated by a narrow middle meatus, which ends abruptly in front of the ethmoid bone. It is difficult to distinguish the middle concha from the ethmoturbinates.

Rat

The ventral meatus is wider and more easily approachable than in the mouse. The ventral meatus continues with a nasopharynx and laryngopharynx of a similar size as that of the mouse.

The soft palate is shorter and thicker than in the mouse.

There are no noticeable differences between the rat and the mouse in the nasal conchae and meatuses. Differences between the illustrations are not representative of species distinctions.

Figure 24A. Mouse. Pharynx (rostral aspect from the open mouth).

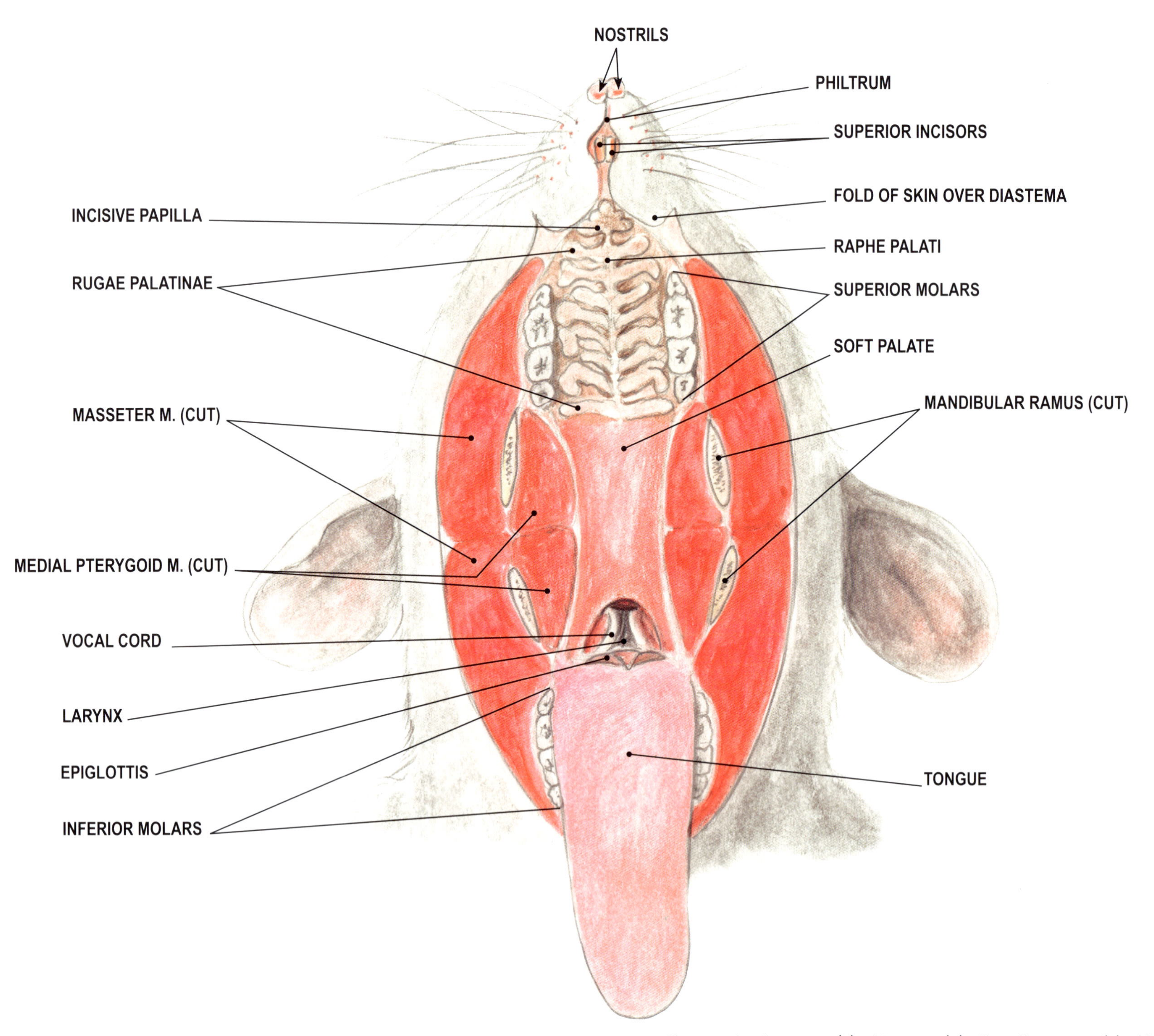

Figure 24B. Rat. Pharynx (rostral aspect from the open mouth).

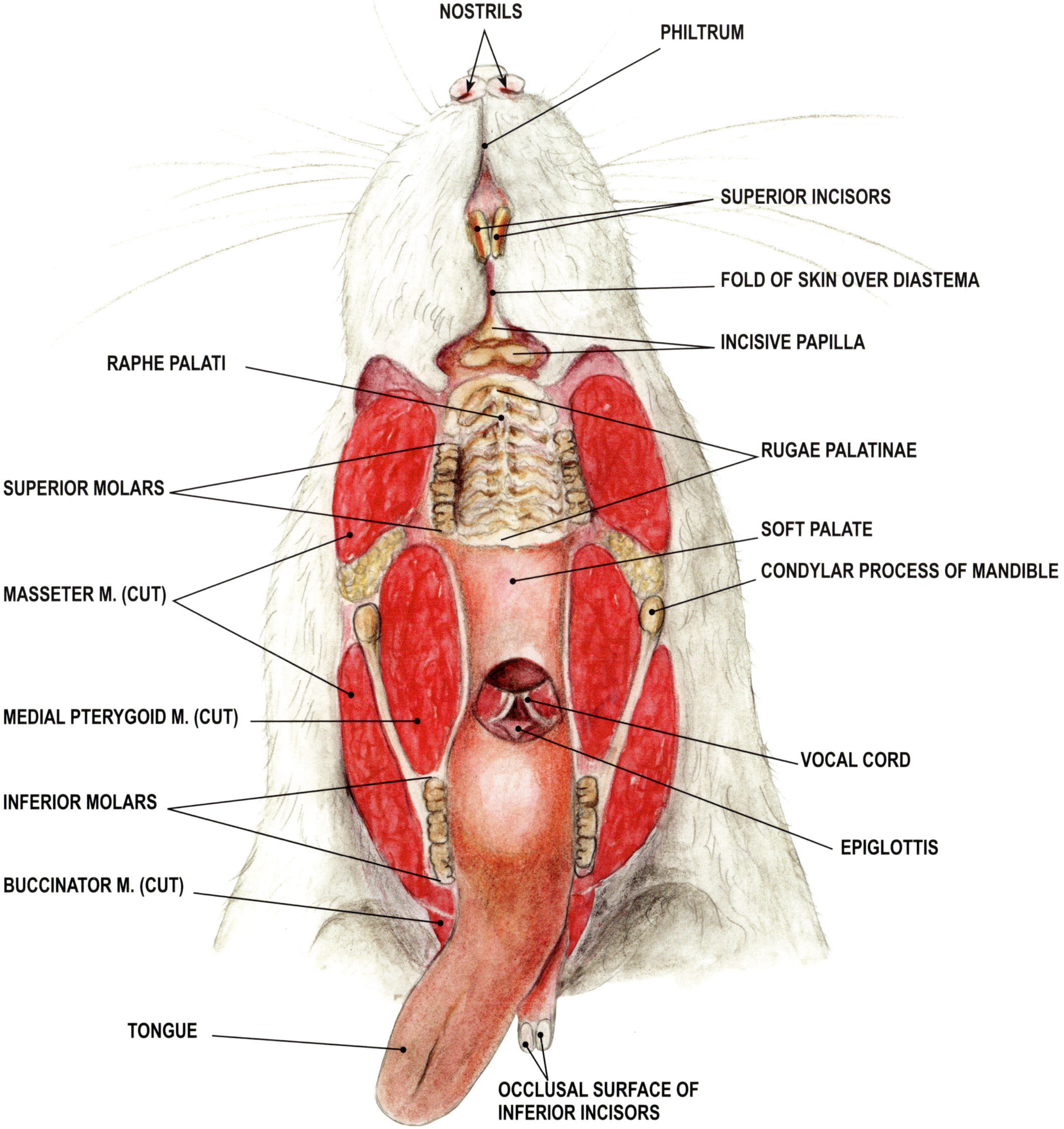

NOSTRILS
PHILTRUM
SUPERIOR INCISORS
FOLD OF SKIN OVER DIASTEMA
INCISIVE PAPILLA
RAPHE PALATI
RUGAE PALATINAE
SUPERIOR MOLARS
SOFT PALATE
CONDYLAR PROCESS OF MANDIBLE
MASSETER M. (CUT)
MEDIAL PTERYGOID M. (CUT)
VOCAL CORD
INFERIOR MOLARS
EPIGLOTTIS
BUCCINATOR M. (CUT)
TONGUE
OCCLUSAL SURFACE OF INFERIOR INCISORS

Figure 24A. Mouse. Pharynx (rostral aspect from the open mouth).

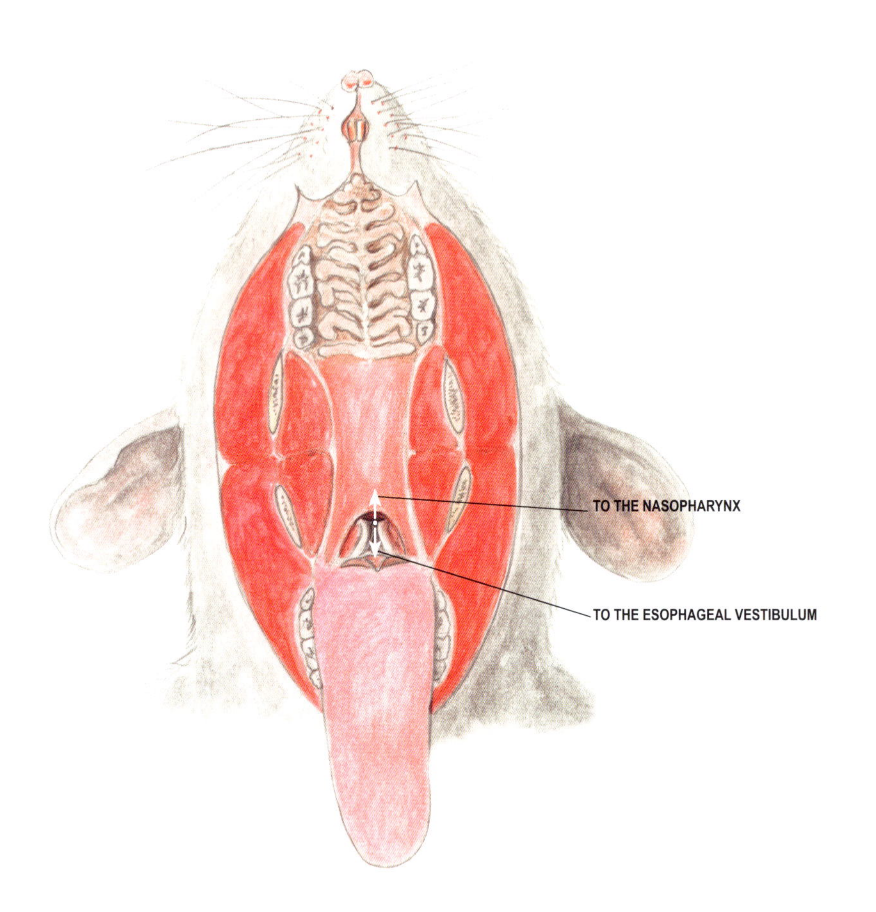

Figure 24B. Rat. Pharynx (rostral aspect from the open mouth).

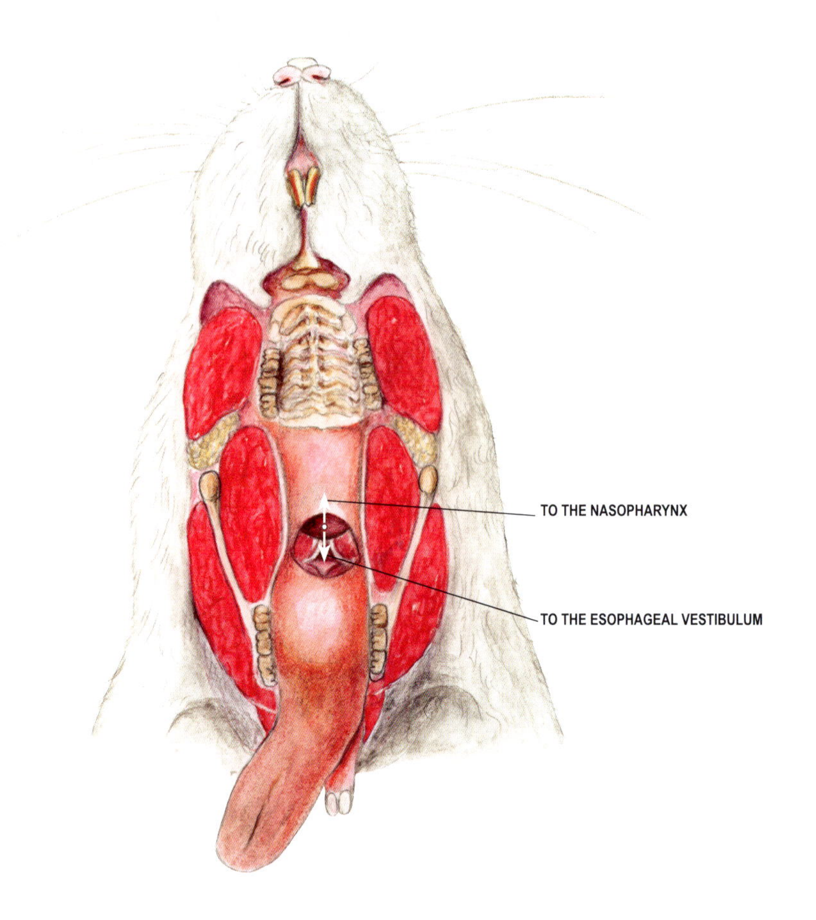

Figure 24. Pharynx (rostral aspect from the open mouth).

The unlabeled figures (opposite) include arrows showing the direction to the nasopharynx and the esophageal vestibulum.

Mouse

The incisive papilla has a lobulated contour; the papilla is connected caudally by a pedicle to the median raphe and the first pair of rugae palatinae.

The rugae palatinae are prominent and have an intricate shape. The soft palate is proportionately longer than in the rat.

There is a common opening at the caudal border of the soft palate, which leads rostrally into the nasopharynx and caudally into the esophageal vestibulum.

The vocal cords are prominent.

Rat

The incisive papilla consists of three prominences:

- one rostral, triangularly shaped; and
- two caudal, elliptical and side by side, completely separated from the rugae palatinae, unlike in the mouse.

The rugae palatinae are alike, having a low height and being almost symmetrical.

The soft palate is short and exposes a common passage, which is rostrally directed to the nasopharynx and caudally directed to the esophageal vestibulum.

The esophageal vestibulum is well outlined and is easily visualized for passing a gastric tube or an endoscope. The glottis is easily accessible for intubation or exploratory surgery.

The vocal cords are proportionately smaller than in the mouse.

Figure 25A. Mouse. Head and neck to upper thorax, including the heart and great vessels (ventral aspect).

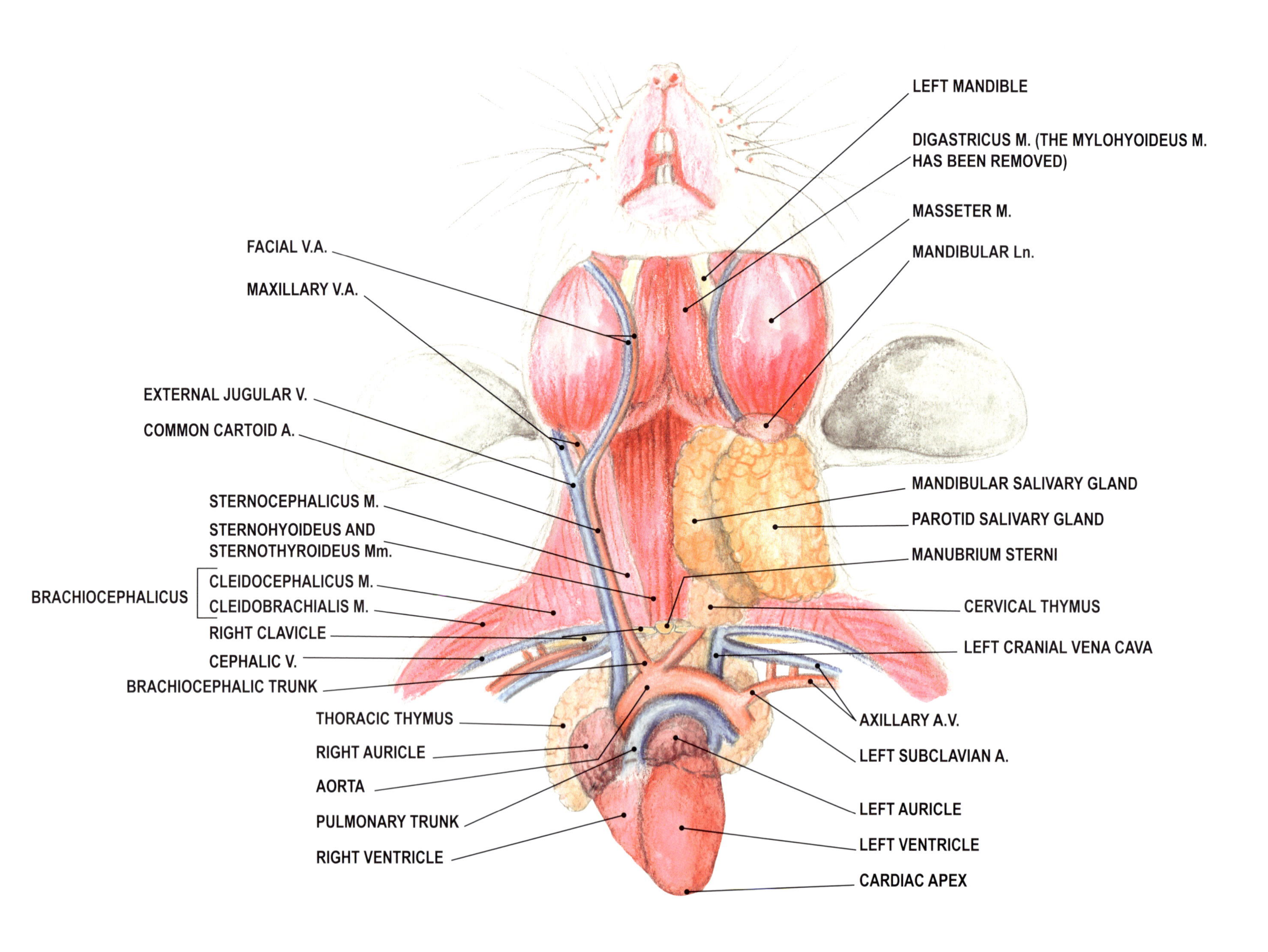

Figure 25B. Rat. Head and neck to upper thorax, including the heart and great vessels (ventral aspect).

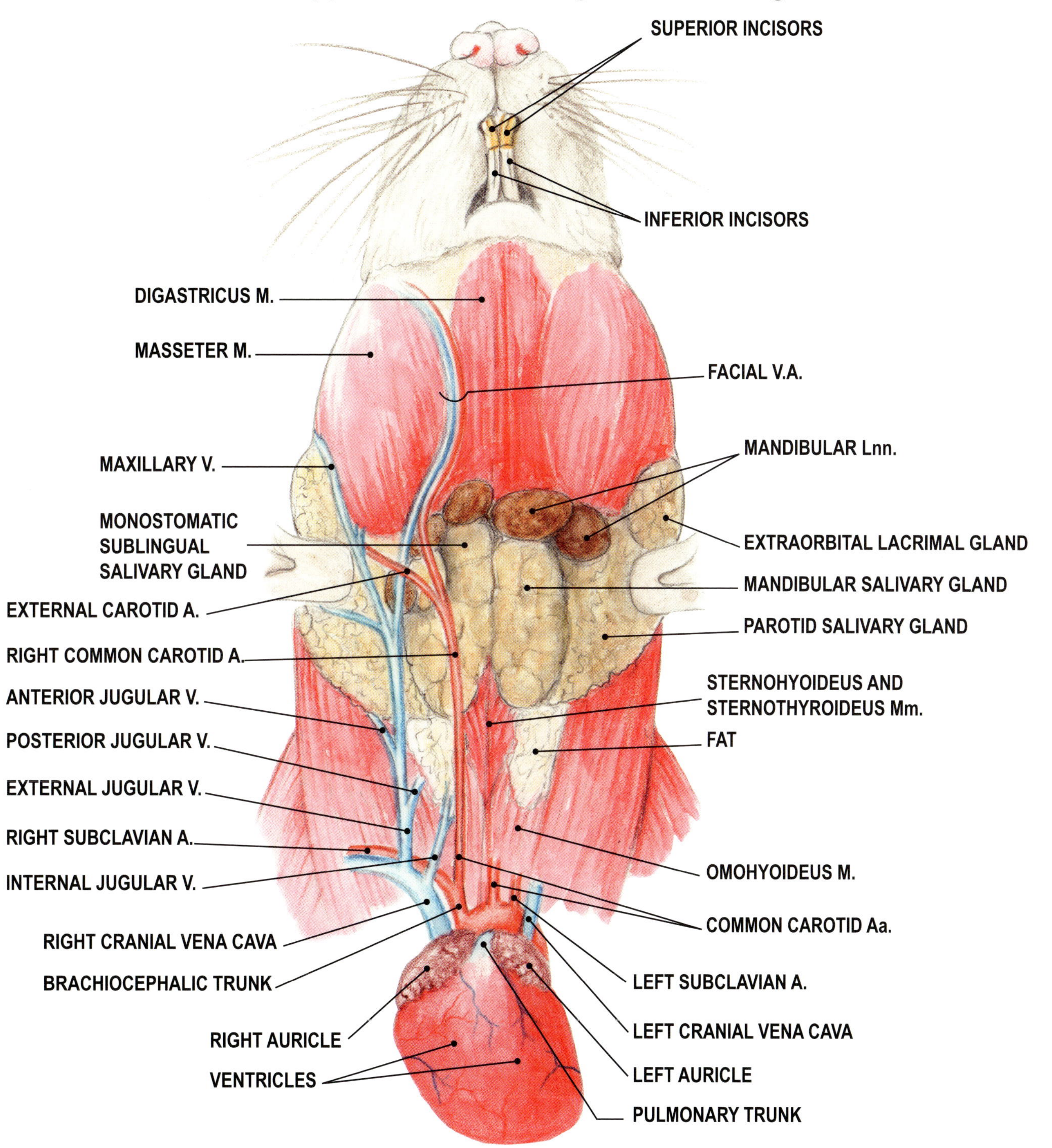

Figure 25A. Mouse. Head and neck to upper thorax, including the heart and great vessels (ventral aspect).

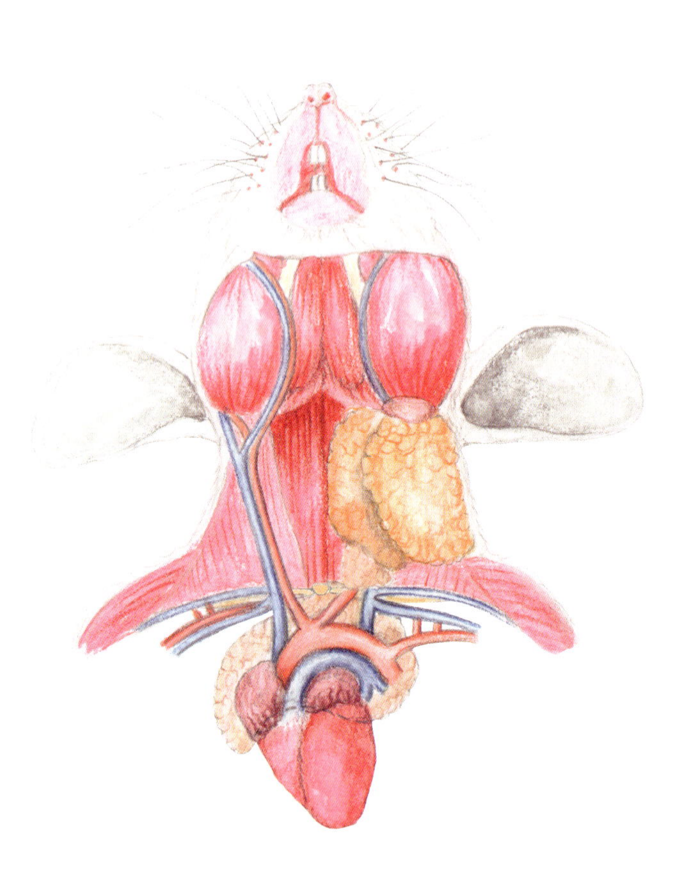

Figure 25B. Rat. Head and neck to upper thorax, including the heart and great vessels (ventral aspect).

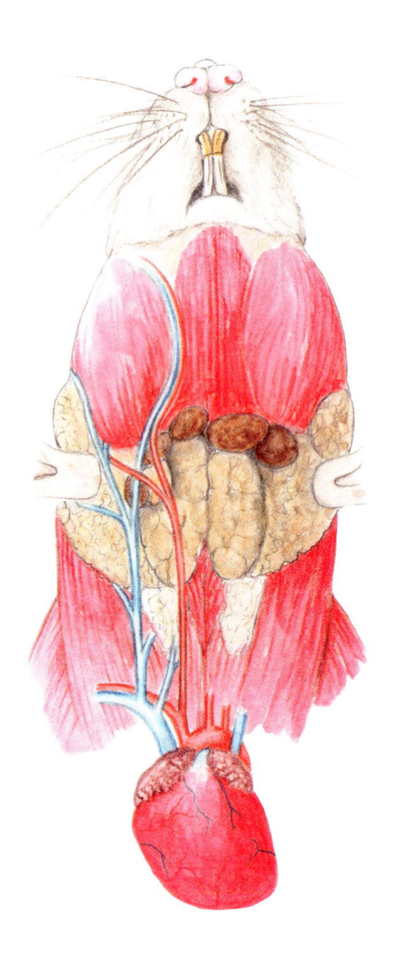

Figure 25. Head and neck to upper thorax, including the heart and great vessels (ventral aspect).

Mouse

The parotid and mandibular salivary glands, the mandibular lymph nodes, and the cervical thymus (and sometimes multilocular adipose tissue) cover the lateral half of the ventral neck.

The external jugular vein and the common carotid artery are overlapped by the salivary glands, lymph nodes, and cervical thymus, which is opposite of the relationships in the rat.

On the right ventral aspect, the external jugular vein and the common carotid artery are shown after removing the parotid and mandibular salivary glands and the cervical thymus; these vessels run parallel to each other between the sternocephalicus and the cleidocephalicus muscles.

The cephalic vein joins the external jugular vein outside of the thoracic cavity, cranial to the clavicle.

The axillary vein joins the external jugular vein caudal to the clavicle.

The right and left cranial venae cavae enter the right atrium in a dorsal position to the aortic arch.

The left common carotid artery (not labeled) originates very close to the brachiocephalic trunk.

The left subclavian artery originates further from the left common carotid artery in the mouse than in the rat.

The thoracic thymus is close to the base of the heart.

The auricular aspect of the heart is shown in this illustration.

Rat

The parotid, mandibular, and monostomatic sublingual salivary glands, the mandibular lymph nodes, and multilocular adipose tissue cover the entire ventral neck. The symmetrical structures (except the fat) come in contact with each other.

The external jugular vein and its affluents run between the medial border of the parotid salivary gland and the lateral border of the mandibular salivary gland. The anterior and posterior jugular veins join the external jugular vein (not shown), as in the mouse.

The external carotid artery originates from the common carotid artery and crosses the facial vein superficially.

The facial vein originates more caudally in the rat than in the mouse; it arises from the external jugular vein in the vicinity of, and dorsal to, the external carotid artery. The facial vein lies over one of the mandibular lymph nodes.

The left common carotid artery originates further away from the brachiocephalic trunk, but closer to the left subclavian artery, in the rat than in the mouse.

The right cranial vena cava joins the right atrium in a dorsal position to the aorta, and the left cranial vena cava is in a ventral position; in the mouse, both venae cavae are dorsal to the aorta.

Even though the thymus is not shown in the rat, it has a similar position as in the mouse, with most of it on the left side.

The auricular aspect of the heart is shown, mostly oriented to the right, which is why the pulmonary trunk is not fully exposed as in the illustration of the mouse.

The deep ventral cervical muscles shown in the rat, and not labeled, are similar in the mouse.

Heart, Vascular Tree, and Respiratory Tract

Figure 26A. Mouse. Projection of the thoracic viscera (left aspect).

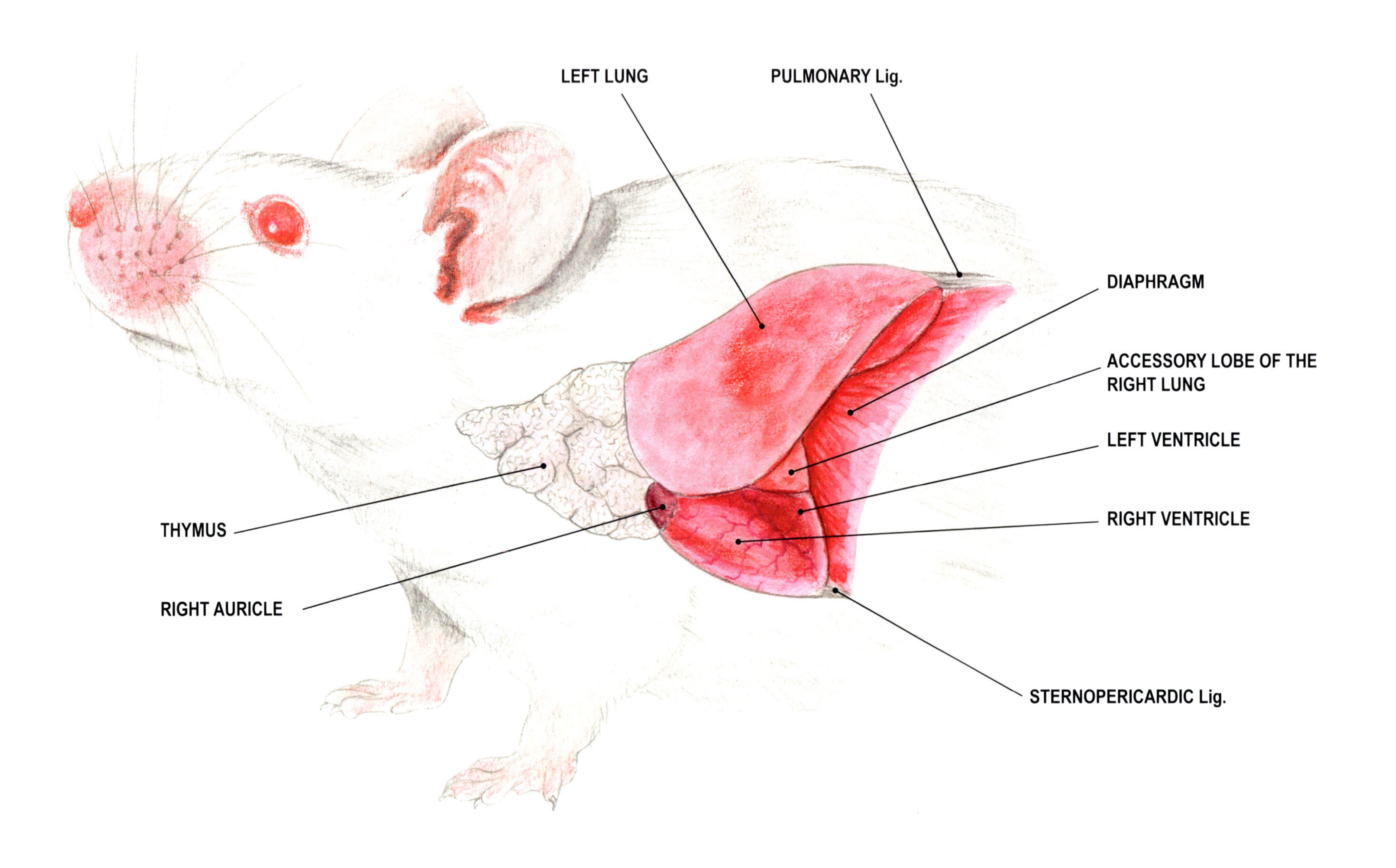

Figure 26B. Rat. Projection of the thoracic viscera (left aspect).

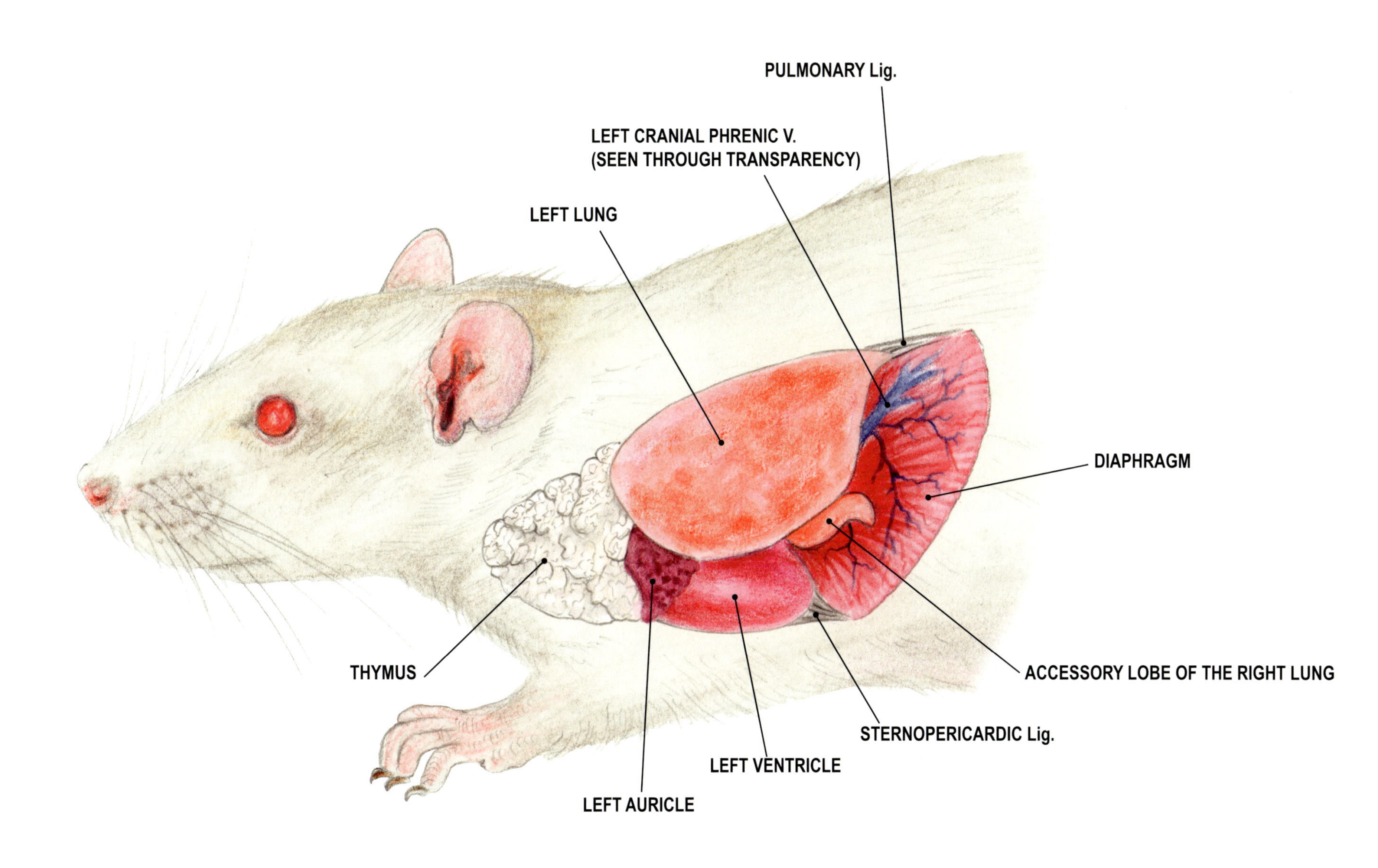

Figure 26A. Mouse. Projection of the thoracic viscera (left aspect).

Figure 26B. Rat. Projection of the thoracic viscera (left aspect).

Figure 26. Projection of the thoracic viscera (left aspect).

The size and shape of the thoracic viscera are illustrated as in life (the heart filled with blood and the lungs inflated with air). The position and location of the heart is shown in Figure 32. The cranial extent of the diaphragm and abdominal viscera is related to multiple factors, including the position of the body (whether flexed, extended, or straight).

Because these specimens are young adults, a large thymus is shown. The size of the thymus is related to the age of the individual, being larger in juvenile animals. Following sexual maturity, the thymus declines in size through involution.

Mouse

The thoracic cavity is proportionately longer in the mouse than in the rat (see Figure 59), which explains why in the mouse the lungs have a greater extent within the ribcage.

The left lung is not divided into lobes.

The right lung has 4 lobes (cranial, middle, caudal, and accessory of caudal lobe). The accessory lobe covers the left auricle of the heart.

The thymus fills the rest of the space cranial to the heart and the lung.

Rat

The left lung is not divided into lobes, and it covers the base of the left ventricle.

The right lung has 4 lobes (cranial, middle, caudal, and accessory of caudal lobe). A portion of the accessory lobe is shown caudal to the heart.

The thymus fills the rest of the space cranial to the heart and the lung.

Figure 27A. Mouse. Projection of the rib cage and the thoracic viscera (left aspect).

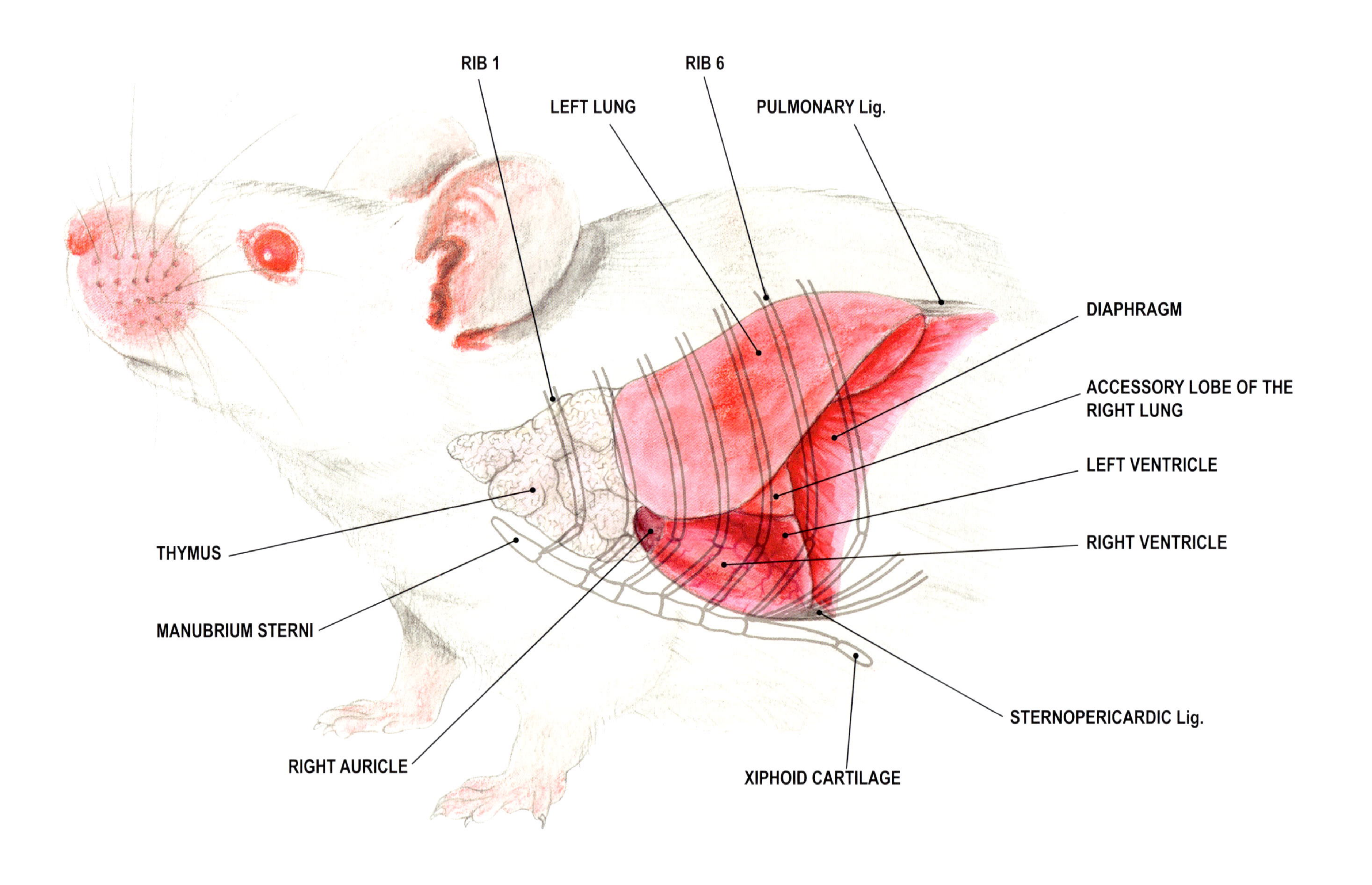

Figure 27B. Rat. Projection of the rib cage and the thoracic viscera (left aspect).

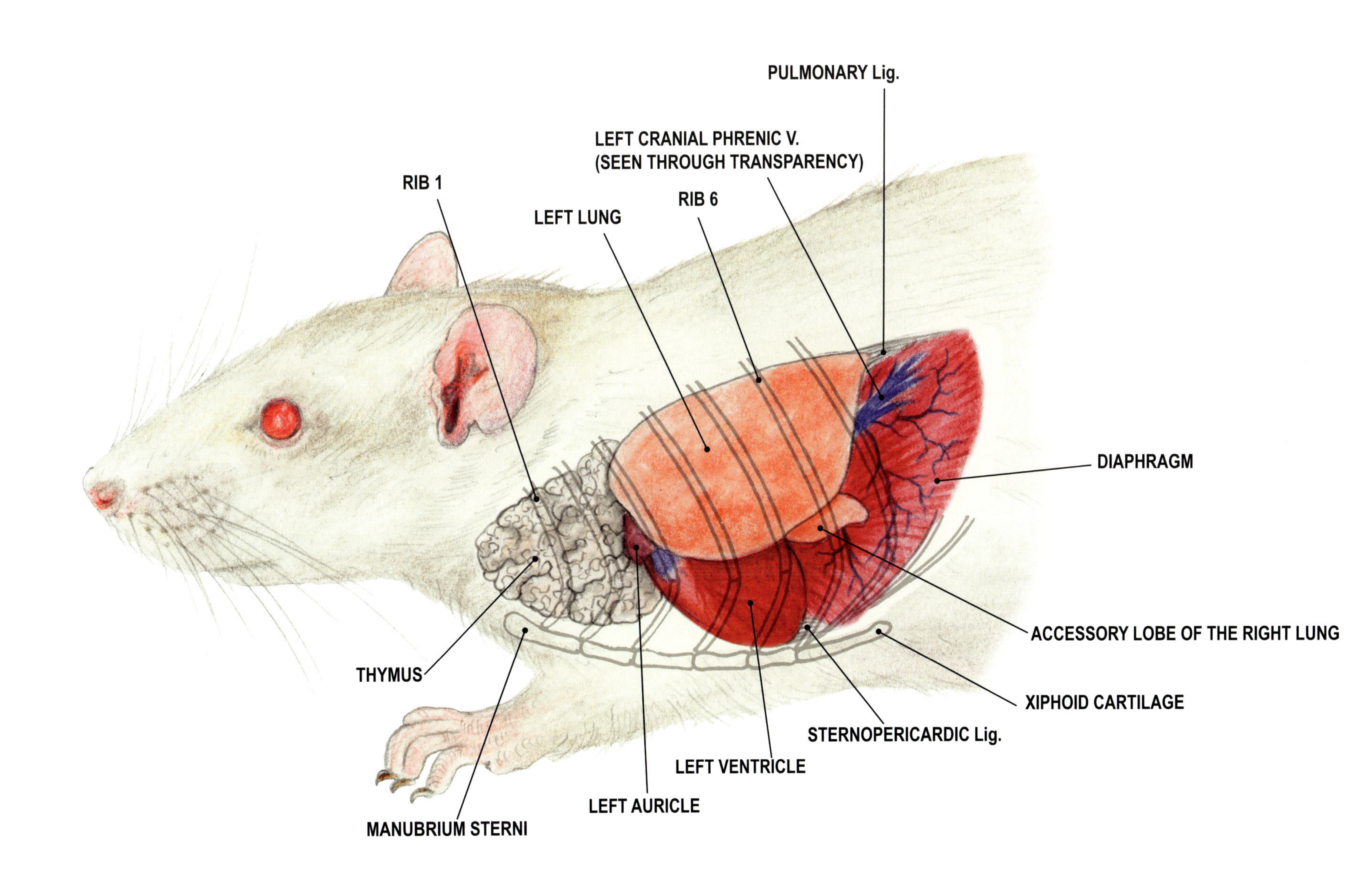

Figure 27A. Mouse. Projection of the rib cage and the thoracic viscera (left aspect).

Figure 27B. Rat. Projection of the rib cage and the thoracic viscera (left aspect).

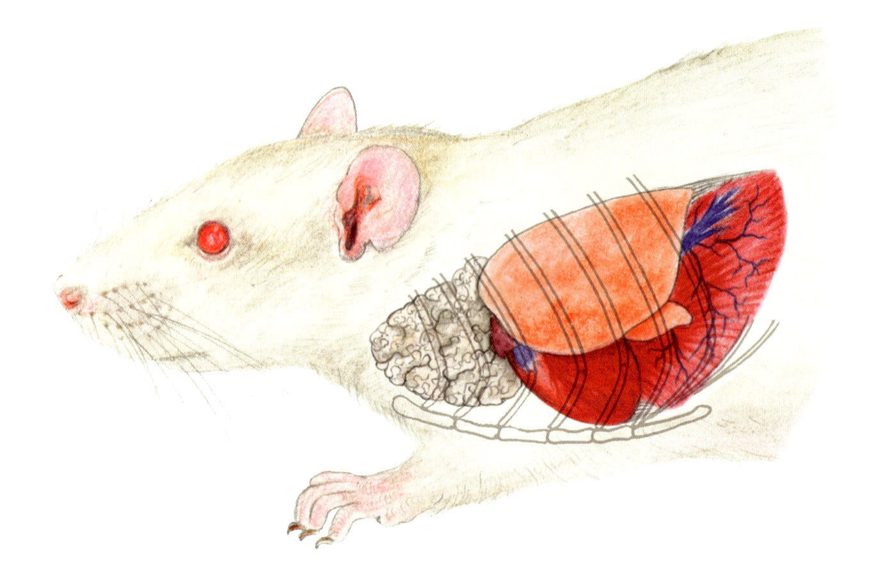

Figure 27. Projection of the rib cage and the thoracic viscera (left aspect).

In both figures, ribs 1 to 7 are shown in their entirety; rib cartilages are added for ribs 8 to 9; the remaining ribs are omitted. See page 103 for more information about these figures.

Mouse

The left lung is not divided into lobes. The left lung extends to the 2nd rib cranially, up to the coronary groove of the heart, and caudally to the 11th rib.

The right lung has 4 lobes (cranial, middle, caudal, and accessory of caudal lobe). Only the accessory lobe can be seen from this perspective, covering the left auricle of the heart.

The caudal end of the thoracic cavity extends up to the 13th rib. Even in maximal inspiration, the lung does not fill the entire capacity of the thoracic cavity caudally. The diaphragm and the liver (not shown) occupy the caudal region of the rib cage. The viscera under the ribs extend cranial to the cupula of the diaphragm, which is subject to move with the respiration.

The thymus fills the rest of the space cranial to the heart and the lung.

Rat

The left lung is not divided into lobes. The left lung fills the space between the 3rd rib cranially and the 8th rib caudally. It covers the base of the left ventricle.

The right lung has 4 lobes (cranial, middle, caudal, and accessory of caudal lobe). Only the accessory lobe can be seen caudal to the heart, in the 5th intercostal space and half of the 6th intercostal space.

The thymus fills the rest of the space cranial to the heart and the lung.

Figure 28A. Mouse. Projection of the thoracic viscera (right aspect).

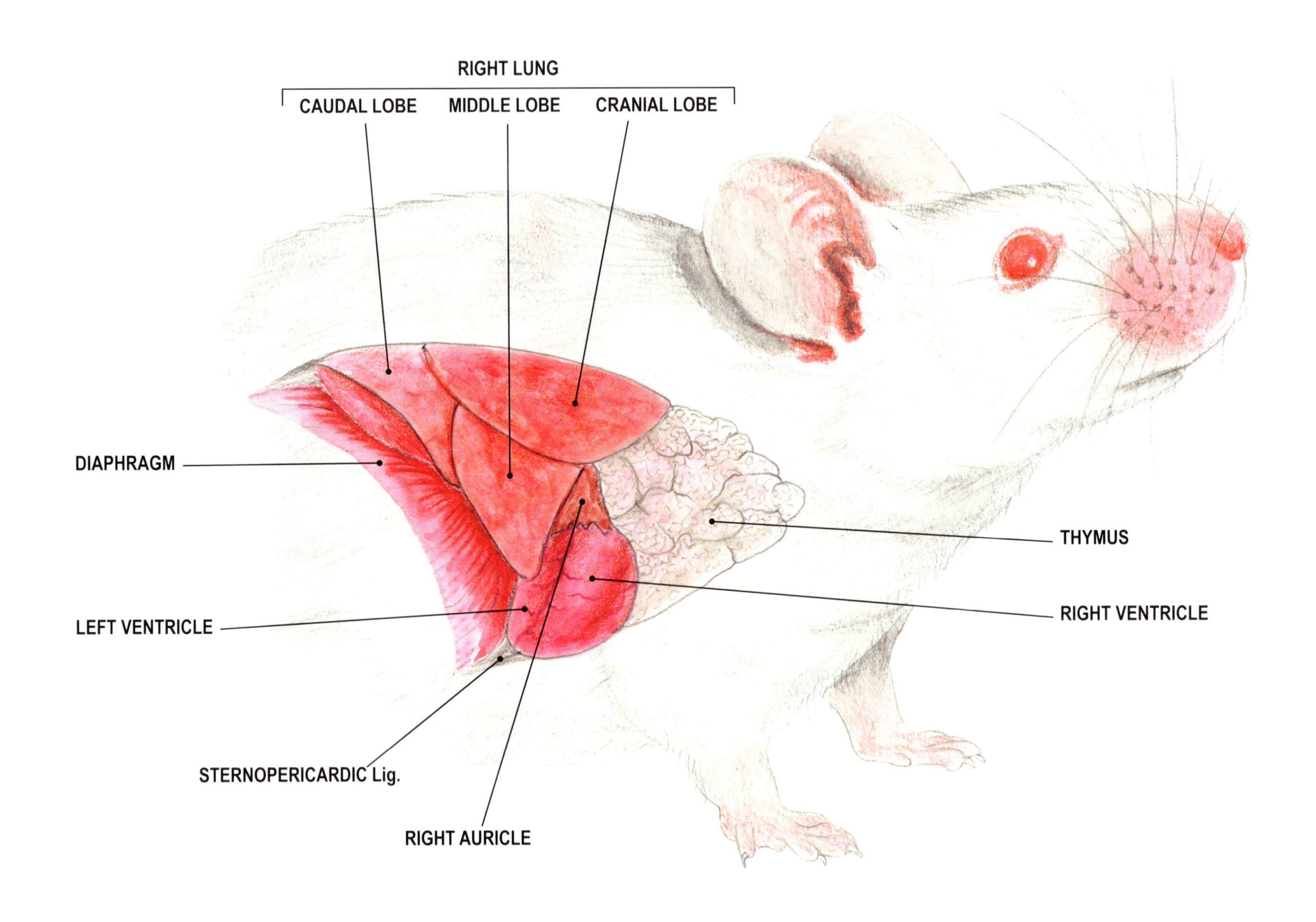

Figure 28B. Rat. Projection of the thoracic viscera (right aspect).

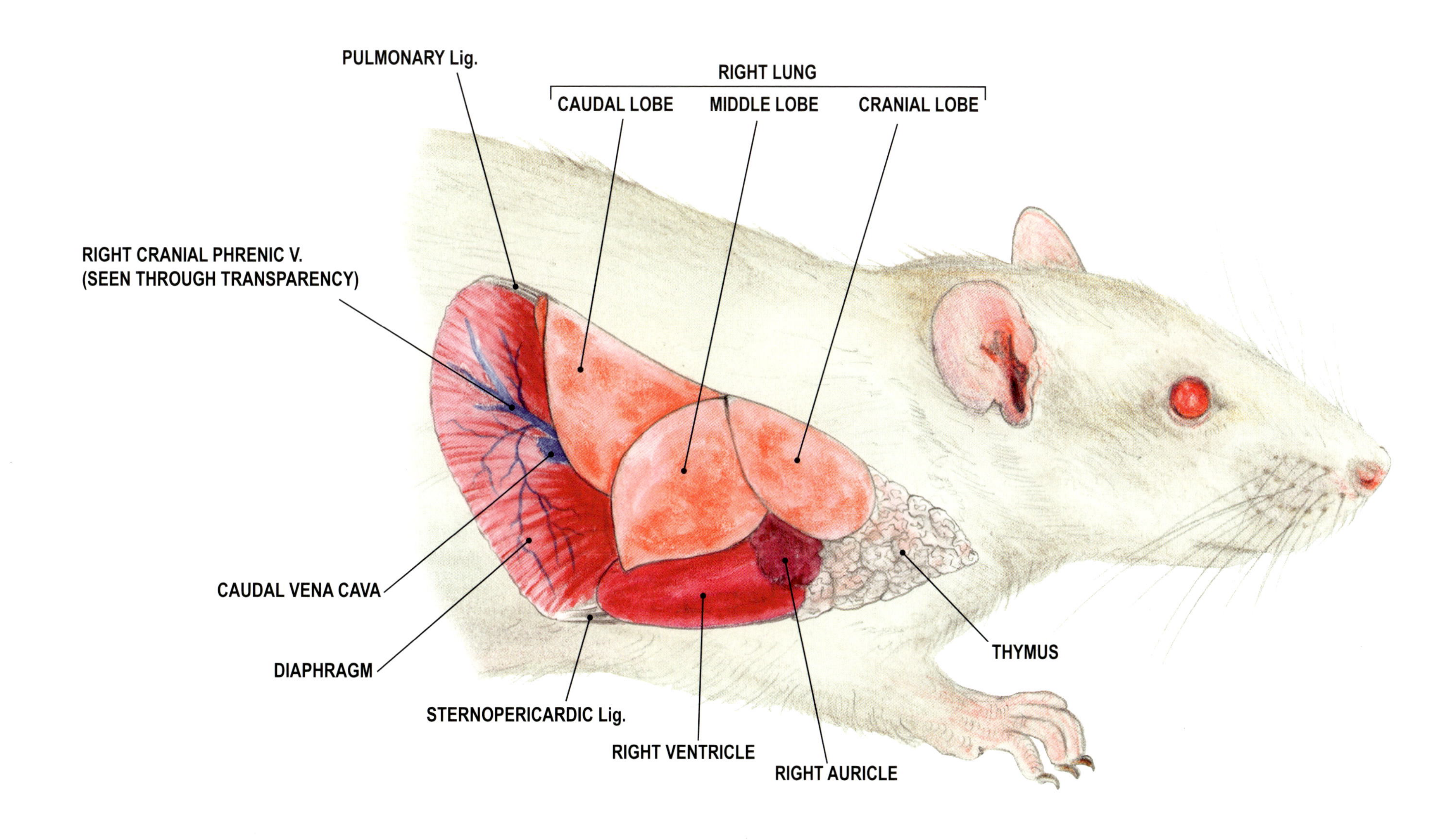

Figure 28A. Mouse. Projection of the thoracic viscera (right aspect).

Figure 28B. Rat. Projection of the thoracic viscera (right aspect).

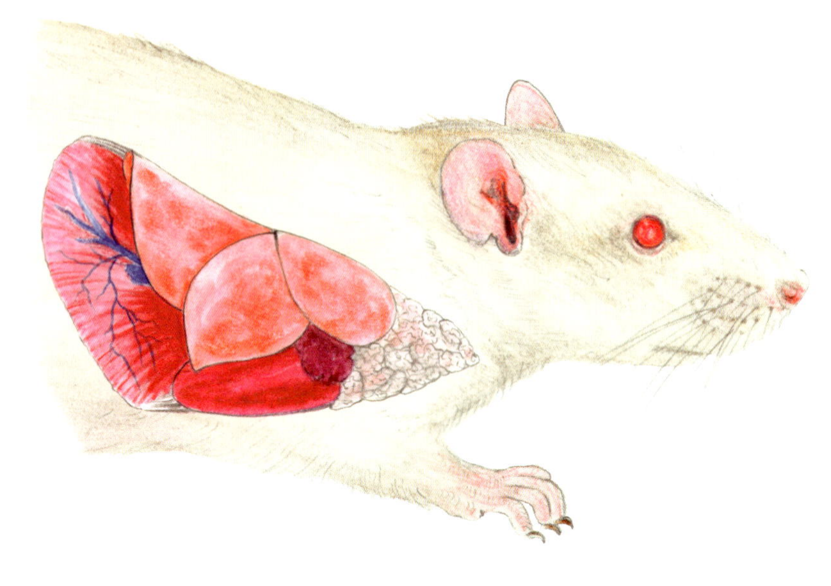

Figure 28. Projection of the thoracic viscera (right aspect).

See page 103 for information about these figures.

Mouse

Most of the right ventricle and part of the left ventricle are exposed.

The right lung has 4 lobes (cranial, middle, caudal, and accessory of caudal lobe).

- The cranial lobe is located in the dorsal third of the thoracic cavity.
- The middle lobe is located ventral to the cranial lobe; it is triangularly shaped, orienting the base dorsally and the tip ventrally. The tip extends to the middle of the left ventricle.
- The caudal lobe fills the space between the caudal extents of the cranial and middle lobes and the diaphragm.
- Only the accessory lobe is not visible. See Figures 26 and 30, where it is seen on the left and ventral aspects.

The thymus fills the space cranial to the heart and the lung.

The pulmonary ligament is not shown in the mouse, as in the rat, because it is covered by the caudal extent of the lung.

Rat

The right ventricle and the apex of the heart are exposed.

The right lung has 4 lobes (cranial, middle, caudal, and accessory of caudal lobe).

- The cranial lobe is located in the dorsal third of the thoracic cavity.
- The middle lobe is located ventral to the previous lobe; it is triangularly shaped, with the base oriented dorsally and the two convex borders meeting in the tip ventrally. The tip extends to the caudal border of the left ventricle.
- The accessory lobe is not shown in this figure. See Figures 26 and 30, where it is seen on the left and ventral aspects.
- The caudal lobe fills the space between the caudal extent of the middle lobe and the diaphragm.

The thymus fills the space cranial to the heart and the cranial lobe of the right lung.

Figure 29A. Mouse. Projection of the rib cage and the thoracic viscera (right aspect).

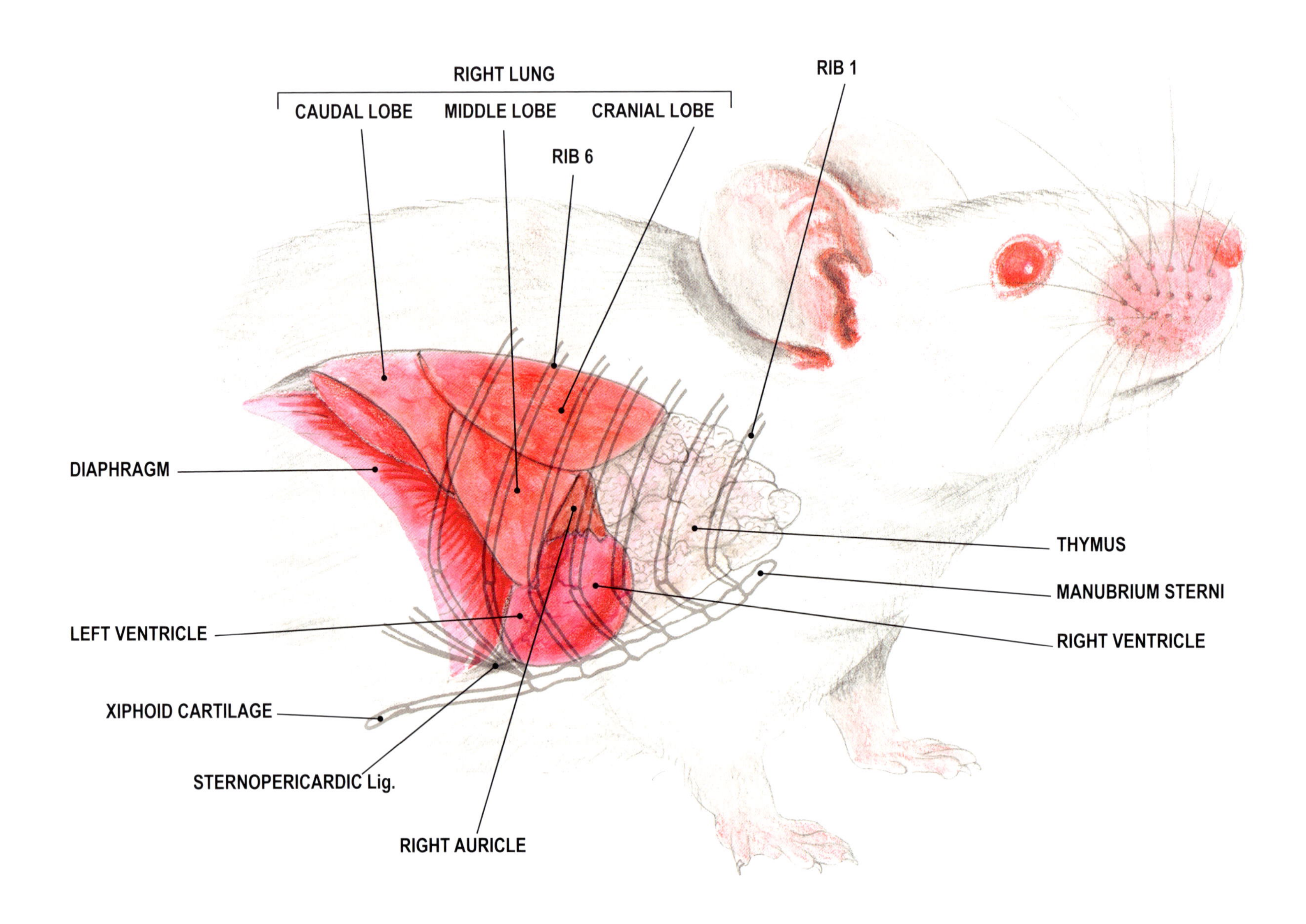

Figure 29B. Rat. Projection of the rib cage and the thoracic viscera (right aspect).

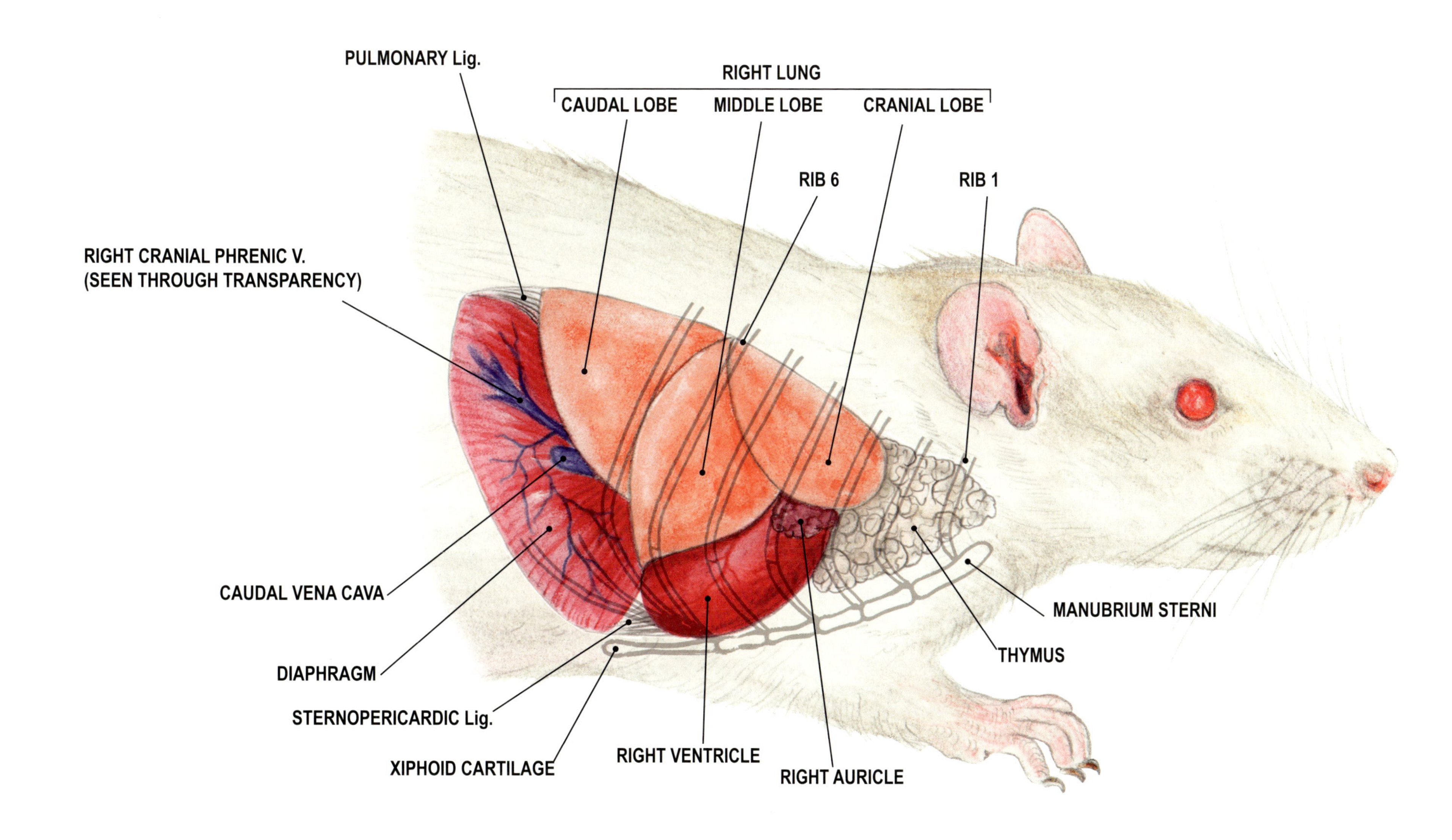

Figure 29A. Mouse. Projection of the rib cage and the thoracic viscera (right aspect).

Figure 29B. Rat. Projection of the rib cage and the thoracic viscera (right aspect).

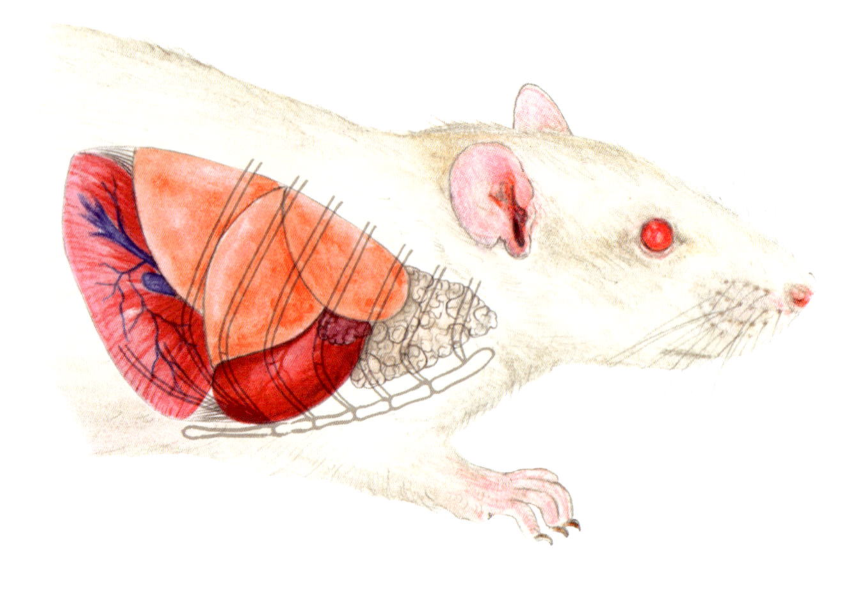

Figure 29. Projection of the rib cage and the thoracic viscera (right aspect).

In both figures, ribs 1 to 7 are shown in their entirety; rib cartilages are drawn for ribs 8 to 9; the remaining ribs are omitted. See page 103 for more information about these figures.

Mouse

Most of the right ventricle and part of the left ventricle are exposed.

The right lung has 4 lobes (cranial, middle, caudal, and accessory of caudal lobe).

- The cranial lobe of the right lung can be outlined in the dorsal third of the thoracic cavity between the roof of the thoracic cavity and a convex line starting from the 3rd rib to the 8th rib.
- The middle lobe is located ventral to the cranial lobe, between the 4th and the 7th ribs. It is triangularly shaped, orienting the base dorsally and the tip ventrally. The tip extends up to the middle of the left ventricle.
- The caudal lobe fills the space between the caudal extents of the cranial and middle lobes and the 12th rib.
- The accessory lobe is not seen in this view. See Figures 27 and 31, where it is seen on the left and ventral aspects.

The thymus fills the space cranial to the heart and the cranial lobe of the lung.

The pulmonary ligament is not shown in the mouse, as in the rat, because it is covered by the caudal extent of the lung.

Rat

The right ventricle, right auricle, and the apex of the heart are exposed.

The right lung has 4 lobes (cranial, middle, caudal, and accessory of caudal lobe).

- The cranial lobe can be outlined in the dorsal third of the thoracic cavity between the roof of the thoracic cavity and a convex line starting from the 2nd intercostal space to the 6th rib.
- The middle lobe is located ventral to the cranial lobe, between the 4th and the 7th ribs. It is triangularly shaped, with the base oriented dorsally and the two convex borders meeting in the tip ventrally. The tip extends to the caudal border of the left ventricle.
- The caudal lobe fills the space between the caudal extent of the middle lobe and the 9th rib.
- The accessory lobe is not shown in this figure. See Figures 27 and 31, where it is seen on the left and ventral aspects.

The thymus fills the space cranial to the heart and the cranial lobe of the lung.

Figure 30A. Mouse. Projection of the thoracic viscera (ventral aspect).

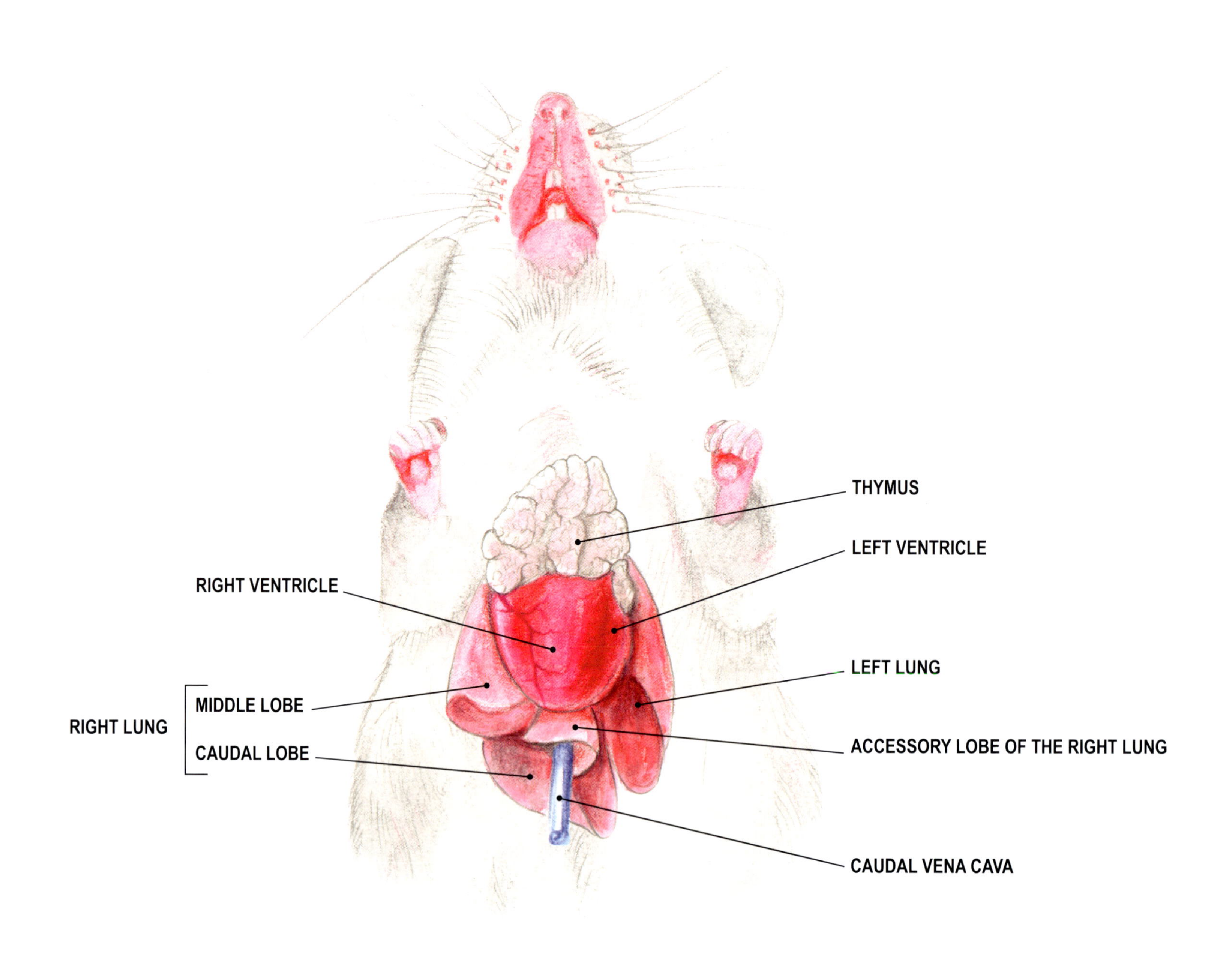

Figure 30B. Rat. Projection of the thoracic viscera (ventral aspect).

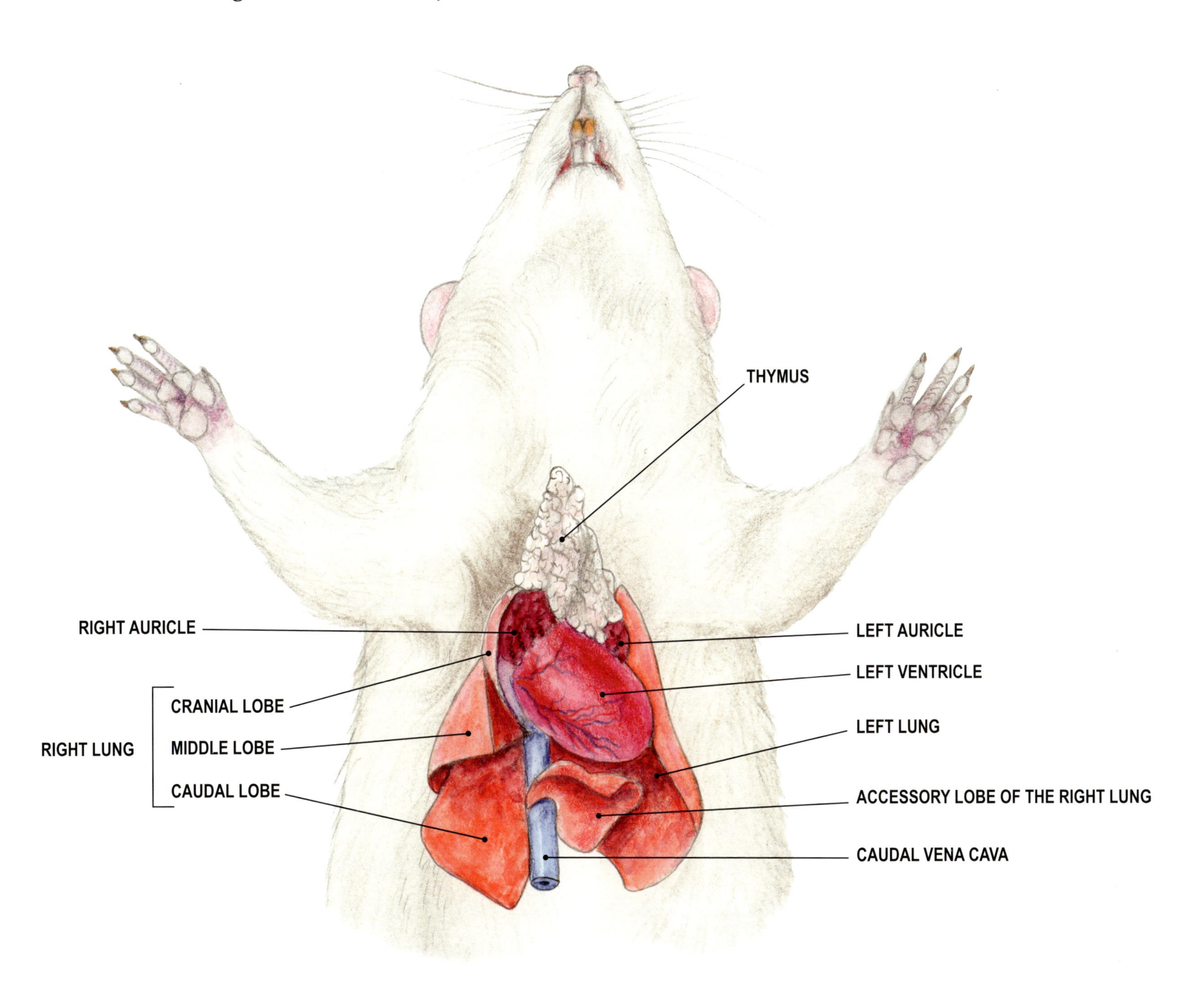

Figure 30A. Mouse. Projection of the thoracic viscera (ventral aspect).

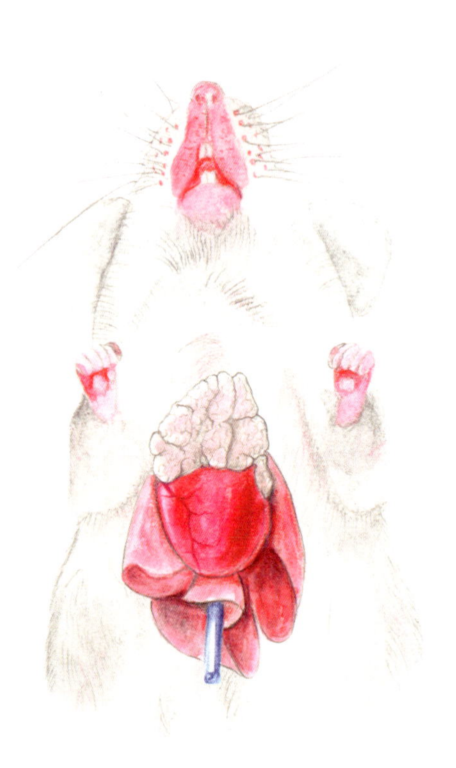

Figure 30B. Rat. Projection of the thoracic viscera (ventral aspect).

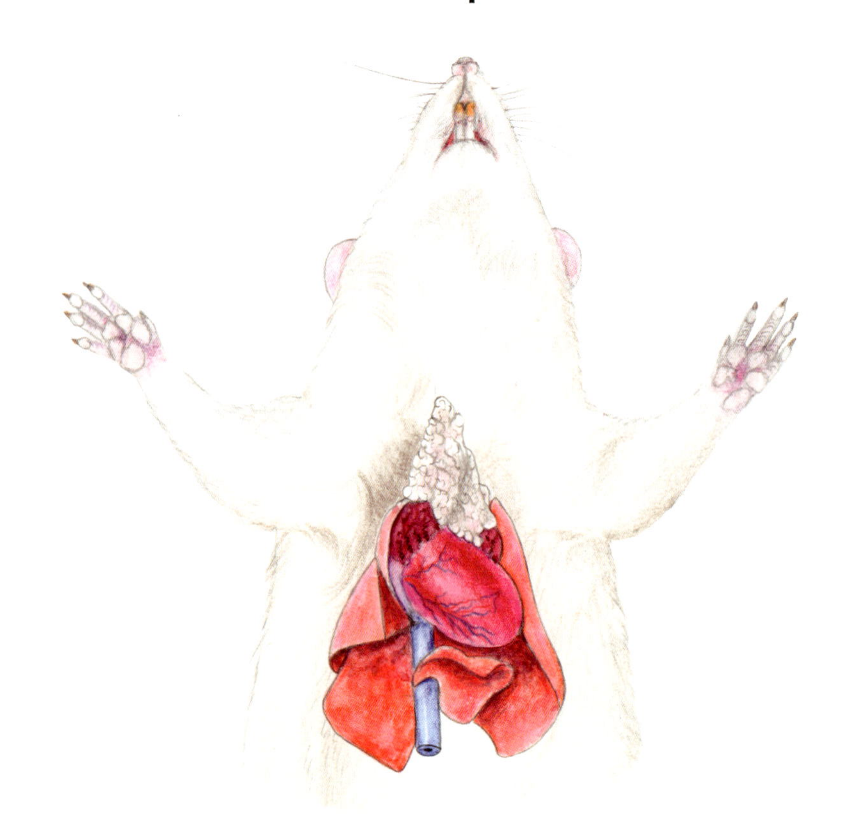

Figure 30. Projection of the thoracic viscera (ventral aspect).

See page 103 for information about these figures.

Mouse

The thymus occupies the cranial third of the thorax and overlaps the base of the heart, concealing much of the left auricle.

The heart extends caudally of the thymus border in the middle of the thoracic cavity.

The left lung extends caudally of the heart and thymus; it is not divided into lobes.

The right lung consists of 4 lobes (cranial, middle, caudal, and accessory of caudal lobe).

- The cranial lobe is concealed by the thymus and heart.
- The middle lobe extends from the right side of the heart.
- The caudal lobe of the right lung extends caudally from the middle lobe and passes into the caudal third of the left half of the thoracic cavity.
- The accessory lobe occupies a small area caudal to the heart and between the middle lobe and the left lung. The accessory lobe curves around the caudal vena cava.

Rat

The thymus occupies the ventral space of the cranial mediastinum. In this figure, the thymus is proportionately narrower in the rat than in the mouse, because of the specimens' difference in age. Caudally, the thymus does not overlap the base of the heart as extensively as in the mouse.

The heart extends caudally from the thymus border to the middle of the thoracic cavity.

The left lung extends caudally from the heart and thymus to the diaphragm (not shown).

The right lung consists of 4 lobes (cranial, middle, caudal, and accessory of caudal lobe).

- Most of the cranial lobe is concealed by the heart.
- The middle lobe extends from the right side of the heart.
- The caudal lobe extends caudally from the middle lobe to the diaphragm (not shown).
- The accessory lobe occupies a small area caudal to the heart and between the middle lobe and the left lung. The accessory lobe curves around the caudal vena cava.

Figure 31A. Mouse. Projection of the rib cage and thoracic viscera (ventral aspect).

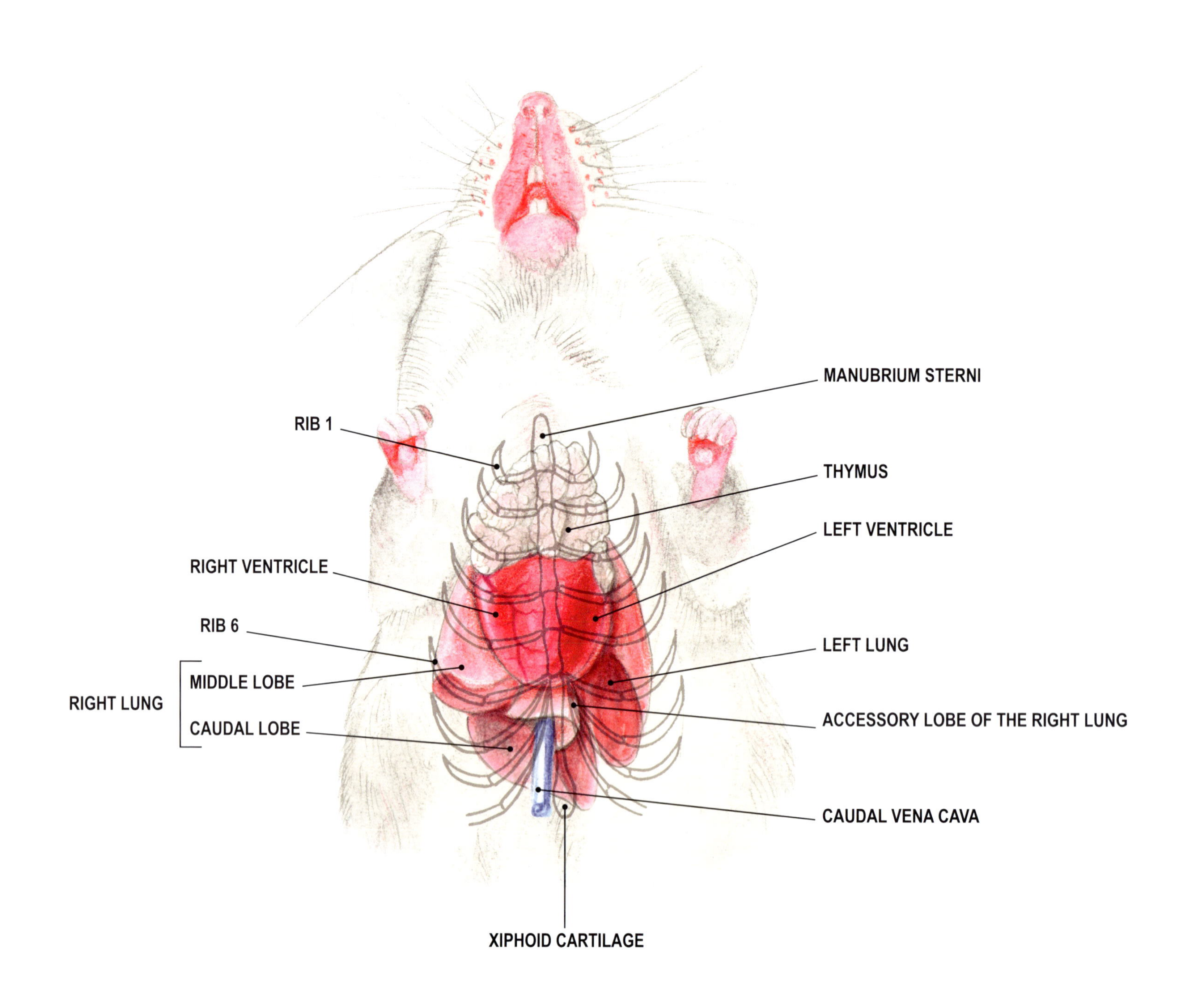

Figure 31B. Rat. Projection of the rib cage and thoracic viscera (ventral aspect).

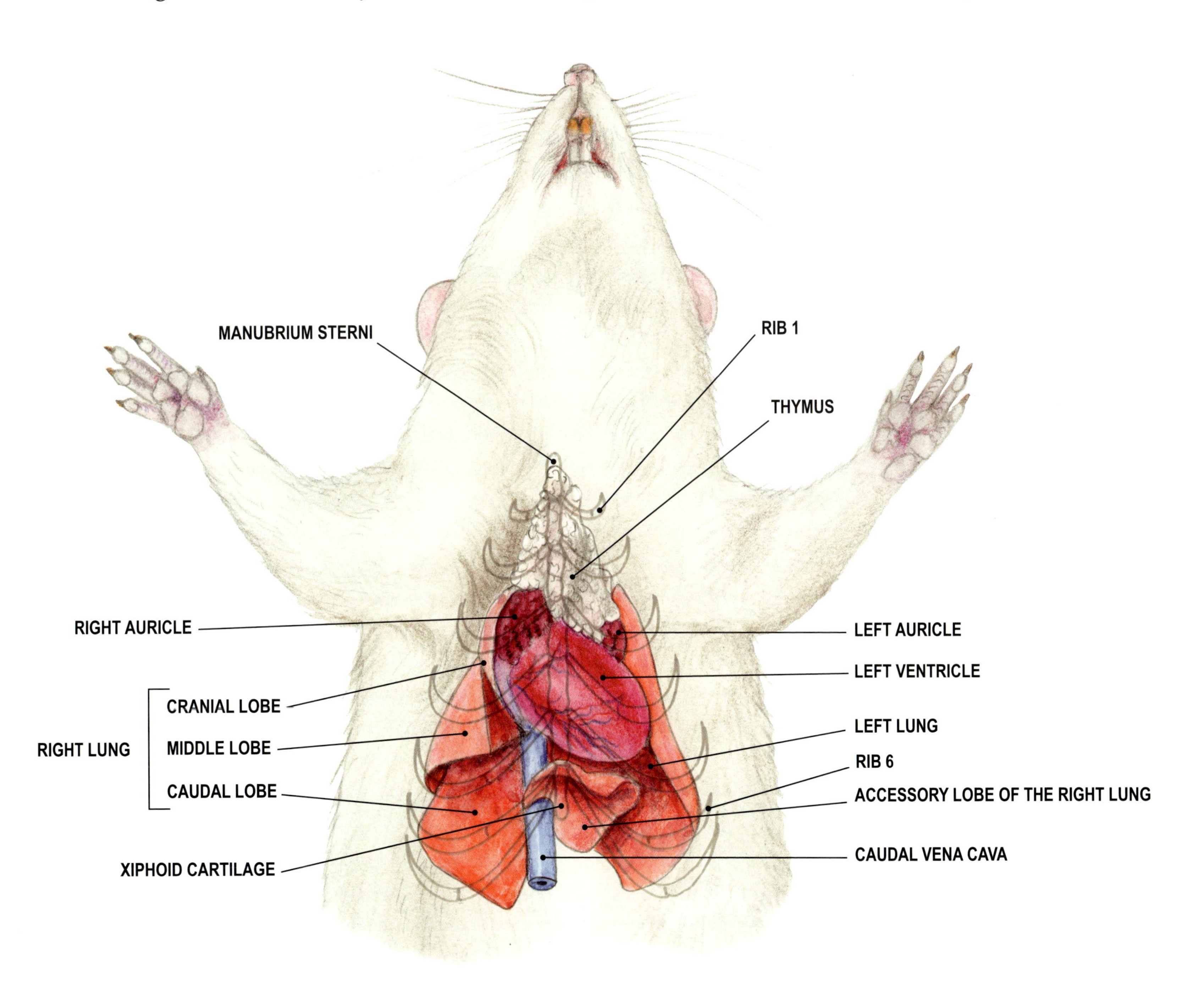

Figure 31A. Mouse. Projection of the rib cage and thoracic viscera (ventral aspect).

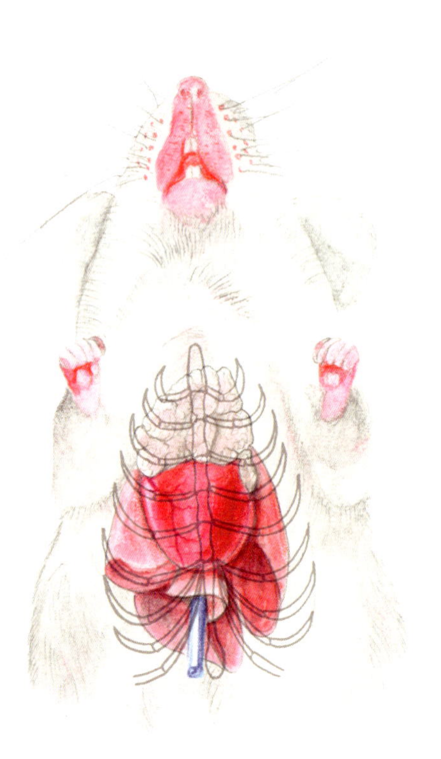

Figure 31B. Rat. Projection of the rib cage and thoracic viscera (ventral aspect).

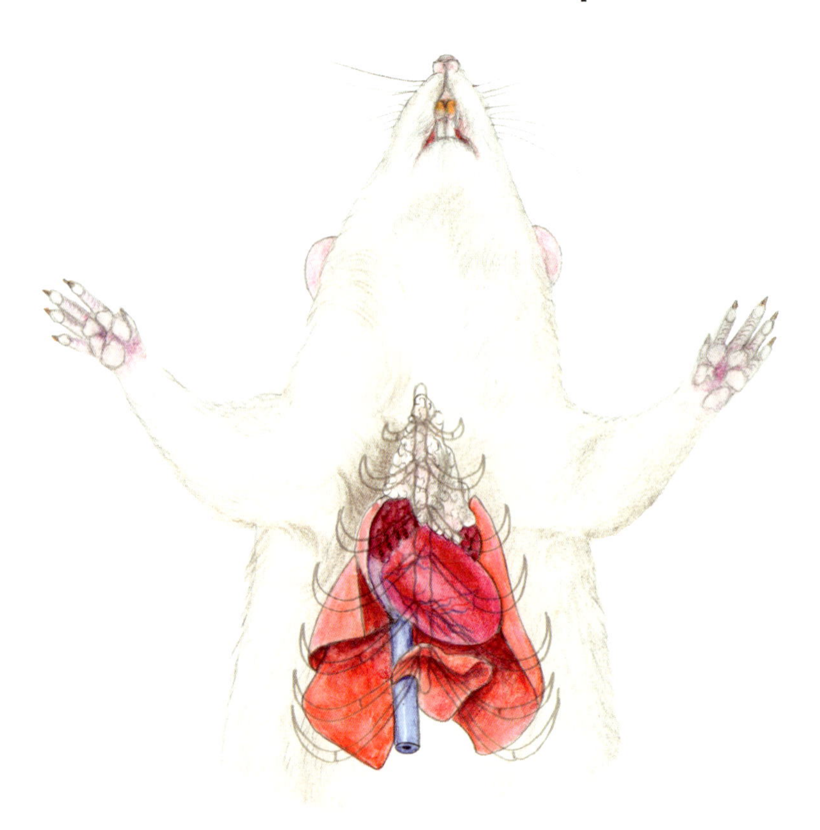

Figure 31. Projection of the rib cage and thoracic viscera (ventral aspect).

In the mouse illustration, ribs 1 to 8 are shown in their entirety; rib 9 is drawn with the rib cartilage and a portion of the rib bone; the remaining ribs are omitted. In the rat illustration, ribs 1 to 7 are shown in their entirety; the remaining ribs are omitted. See page 103 for more information about these figures.

Mouse

The thymus extends from the manubrium sterni to the 3rd pair of ribs and cartilages. It overlaps the cranial lobe of the right lung.

The heart extends from the caudal border of the thymus to the 6th pair of rib cartilages. The heart's axis corresponds to the sternum, and its borders correspond to the costochondral junctions at equal distances right and left from the sternum.

The left lung extends caudally from the heart and thymus to the 8th rib cartilage, where it has a rounded end. Its medial line is parallel and close to the sternum.

The right lung has 4 lobes (cranial, middle, caudal, and accessory of caudal lobe).

- The cranial lobe of the right lung is concealed by the thymus and the heart.
- The middle lobe extends caudally from the right side of the heart to the 6th rib and rib cartilage.
- The caudal lobe extends caudally from the middle lobe and passes into the left half of the thoracic cavity.
- The accessory lobe of the right lung occupies a small area caudal to the heart. This lobe lies between the middle lobe and the left lung. The accessory lobe curves around the caudal vena cava.

Rat

The thymus extends from the manubrium sterni to the 3rd pair of ribs and cartilages. In this figure, the thymus is proportionately narrower in the rat than in the mouse, because of the specimens' difference in age. Caudally, the thymus does not overlap the base of the heart as extensively as in the mouse.

The heart extends caudally from the thymus border to the 5th pair of rib cartilages and laterally to the costochondral junctions.

The left lung extends caudally from the heart and thymus to the 8th rib cartilage, where it has a rounded end. Its medial line is parallel to and near the sternum.

The right lung has 4 lobes (cranial, middle, caudal, and accessory of caudal lobe).

- The cranial lobe of the right lung is mostly concealed by the heart.
- The middle lobe extends from the right side of the heart to the 5th rib and rib cartilage.
- The caudal lobe extends caudally from the middle lobe to the diaphragm (not shown).
- The accessory lobe of the right lung occupies a small area caudally from the heart and between the middle lobe and the left lung. The accessory lobe curves around the caudal vena cava.

Figure 32A. Mouse. Topography of the heart in situ (left aspect).

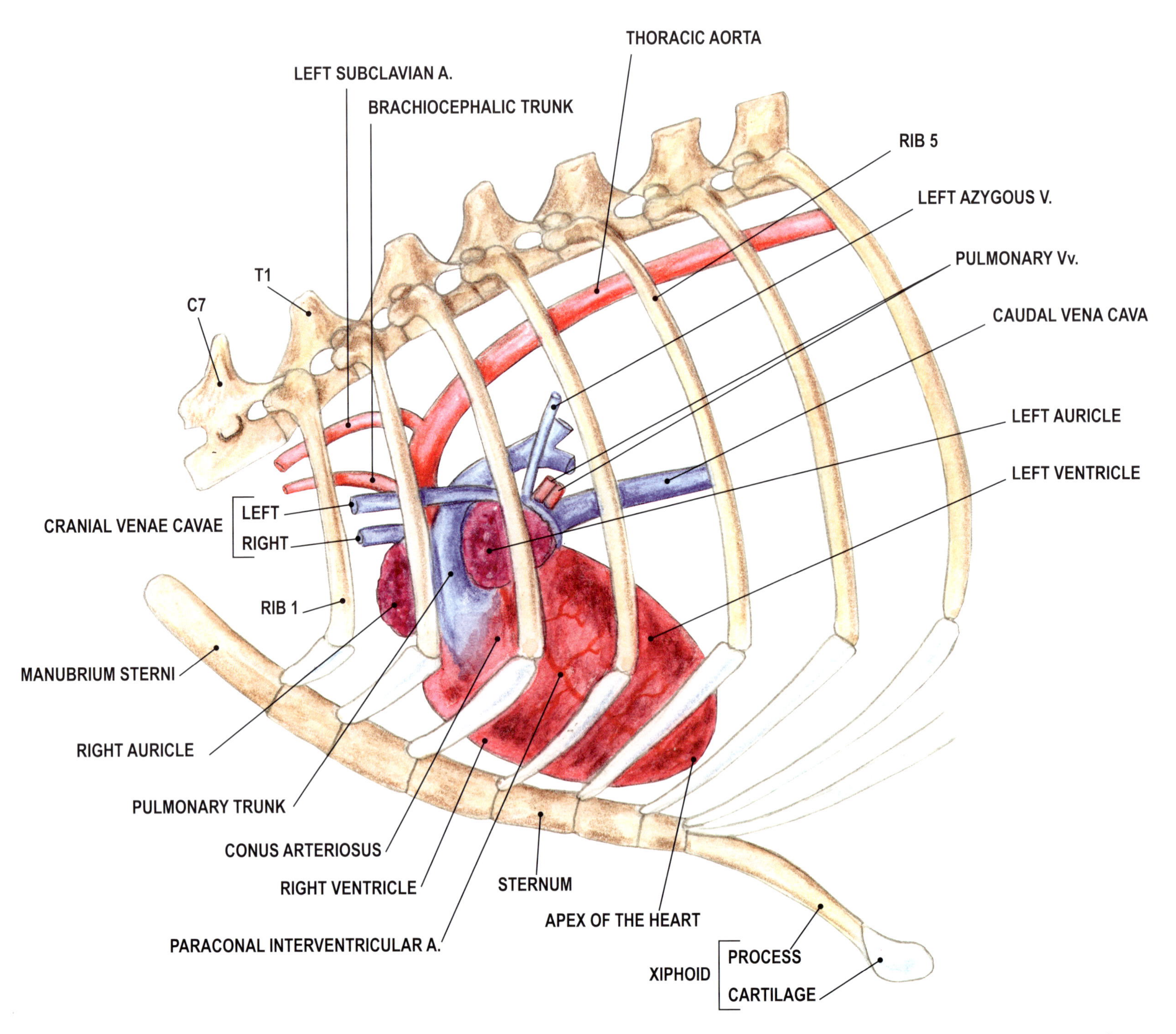

Figure 32B. Rat. Topography of the heart in situ (left aspect).

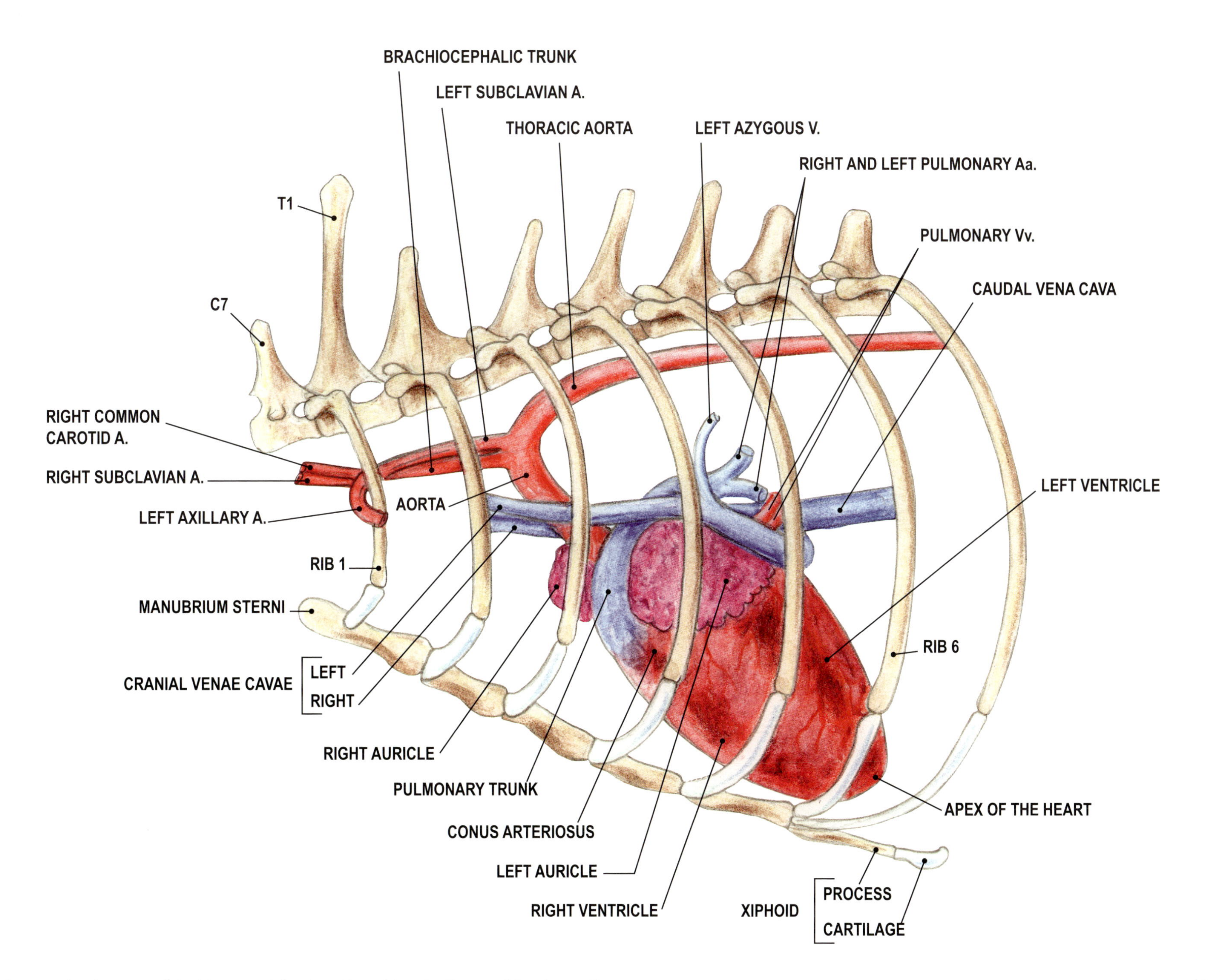

Figure 32A. Mouse. Topography of the heart in situ (left aspect).

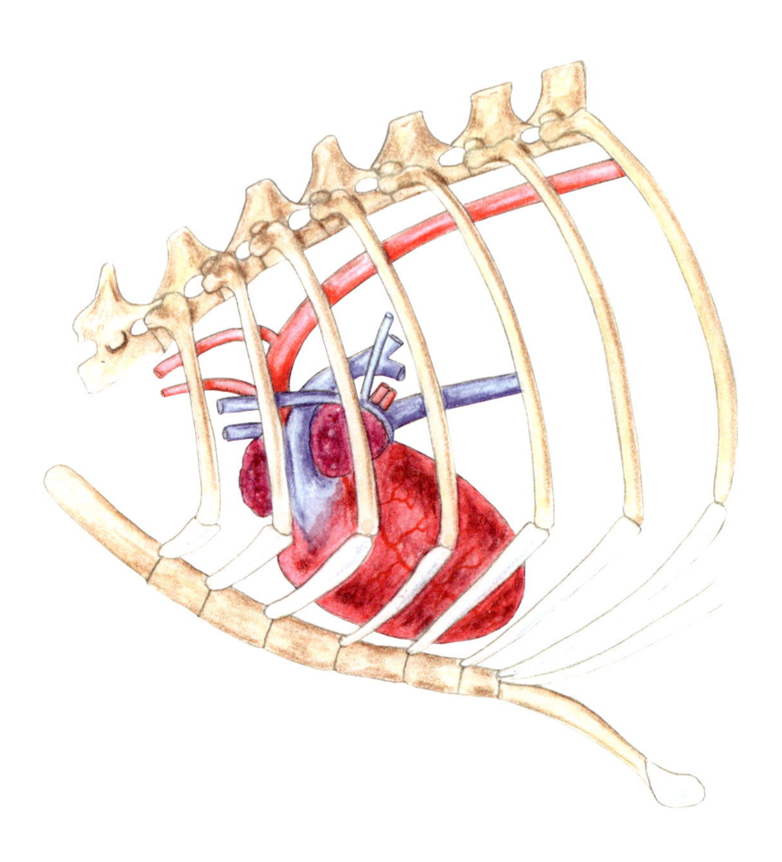

Figure 32B. Rat. Topography of the heart in situ (left aspect).

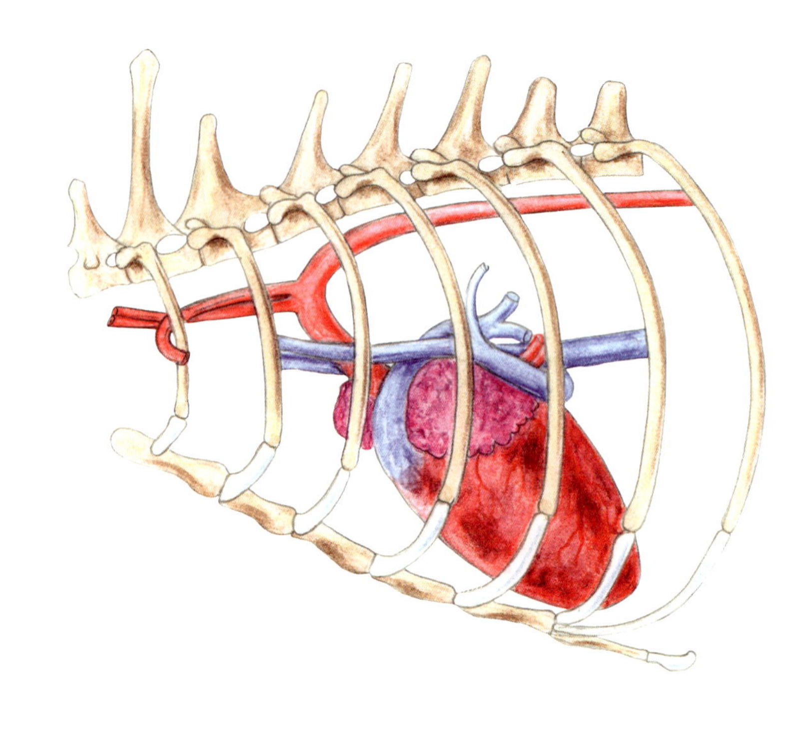

Figure 32. Topography of the heart in situ (left aspect).

The size and shape of the heart are illustrated as in life (filled with blood). The position of the heart varies slightly among specimens due to physiological conditions related to blood distribution, euthanasia, etc. Variations in heart position seen between the left and right (see Figure 33) views of each species reflect the variation observed in specimens from which the illustrations were developed.

Mouse

Two differentiating features can be observed in the topography of the heart:

- The long axis of the heart is parallel with the sternum.
- The heart is located between the 1st intercostal space and 5th intercostal space, almost touching the 6th rib cartilage, and in some specimens extending to the 6th intercostal space.

The right auricle can be found in the 1st intercostal space, dorsal to the 2nd costochondral junction.

The locations of the heart valves are as follows:

- The pulmonary valve can be found in the 2nd intercostal space, dorsal to the 2nd costochondral junction and close to the 2nd rib.
- The aortic valve is approachable in the 2nd intercostal space, dorsal to the 2nd costochondral junction but close to the 3rd rib.
- The bicuspid atrioventricular (mitral) valve can be found in the 3rd intercostal space, dorsal to the costochondral junction.

The shape of the aorta is very different from that of the rat.

Rat

The longitudinal axis of the heart makes an angle of 30°–40° with the sternum.

The heart is located between the 2nd intercostal space and the 6th intercostal space.

The locations of the heart valves are different from those of the mouse:

- The pulmonary valve is located in the middle of the 3rd intercostal space, at the level of the 3rd costochondral junction.
- The aortic valve can be found close to, but caudal, to the pulmonary valve.
- The bicuspid atrioventricular (mitral) valve is located in the 4th intercostal space, close to the 4th rib and dorsal to the costochondral junction.

The shape of the aorta shown in the illustration is typical for the rat.

Rib cage

The spinous processes of the thoracic vertebrae are elongated in the rat. Those in the mouse are diminutive and stubby.

In the mouse, the rib cartilages are angled more cranially. The points of attachment of the cartilages on the sternum are also more cranial, and the sternum is comparatively shorter in the mouse. The mouse has a more pronounced manubrium sterni and xiphoid process than the rat.

Figure 33A. Mouse. Topography of the heart in situ (right aspect).

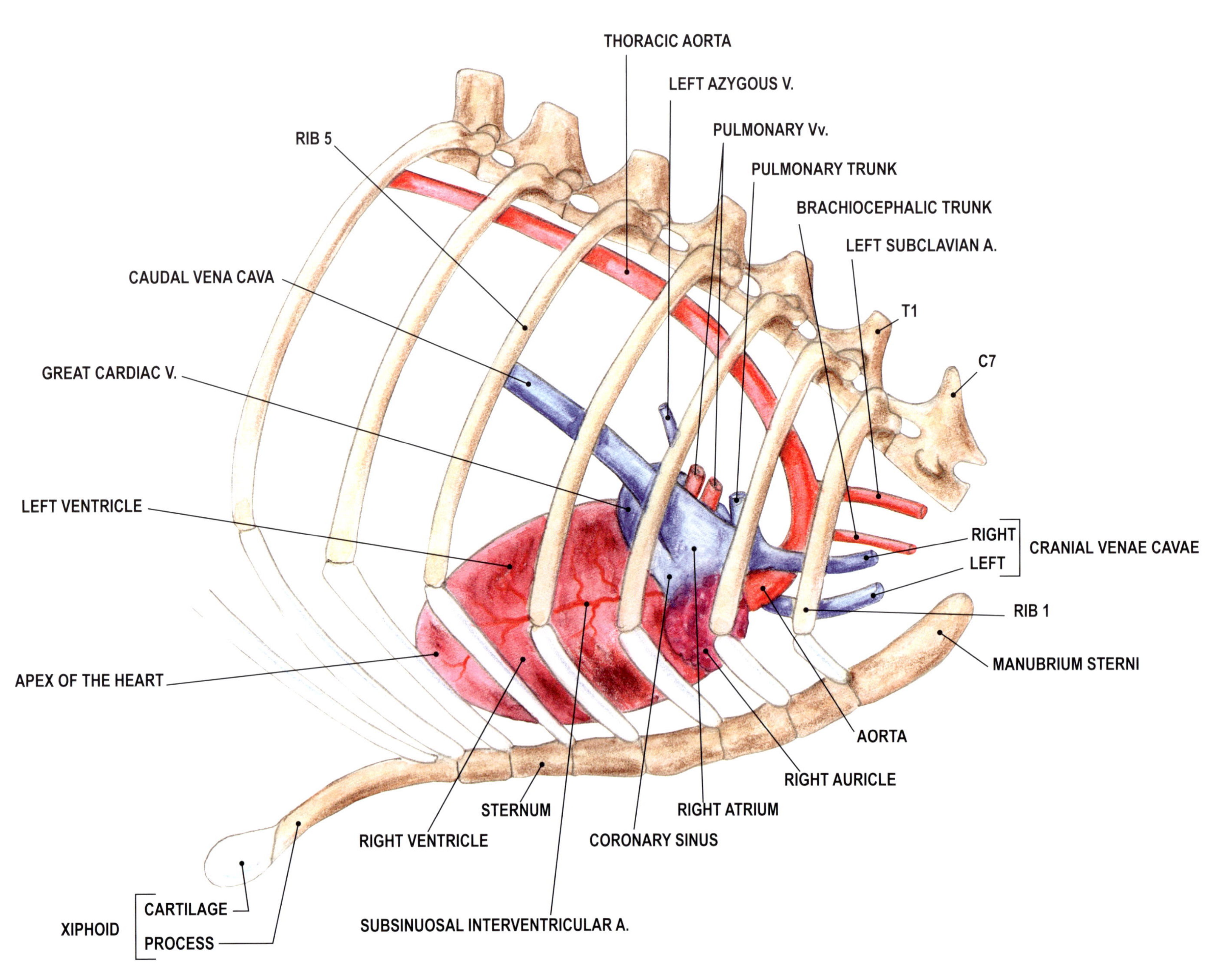

Figure 33B. Rat. Topography of the heart in situ (right aspect).

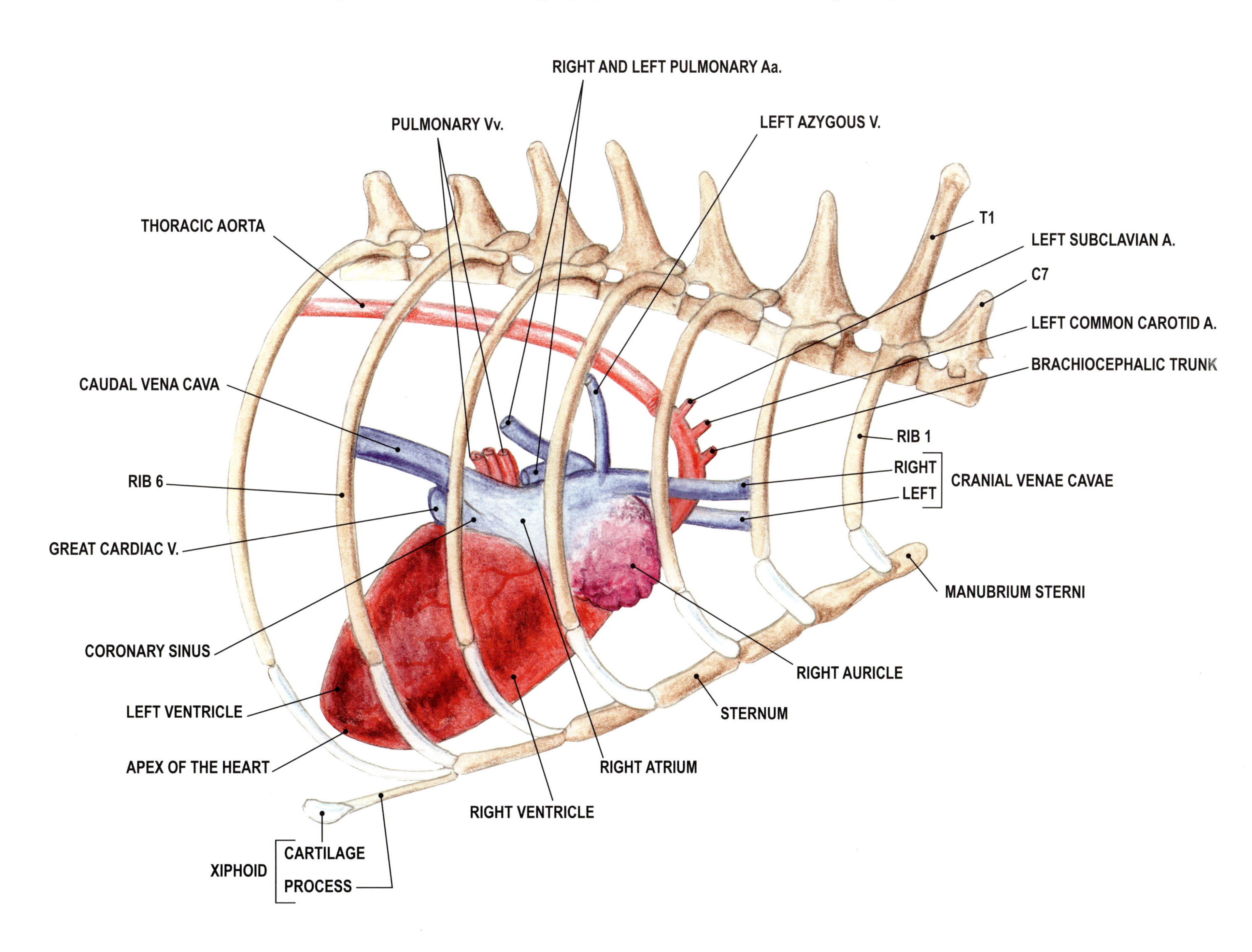

Figure 33A. Mouse. Topography of the heart in situ (right aspect).

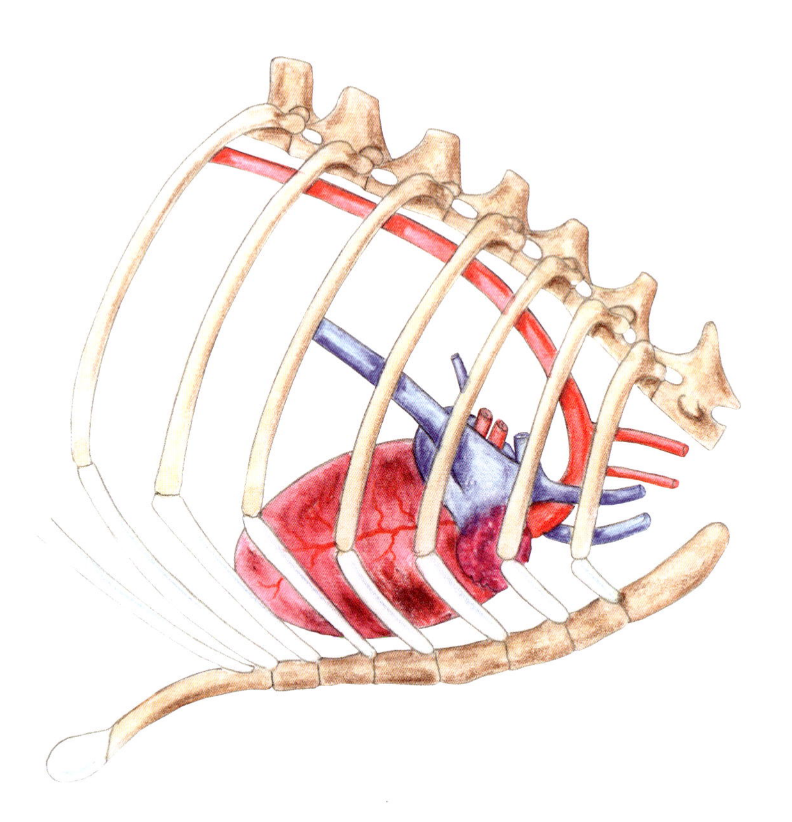

Figure 33B. Rat. Topography of the heart in situ (right aspect).

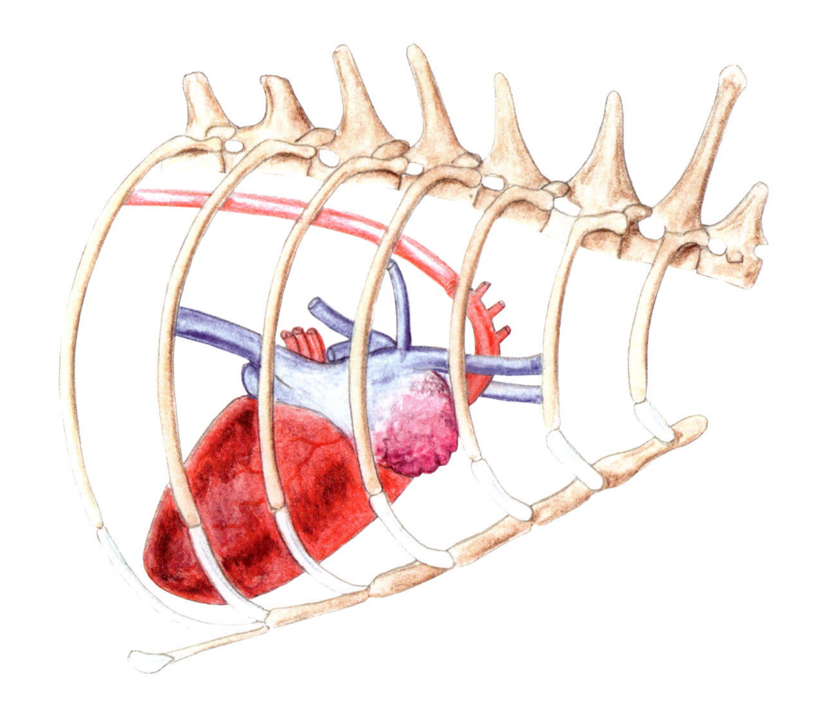

Figure 33. Topography of the heart in situ (right aspect).

See page 127 for information about these figures.

Mouse

The location of the tricuspid atrioventricular valve is in the middle of the 2nd intercostal space, at the level of the 2nd and 3rd costochondral junctions.

Rat

The tricuspid atrioventricular valve can be found in the 3rd intercostal space, close to the 4th rib at the level of the 4th costochondral junction.

Figure 34A. Mouse. Heart (atrial and auricular aspects).

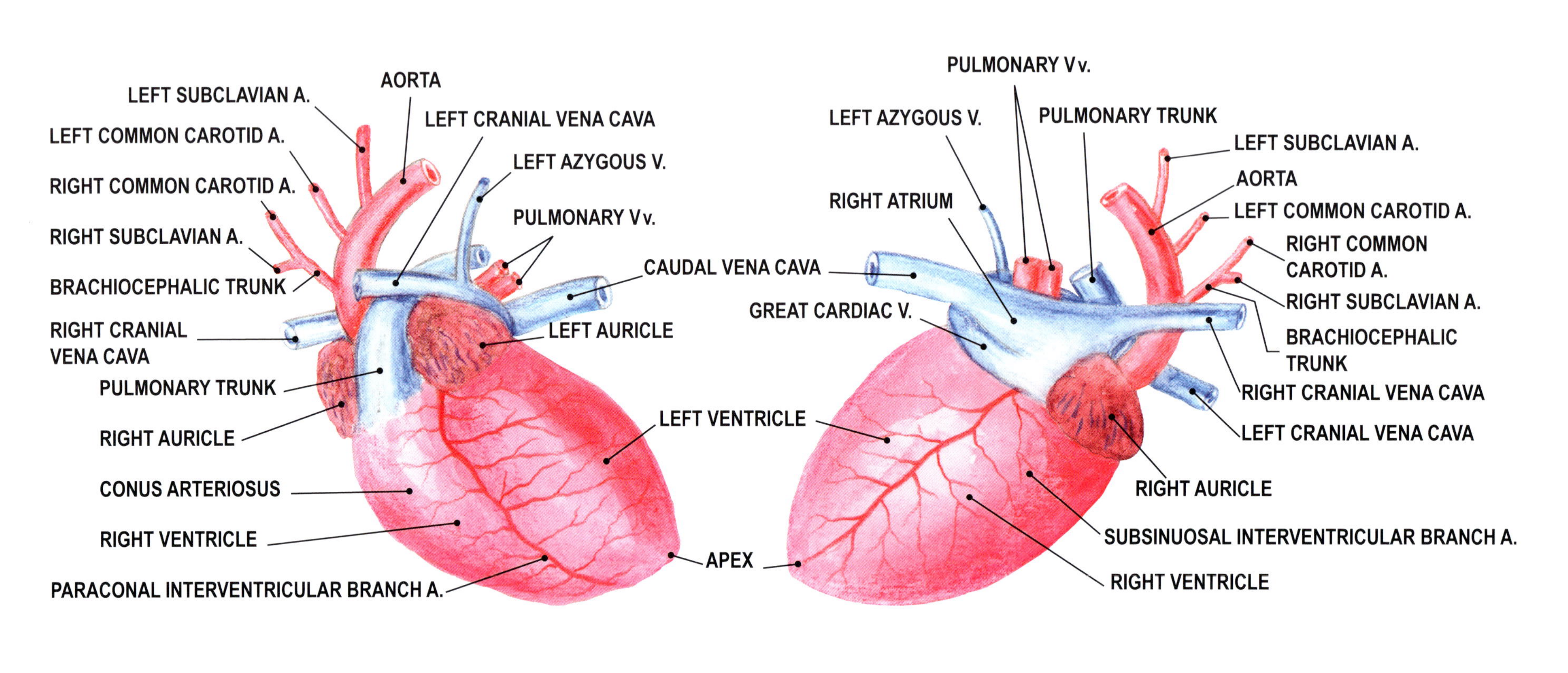

Figure 34B. Rat. Heart (atrial and auricular aspects).

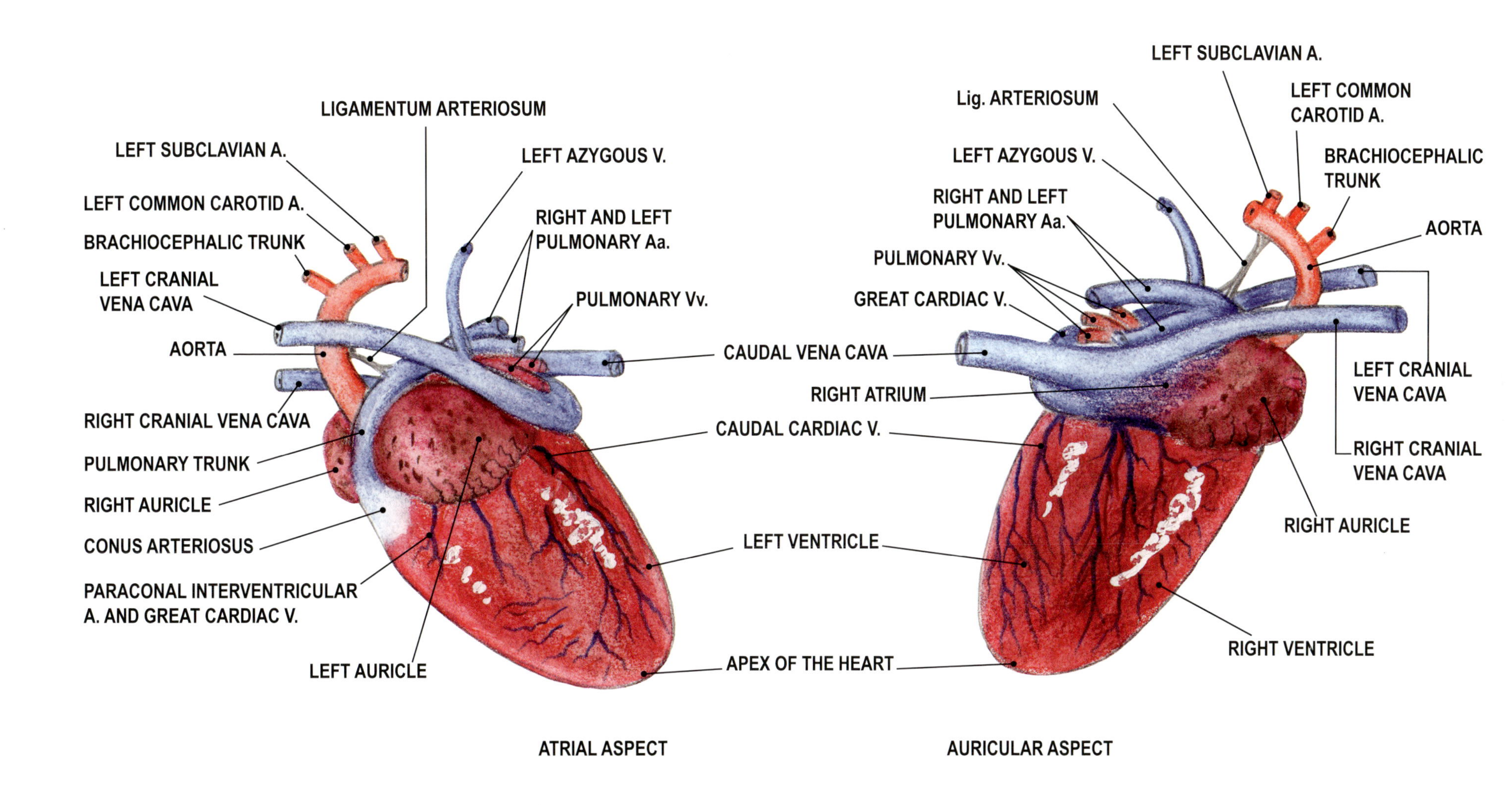

Figure 34A. Mouse. Heart (atrial and auricular aspects). **Figure 34B. Rat. Heart (atrial and auricular aspects).**

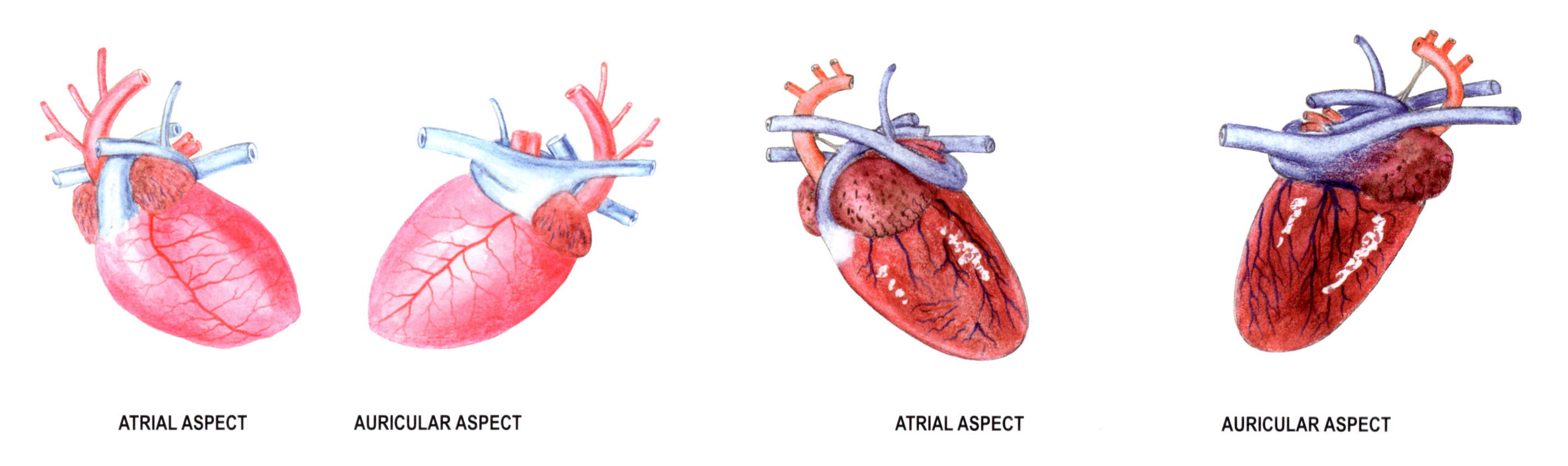

Figure 34. Heart (atrial and auricular aspects).

In this view, there are no appreciable differences in gross anatomy of the heart between the mouse and the rat.

Figure 35A. Mouse. Heart in situ (ventral aspect) and reflected cranially (dorsal aspect).

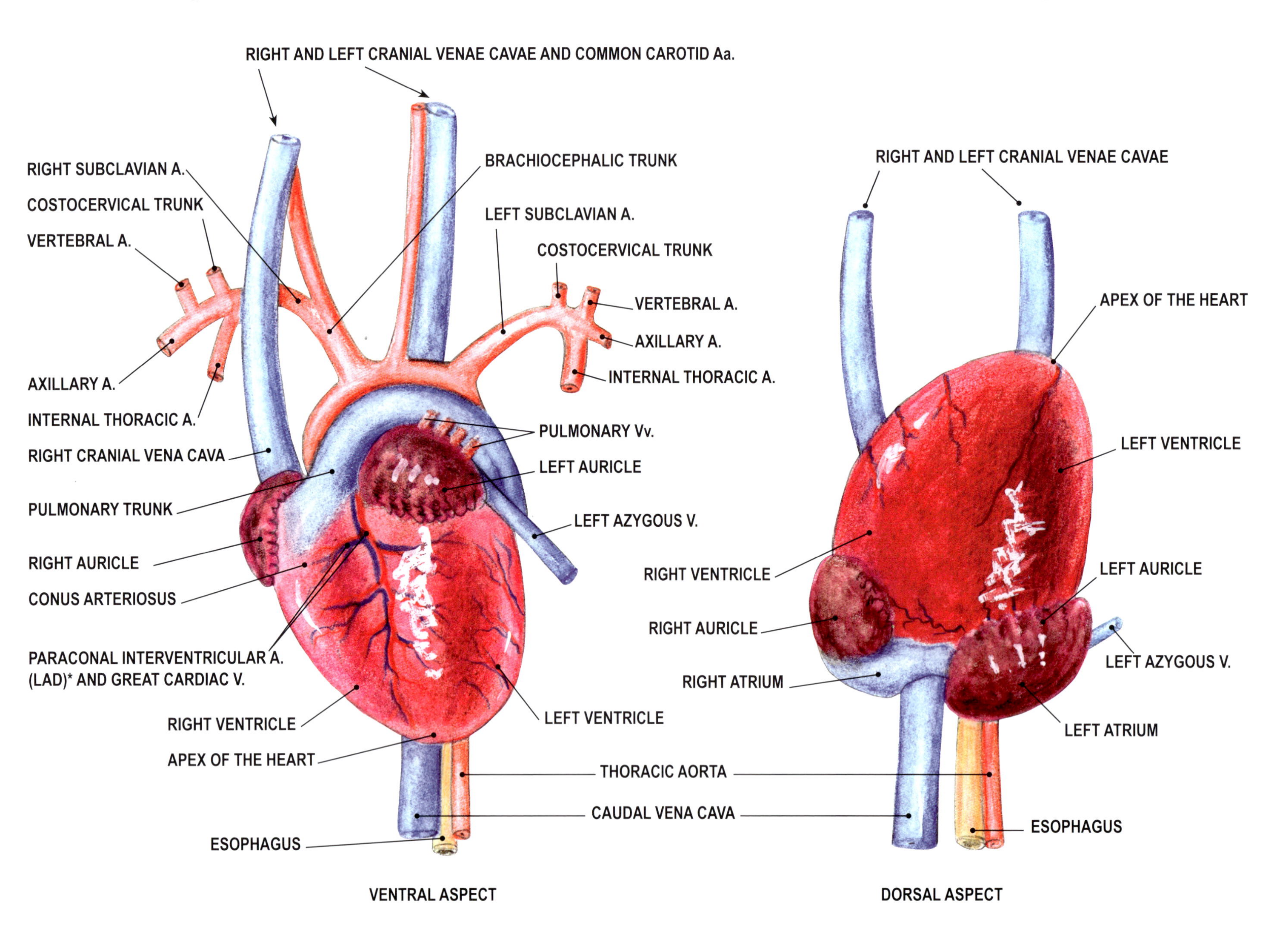

*Left anterior descending artery. Many researchers ligate this artery to induce myocardial infarctions in mice and rats.

Figure 35B. Rat. Heart in situ (ventral aspect) and reflected cranially (dorsal aspect).

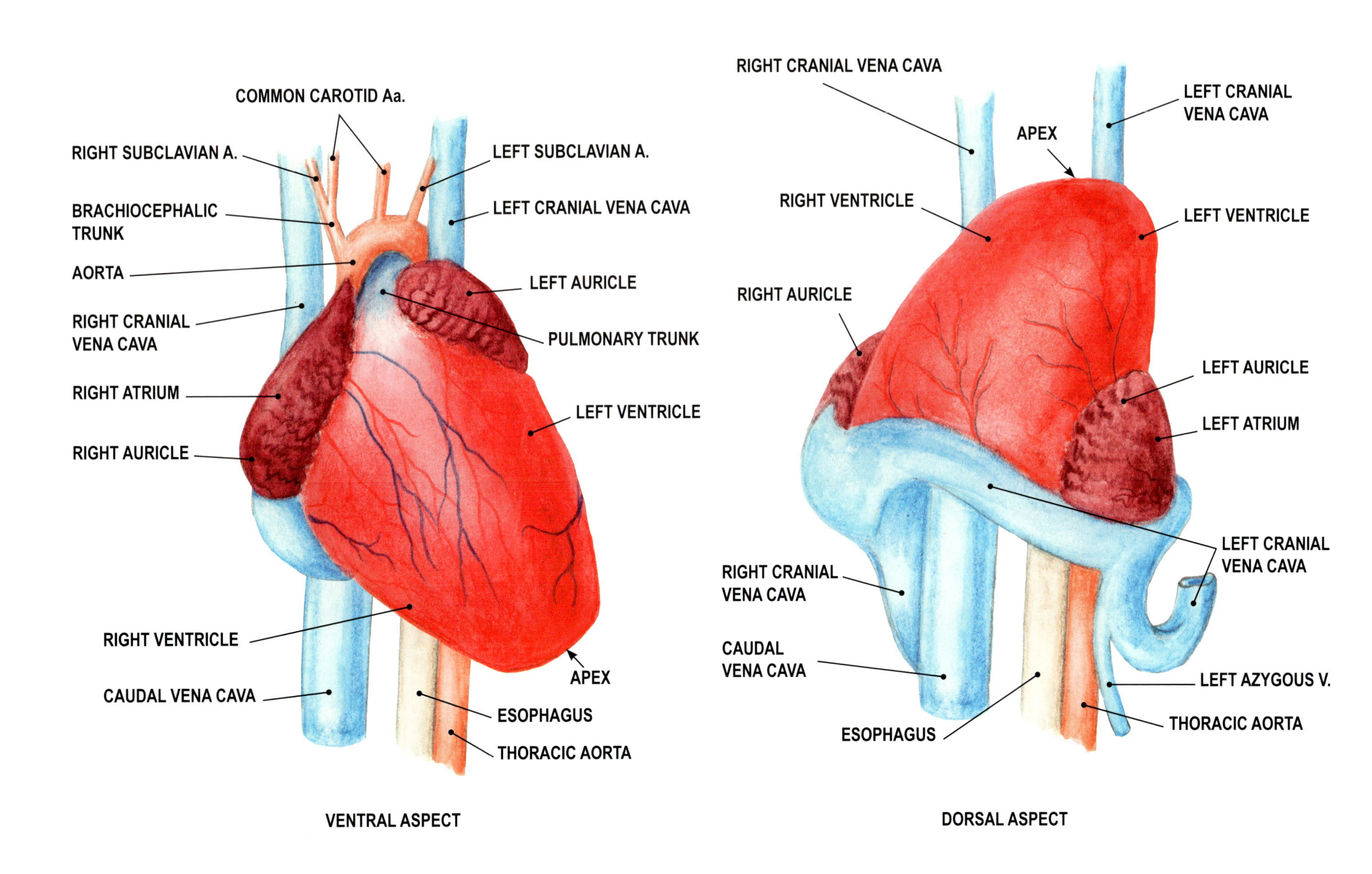

Figure 35A. Mouse. Heart in situ (ventral aspect) and reflected cranially (dorsal aspect).

Figure 35B. Rat. Heart in situ (ventral aspect) and reflected cranially (dorsal aspect).

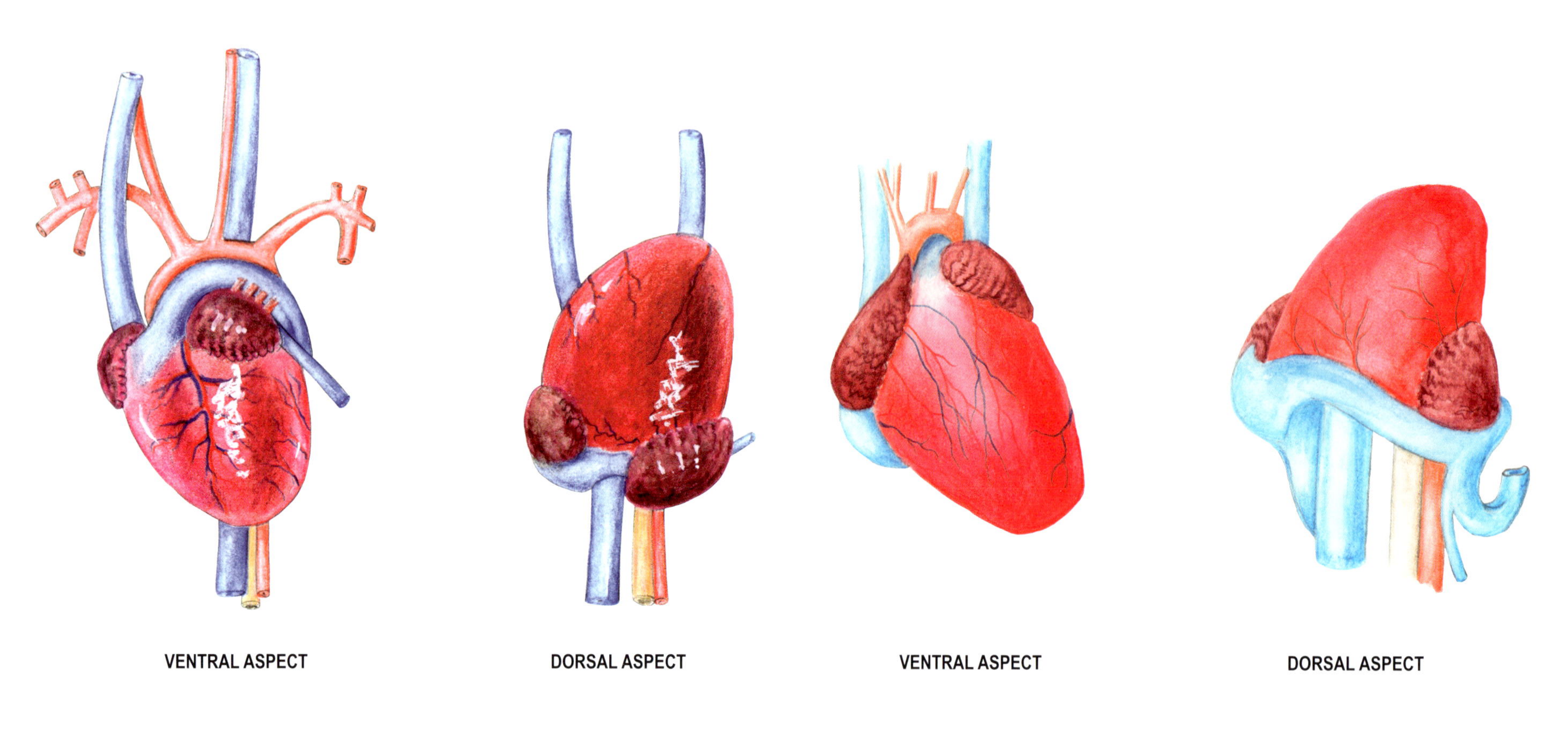

Figure 35. Heart in situ (ventral aspect) and reflected cranially (dorsal aspect).

Mouse

Ventral Aspect

The auricular aspect of the heart is shown in a position similar to how it appears within the thoracic cavity after removing the pericardium, orienting the apex caudally and the base cranially.

The auricles are fully exposed; the limit between the two ventricles is well defined by the paraconal interventricular artery (LAD*) and the great cardiac vein.

The conus arteriosus and the pulmonary trunk up to the bifurcation of the right and left pulmonary arteries are visible.

The aortic arch and the detailed branches of the right and left subclavian arteries are fully exposed because of the position of the heart.

The right and left cranial venae cavae, the left azygous vein, and the caudal vena cava are also visible.

Dorsal Aspect

The reflected heart shows the apex oriented cranially and the base caudally. As reference structures, the cranial vena cava, the caudal vena cava, the thoracic aorta, and the esophagus are in the same place as in the previous figure.

The atrial aspect of the heart is now exposed, with both auricles in continuation of the atria.

Rat

Ventral Aspect

Compared with the mouse figure, the rat heart is slightly turned on the long axis to the right, exposing most of the right ventricle, the full right auricle, and part of the right atrium.

In this view, the left auricle conceals the relationships with the pulmonary veins, and the left azygous vein is behind the left auricle (unlike in the mouse heart). The conus arteriosus is fully exposed, and the pulmonary trunk shows only its origin.

The aorta shows only the origin of the brachiocephalic trunk, the left common carotid artery, and the left subclavian artery.

Dorsal Aspect

The thoracic aorta, the caudal vena cava, and the esophagus are landmarks for the orientation of the heart.

The right and left cranial venae cavae, the caudal vena cava, and the left azygous vein are visible.

This view exposes the left atrium and the entire left auricle.

The right atrium cannot be shown; only a little of the right auricle is visible.

*Left anterior descending artery in human anatomy. Many researchers ligate this artery to induce myocardial infarctions in mice and rats.

Figure 36. Longitudinal section through the heart (atrial aspect), semi-schematic.

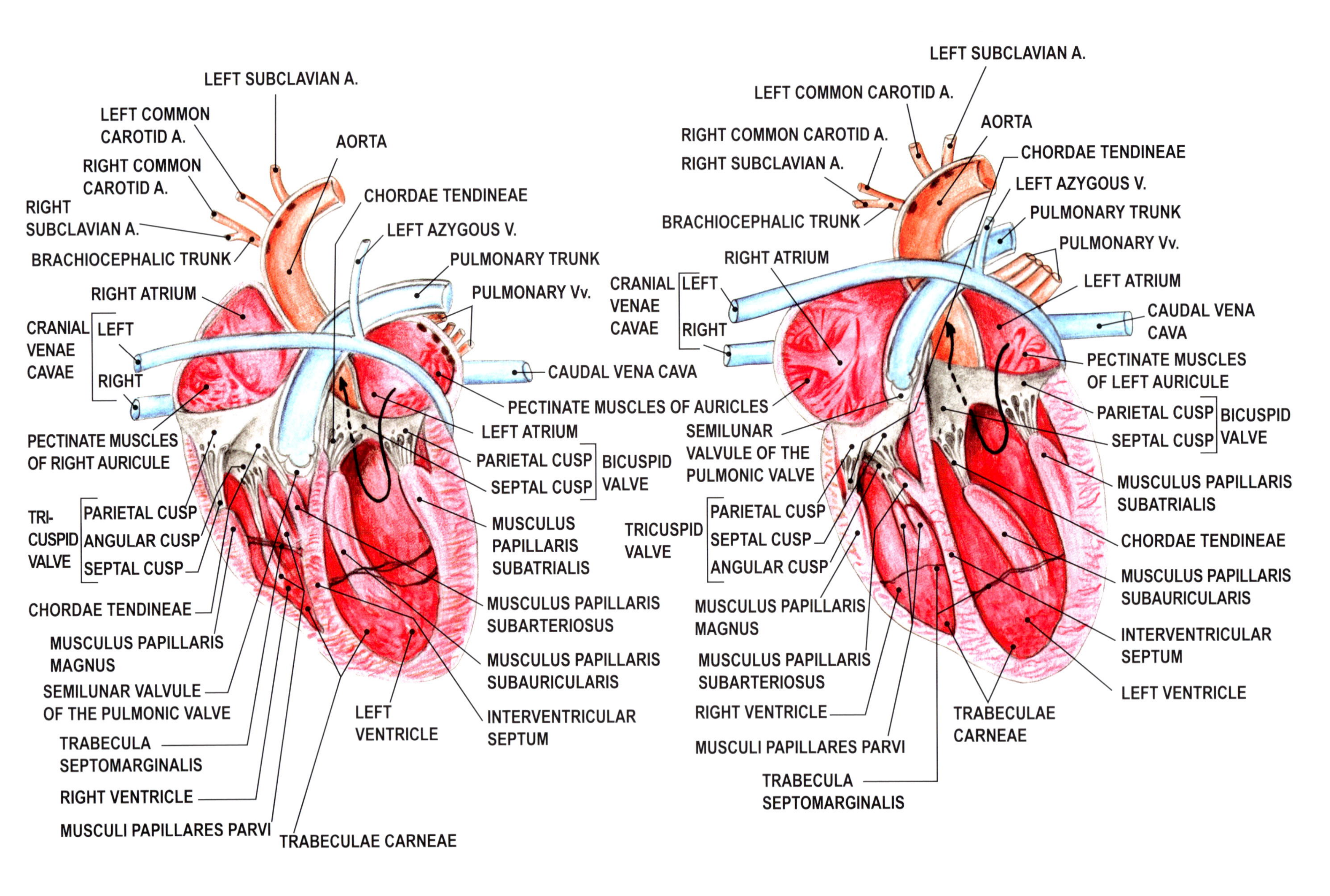

Figure 36. Longitudinal section through the heart (atrial aspect), semi-schematic.

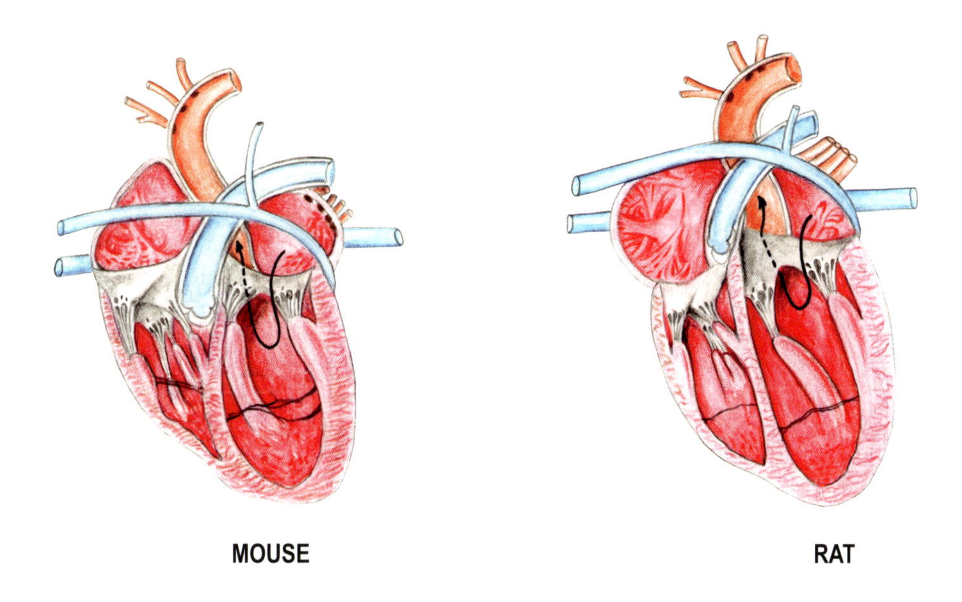

Showing all structures exposed in the figures is not possible in a single sagittal section that is anatomically correct. In order for the reader to know all structures that can be identified in parallel longitudinal sections of the heart, each figure was developed as a semi-schematic, bringing together in one figure more structures than would be viewed in any single section.

The arrows shown in the left heart indicate the direction of blood flow from the left atrium to the left ventricle and to the aorta. In the right heart, the direction of blood flow is from the left atrium to the right ventricle and to the pulmonary trunk.

This section exposes the atria, the ventricles, the right atrioventricular (tricuspid) valve and the left atrioventricular (bicuspid, mitral) valve, the aorta and the pulmonary trunk, the cusps, the chordae tendineae, the papillary muscles, the trabeculae carneae, and the trabeculae septomarginales.

Only the pulmonary valve can be seen in these figures, with the three valvules in each species.

The position and shape of the papillary muscles are slightly different in the mouse than in the rat.

The pectinate muscles of the right atrium are larger than those of the left atrium in both species. The pectinate muscles are proportionately larger in the rat than in the mouse.

Figure 37A. Mouse. Cervicothoracic organs (ventral aspect).*

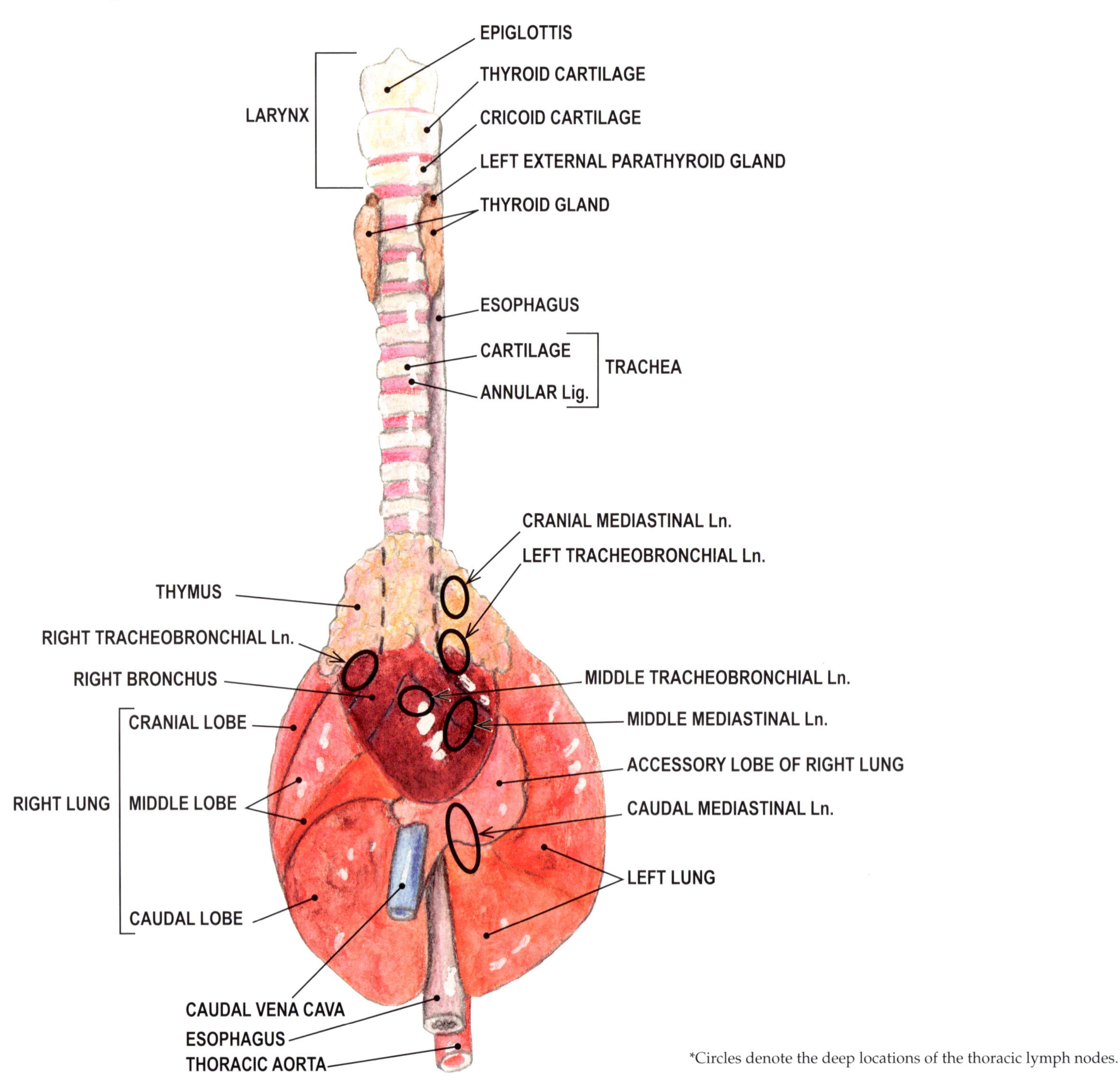

*Circles denote the deep locations of the thoracic lymph nodes.

Figure 37B. Rat. Cervicothoracic organs (ventral aspect).*

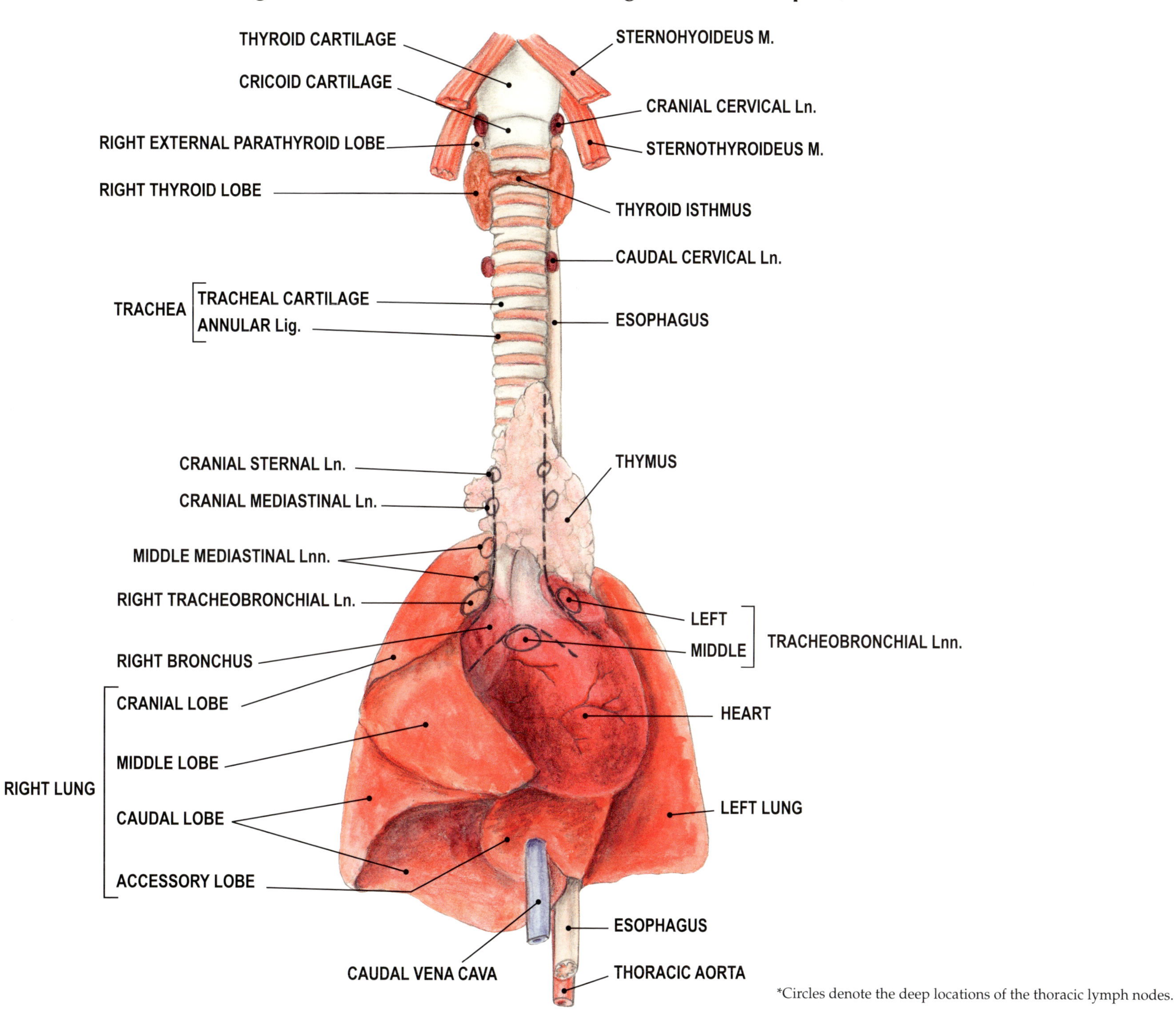

*Circles denote the deep locations of the thoracic lymph nodes.

Figure 37A. Mouse. Cervicothoracic organs (ventral aspect).*

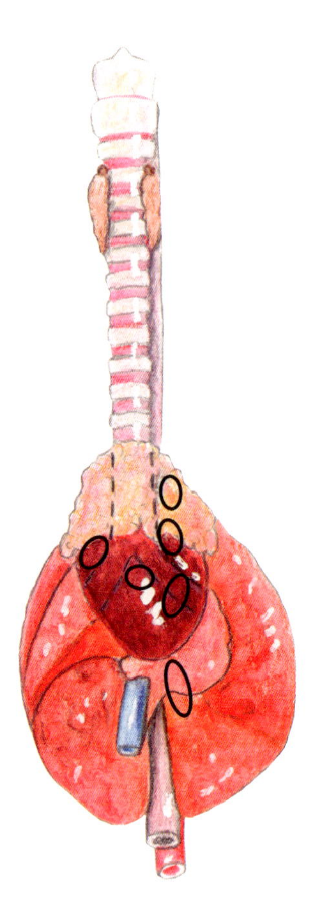

Figure 37B. Rat. Cervicothoracic organs (ventral aspect).*

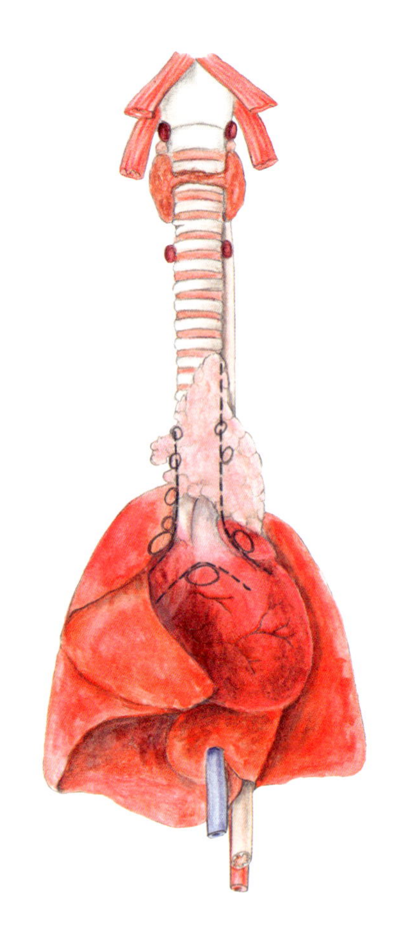

*Circles denote the deep locations of the thoracic lymph nodes.

Figure 37. Cervicothoracic organs (ventral aspect).

In both figures, the lungs are deflated, as seen in an opened thorax. Because these specimens are young adults, a large thymus is shown. The size of the thymus is related to the age of the individual, being larger in juvenile animals. Following sexual maturity, the thymus declines in size through involution.

Mouse

The cervical area includes the cartilages of the larynx, the tracheal cartilages and the annular ligaments, the thyroid and parathyroid glands, and the esophagus.

The thoracic part of the trachea and the bifurcation into the right and left primary bronchi as well as the ventral aspect of the lungs are illustrated in relationship to the heart. The caudal vena cava runs between the right caudal lobe and the accessory lobe, the esophagus, the thoracic aorta, and several lymph nodes.

Thymic involution occurs rapidly between 35 and 80 days, and then continues at a slower pace until the thymus becomes thin and leaf-like in the older adult.[8] Athymic animals lack a thymus; immunocompromised animals may have a smaller thymus.

Rat

Shown are the larynx, tracheal cartilages and annular ligaments, the thyroid gland (with the right and left lobes connected by an isthmus, which is not present in the mouse), the external parathyroid glands, and the esophagus.

The sternohyoideus and sternothyroideus muscles are cut close to their attachments, respectively, to the hyoid bone and thyroid cartilage. These muscles are similar to those in the mouse (not shown).

The thoracic part of the trachea, the right and left bronchi, and the ventral aspect of the lungs are the structures of the respiratory apparatus in the thoracic cavity.

The heart shows the auricular aspect. The thymus and the lymph nodes are shown in relationship to the heart.

In rats, thymus weight peaks at sexual maturity and declines slowly thereafter. By 18 months of age, the rat lacks a grossly visible thymus.[1]

Figure 38A. Mouse. Heart and vascular tree (ventral aspect).

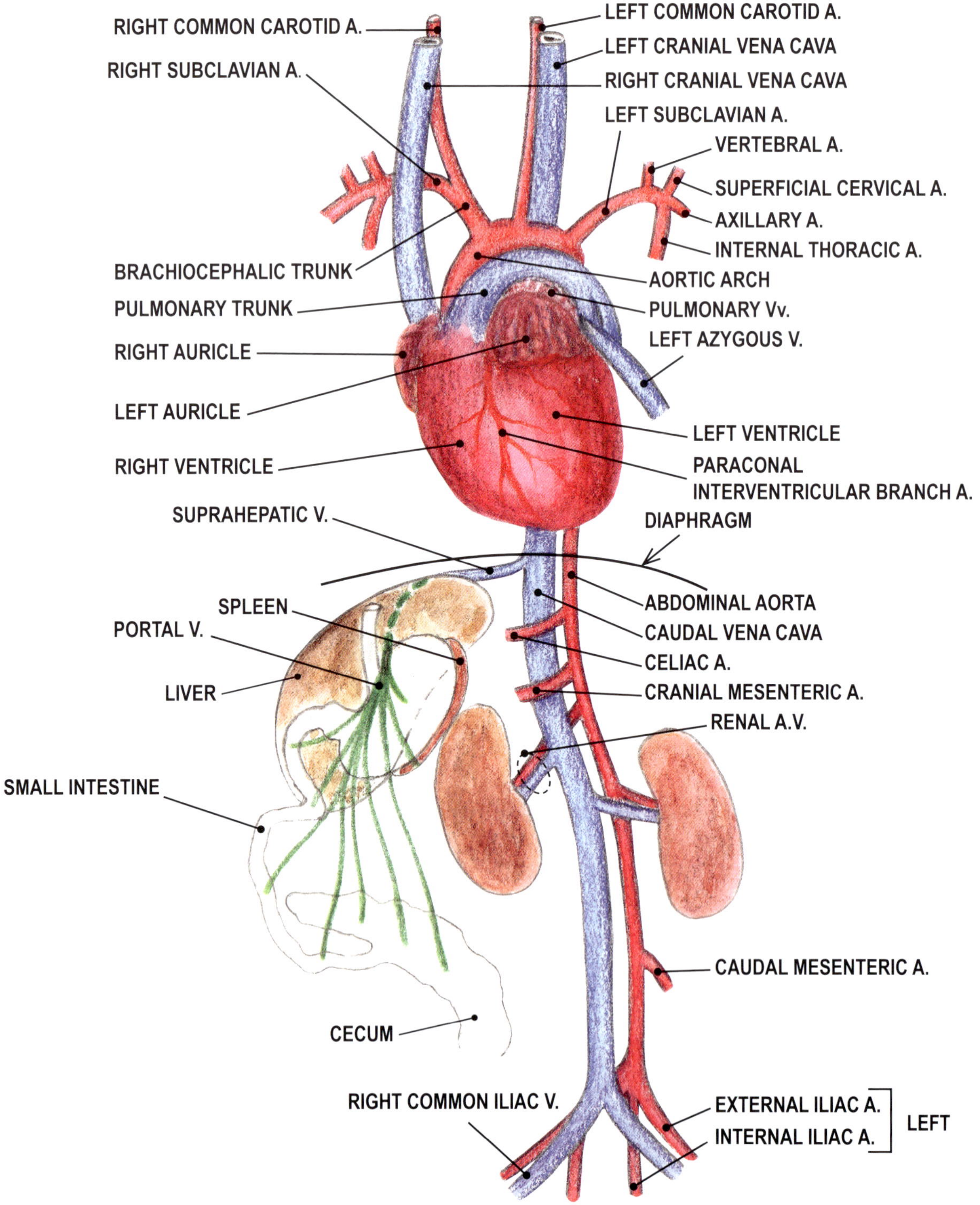

RIGHT COMMON CAROTID A.
LEFT COMMON CAROTID A.
LEFT CRANIAL VENA CAVA
RIGHT SUBCLAVIAN A.
RIGHT CRANIAL VENA CAVA
LEFT SUBCLAVIAN A.
VERTEBRAL A.
SUPERFICIAL CERVICAL A.
AXILLARY A.
INTERNAL THORACIC A.
BRACHIOCEPHALIC TRUNK
AORTIC ARCH
PULMONARY TRUNK
PULMONARY Vv.
LEFT AZYGOUS V.
RIGHT AURICLE
LEFT AURICLE
LEFT VENTRICLE
RIGHT VENTRICLE
PARACONAL INTERVENTRICULAR BRANCH A.
SUPRAHEPATIC V.
DIAPHRAGM
ABDOMINAL AORTA
SPLEEN
CAUDAL VENA CAVA
PORTAL V.
CELIAC A.
LIVER
CRANIAL MESENTERIC A.
RENAL A.V.
SMALL INTESTINE
CAUDAL MESENTERIC A.
CECUM
RIGHT COMMON ILIAC V.
EXTERNAL ILIAC A.
INTERNAL ILIAC A.
LEFT

Figure 38B. Rat. Heart and vascular tree (ventral aspect).

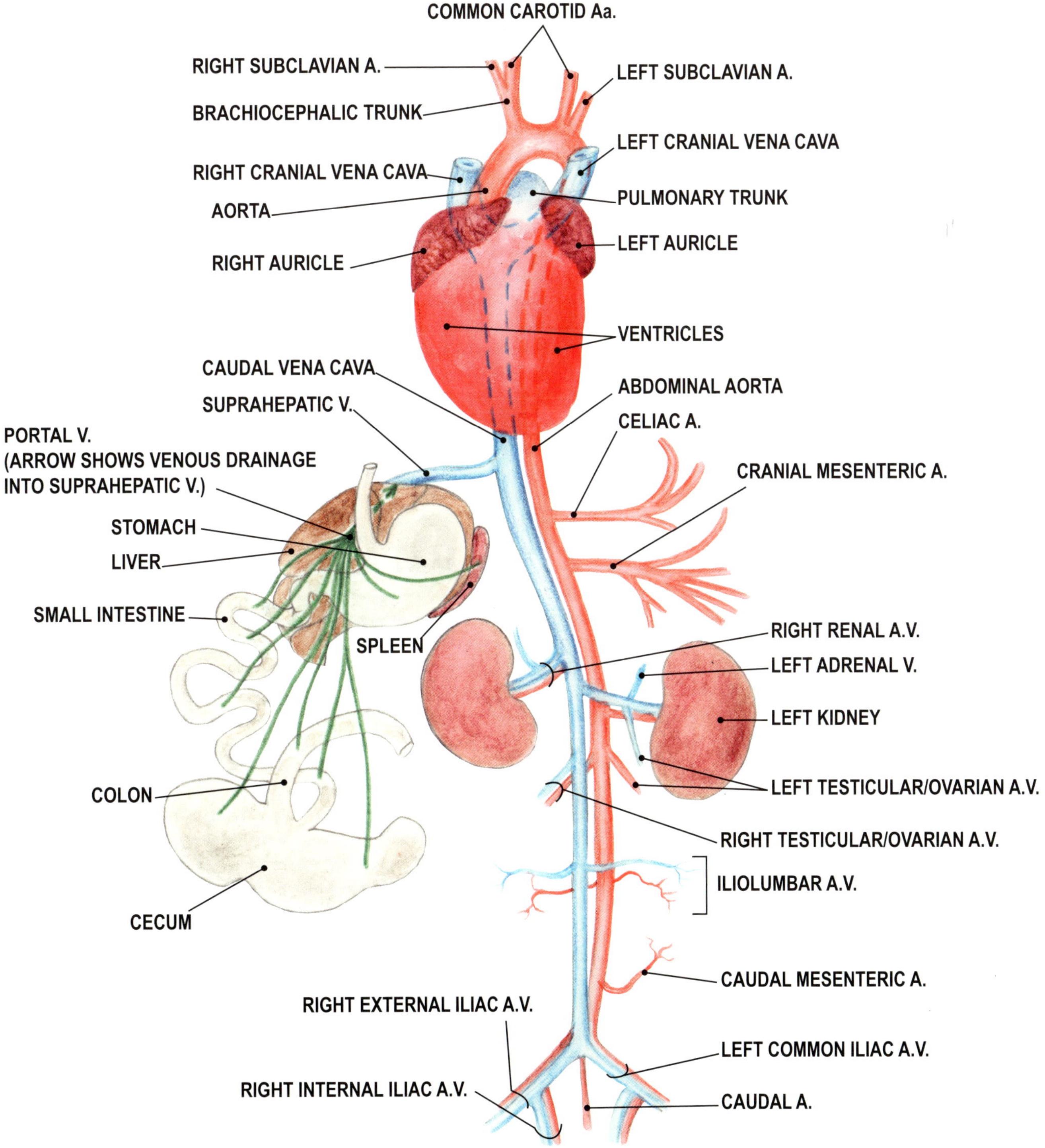

COMMON CAROTID Aa.
RIGHT SUBCLAVIAN A.
LEFT SUBCLAVIAN A.
BRACHIOCEPHALIC TRUNK
LEFT CRANIAL VENA CAVA
RIGHT CRANIAL VENA CAVA
PULMONARY TRUNK
AORTA
LEFT AURICLE
RIGHT AURICLE
VENTRICLES
CAUDAL VENA CAVA
ABDOMINAL AORTA
SUPRAHEPATIC V.
CELIAC A.
PORTAL V.
(ARROW SHOWS VENOUS DRAINAGE
INTO SUPRAHEPATIC V.)
CRANIAL MESENTERIC A.
STOMACH
LIVER
SMALL INTESTINE
SPLEEN
RIGHT RENAL A.V.
LEFT ADRENAL V.
LEFT KIDNEY
COLON
LEFT TESTICULAR/OVARIAN A.V.
RIGHT TESTICULAR/OVARIAN A.V.
ILIOLUMBAR A.V.
CECUM
CAUDAL MESENTERIC A.
RIGHT EXTERNAL ILIAC A.V.
LEFT COMMON ILIAC A.V.
RIGHT INTERNAL ILIAC A.V.
CAUDAL A.

Figure 38A. Mouse. Heart and vascular tree (ventral aspect).

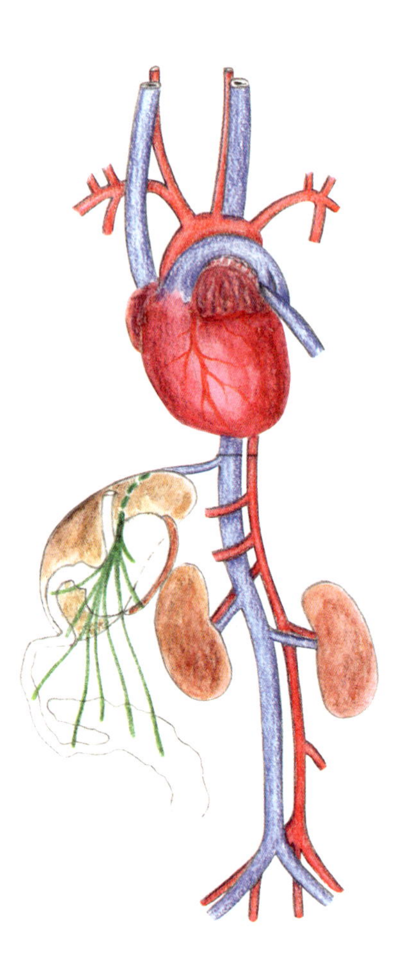

Figure 38B. Rat. Heart and vascular tree (ventral aspect).

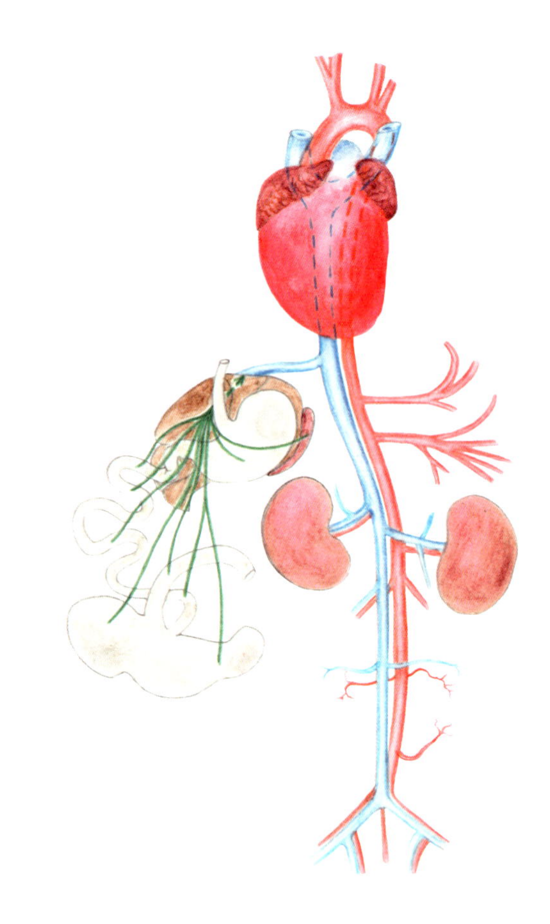

Figure 38. Heart and vascular tree (ventral aspect).

The atria are not visible from this perspective (see Figure 35). The abdominal aorta with the renal arteries; the origins of the celiac, cranial, and caudal mesenteric arteries; and the quadrifurcation of the aorta into the symmetrical external and internal iliac arteries are the main arterial branches in the abdominal cavity.

Mouse

The auricular aspect of the heart is exposed (see details on Figure 35).

The aortic arch, brachiocephalic trunk, left common carotid artery, left subclavian artery, and the pulmonary trunk are the main arteries at the base of the heart.

The right and left cranial venae cavae, the caudal vena cava, and the last segments of the pulmonary veins are the veins at the base of the heart.

Caudal to the diaphragm, the main viscera inside of the abdominal cavity are outlined.

The caudal vena cava originating from the common iliac veins, the renal veins, and the portal vein emptying into the suprahepatic vein are the main veins in the abdominal cavity.

Rat

The heart shows the auricular aspect, with most of the right ventricle and right auricle exposed (see details on Figure 35).

Shown at the base of the heart are the aortic arch; the origins of the brachiocephalic trunk, left common carotid artery, and left subclavian artery; the pulmonary trunk; and the right and left cranial venae cavae.

The abdominal aorta is shown with the renal arteries and the origins of the celiac, cranial mesenteric, testicular/ovarian, iliolumbar, and caudal mesenteric arteries. The aorta ends at the caudal artery and the external iliac arteries; the internal iliac arteries originate from the latter.

The caudal vena cava originates from the external iliac veins, and receives the iliolumbar, testicular/ovarian, renal and hepatic veins.

The portal vein collects blood from the abdominal viscera and connects with the suprahepatic vein.

Abdominal Structures

Figure 39A. Mouse, male. Abdominal topography (left aspect).

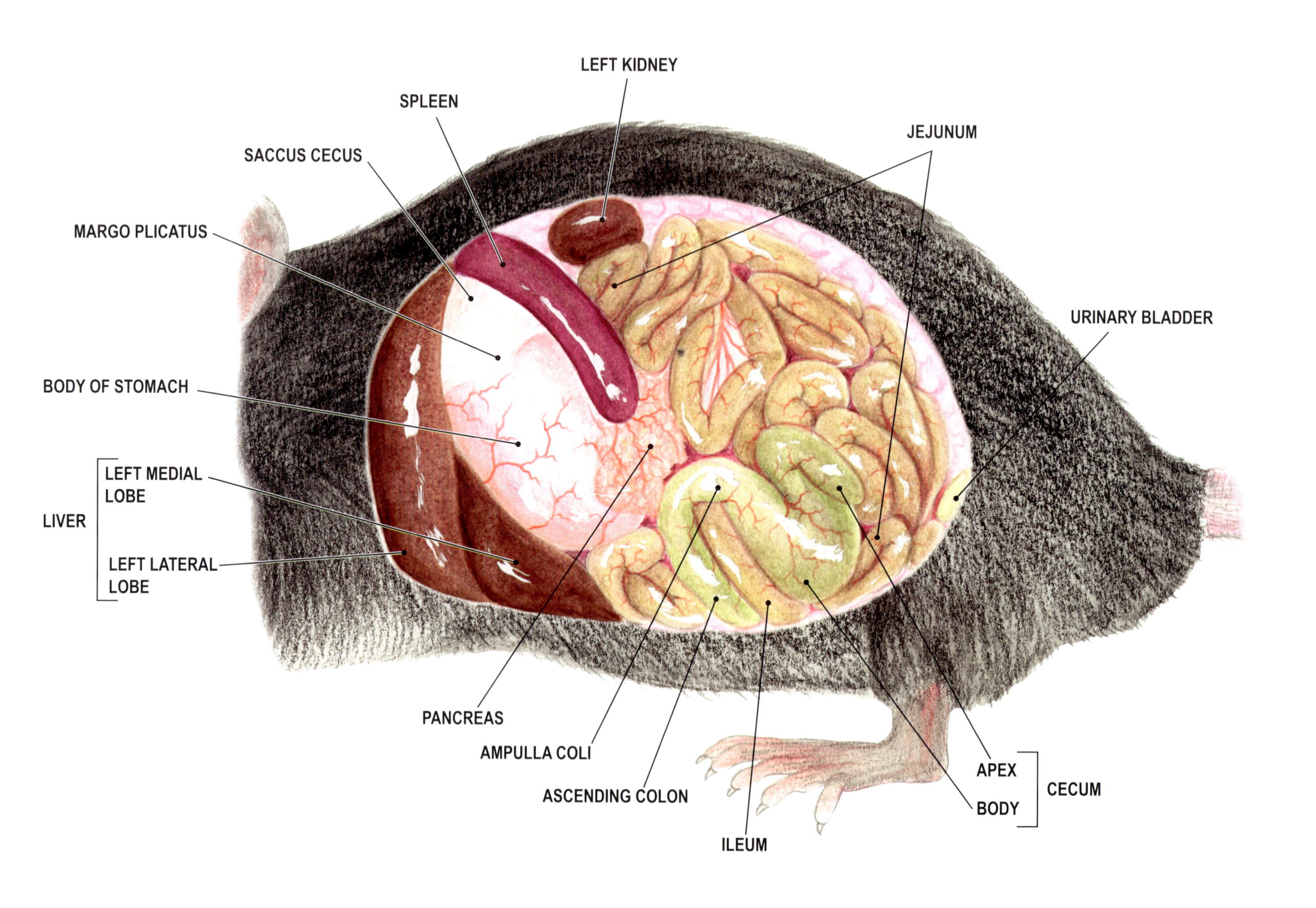

Figure 39B. Rat, female. Abdominal topography (left aspect).

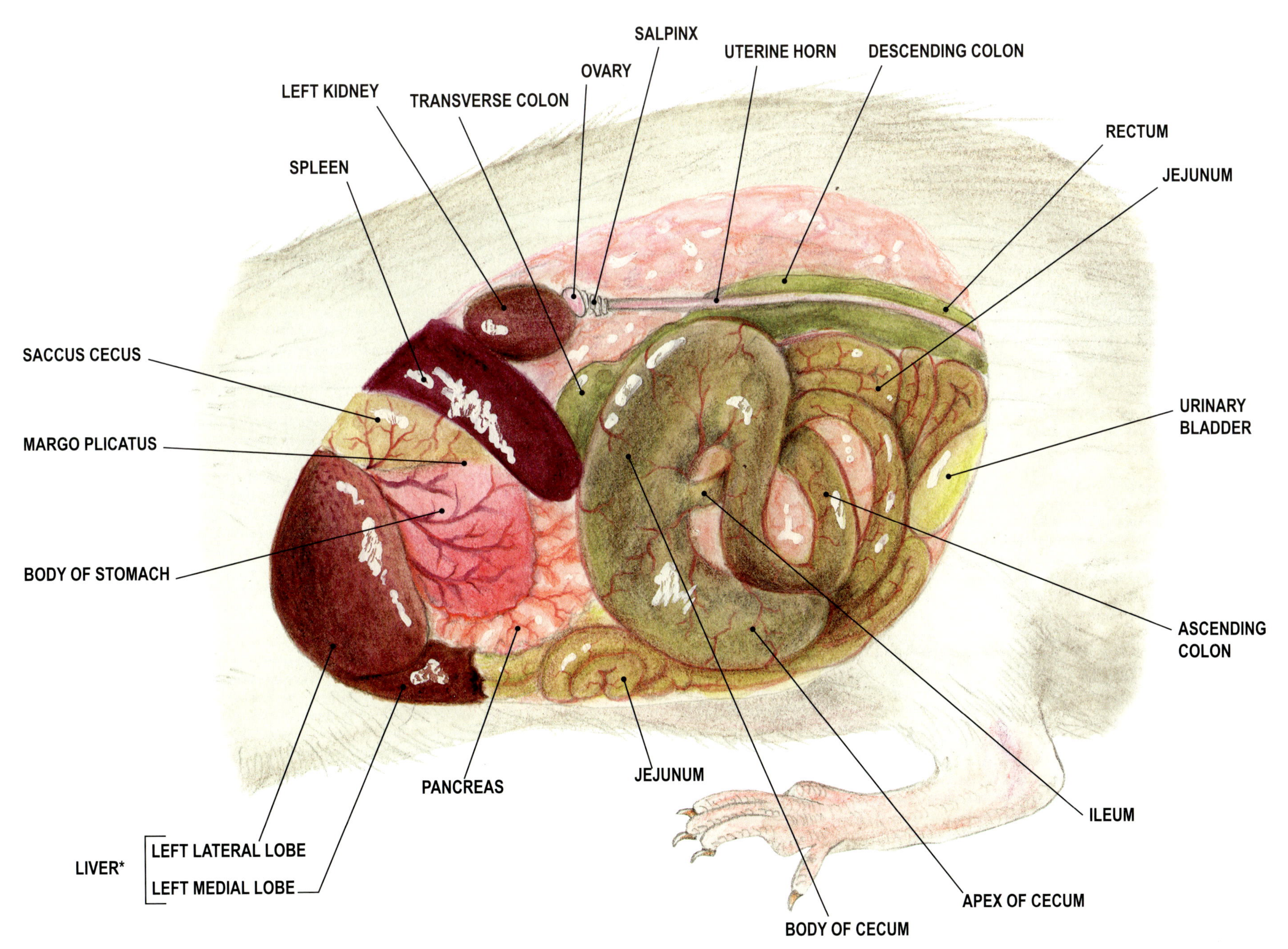

*The rat lacks a gallbladder.

Figure 39A. Mouse, male. Abdominal topography (left aspect).

Figure 39B. Rat, female. Abdominal topography (left aspect).

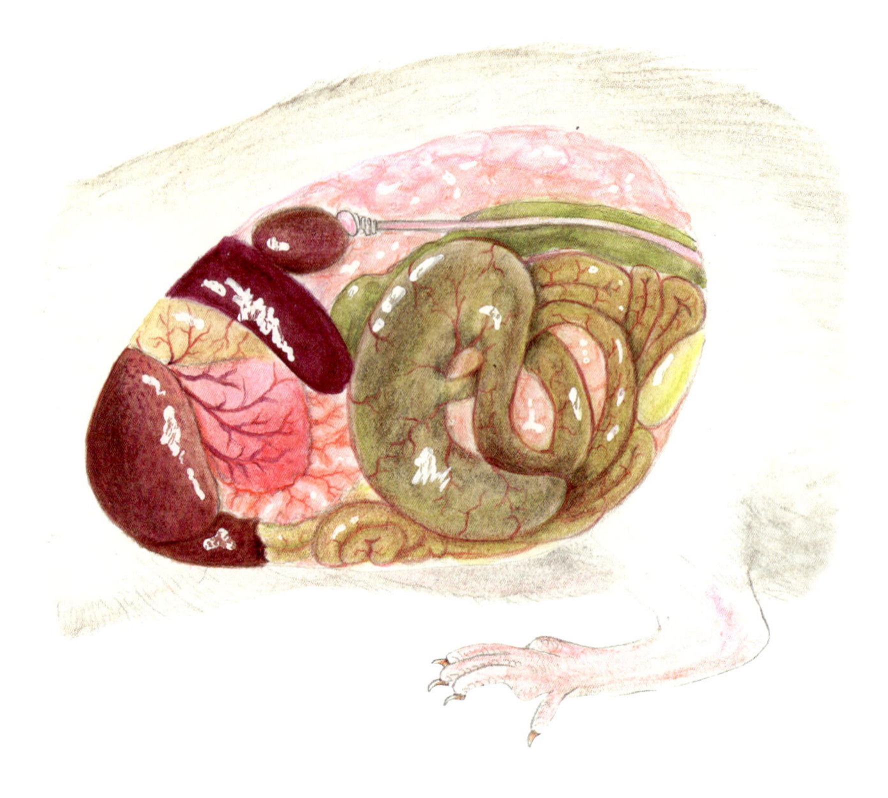

Figure 39. Abdominal topography (left aspect).

Mouse

In a craniocaudal direction (from left to right), the left lateral and the left medial lobes of the liver are in intimate contact with the diaphragm (cranially, not shown) and the stomach (caudally). The stomach body and saccus cecus are separated by the margo plicatus (seen due to tissue transparency). The spleen and the left kidney follow dorsally; the pancreas and part of the jejunum are seen ventrally. The largest part of the jejunum, the ileum, the cecum, and the ascending colon fill the remainder of the left aspect of the abdominal cavity.

Rat

In a craniocaudal direction (from left to right), the left medial and left lateral lobes of liver, the stomach body and saccus cecus separated by the margo plicatus (seen due to tissue transparency), the spleen, and the left kidney are shown. The jejunum (on the floor of the abdominal cavity) and the pancreas are located between the liver and stomach cranially and the body of the cecum caudally. The transverse and descending colons, the rectum, then the cecum body and apex, the ileum, and the jejunum fill the rest of the space as digestive structures. When full with urine, the urinary bladder drops in the abdominal cavity far ventral from the pelvic inlet. In contact with the caudal pole of the left kidney, the left ovary, salpinx (fallopian or uterine tube), and uterine horn are also visible.

Figure 40A. Mouse, male. Abdominal topography (right aspect).

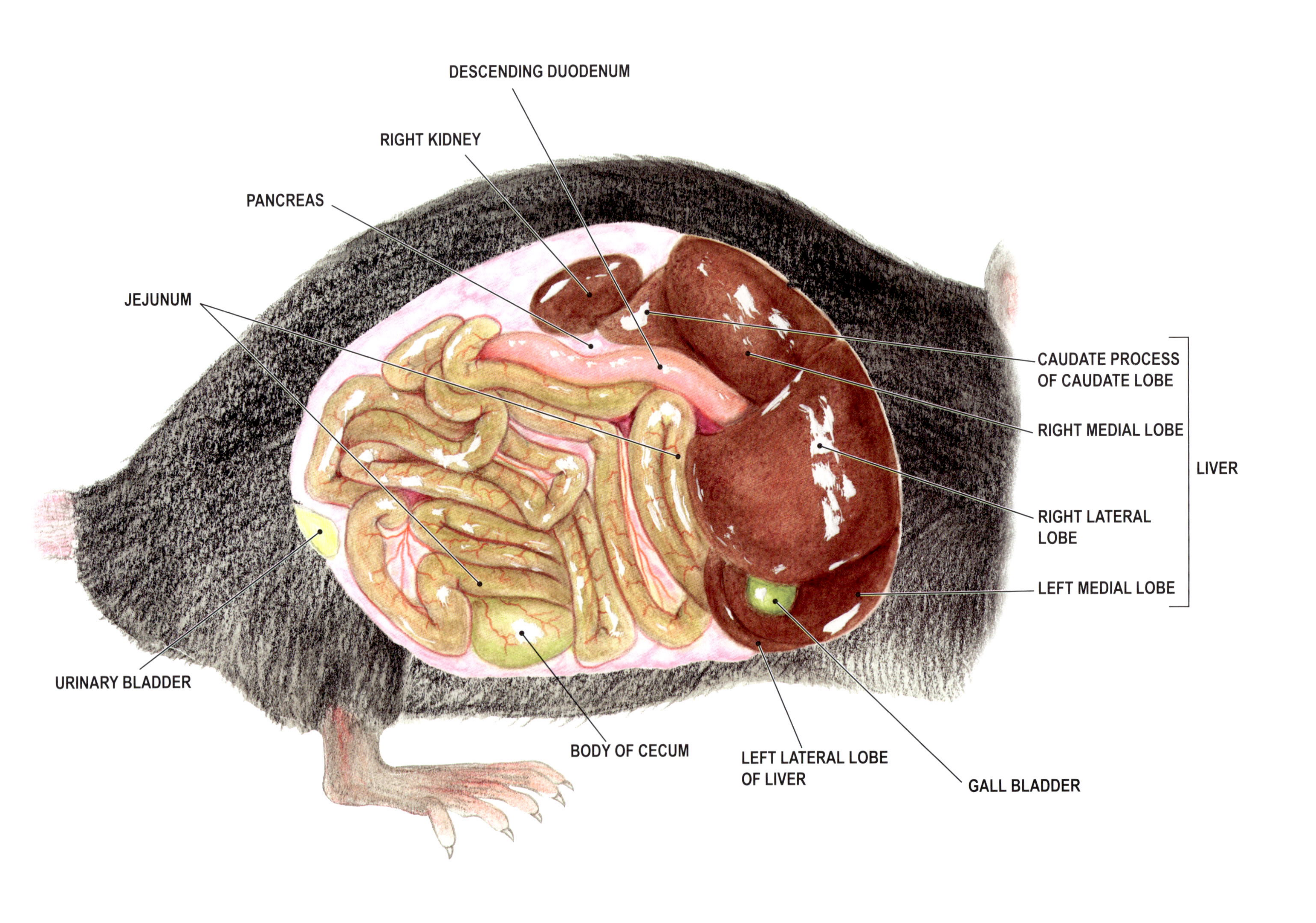

Figure 40B. Rat, female. Abdominal topography (right aspect).

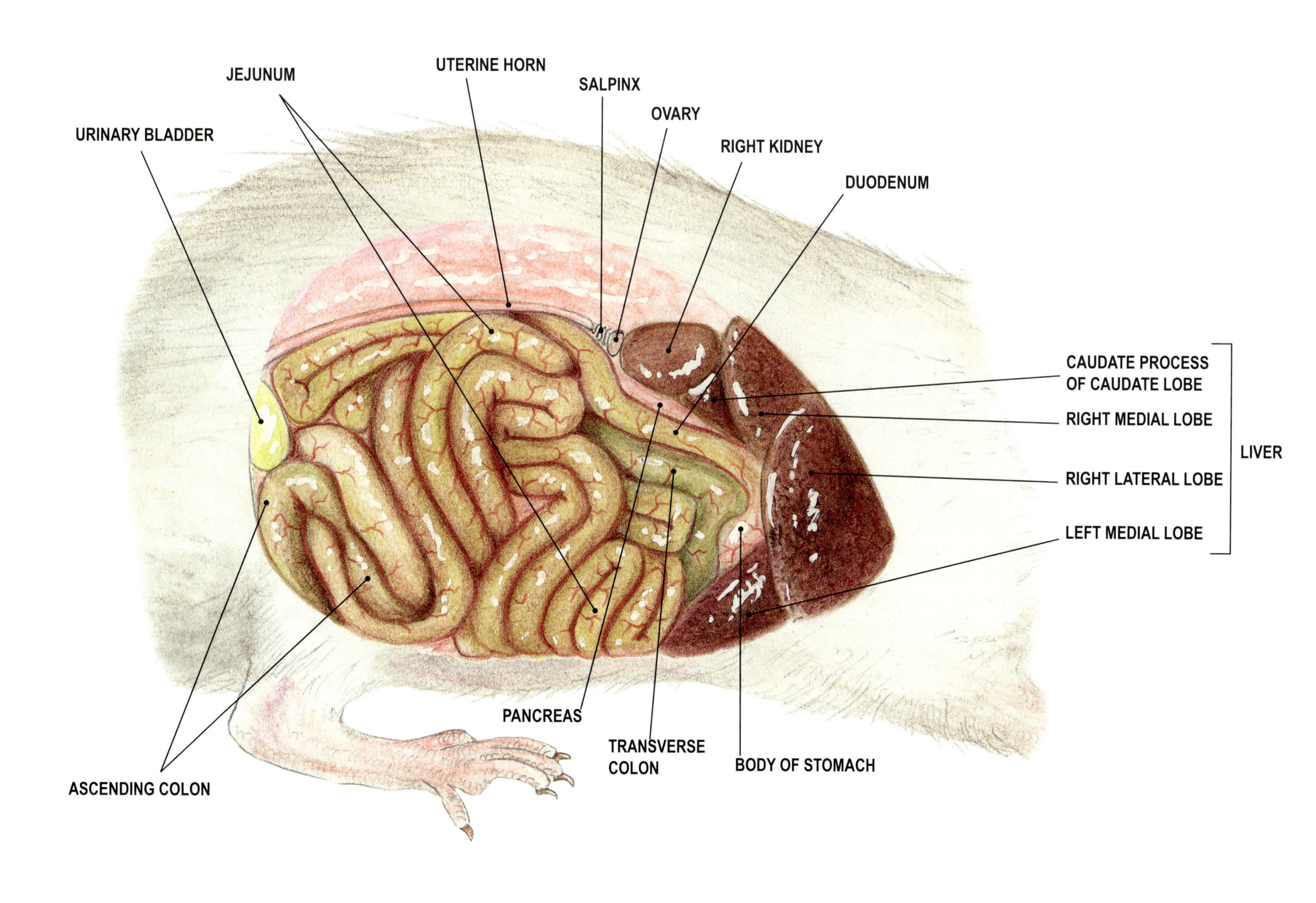

Figure 40A. Mouse, male. Abdominal topography (right aspect).

Figure 40B. Rat, female. Abdominal topography (right aspect).

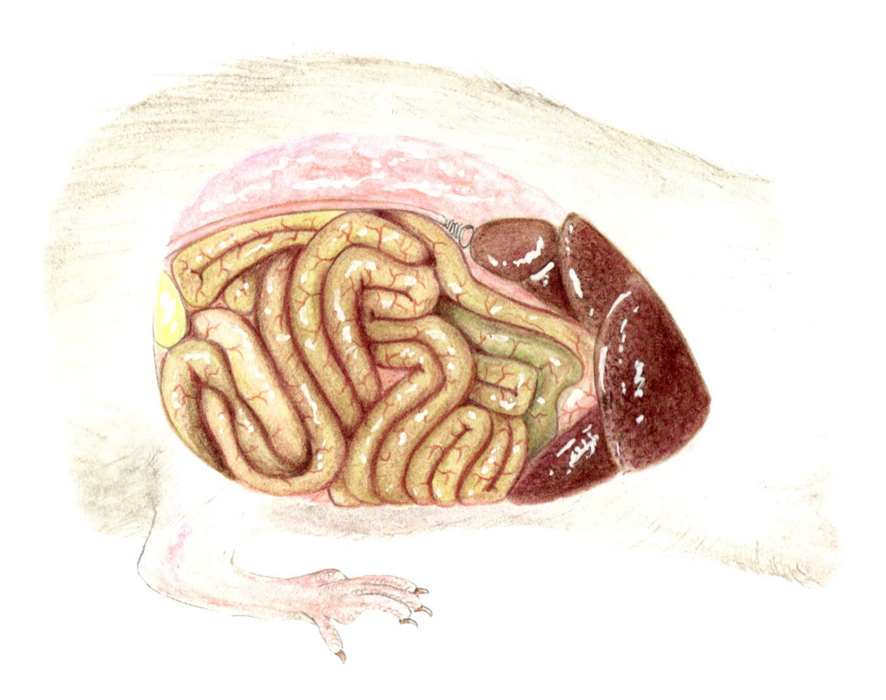

Figure 40. Abdominal topography (right aspect).

Mouse

In a craniocaudal direction (from right to left) and in a dorsoventral direction, the right medial, right lateral, left medial, and left lateral lobes of the liver are in intimate contact with the diaphragm (not shown). The caudate process of the caudate lobe of the liver is intercalated between the right medial lobe of the liver and the right kidney; the gallbladder is shown at the ventral border of the right lateral lobe of the liver. Ventral to the right medial lobe and caudate process of the caudate lobe of the liver and the right kidney, the descending duodenum runs in a craniocaudal direction. With the exception of a small area on the ventral aspect of the abdomen filled by the cecum, the rest of the abdominal cavity is covered by the jejunum. When full with urine, the urinary bladder can be seen as the most caudal structure, in contact with the jejunum.

Rat

In a craniocaudal direction (from right to left) and in a dorsoventral direction, the right medial, right lateral, and the left medial lobes of the liver are in intimate contact with the diaphragm (not shown). The caudate process of the caudate lobe of the liver and the right kidney follow on the dorsal aspect. The body of the stomach, the duodenum, and the pancreas fill the space caudal to the liver. The remainder of the abdominal cavity is filled by the transverse colon cranially, jejunal loops, the ascending colon, and the urinary bladder (visible when full). Starting from the caudal pole of the right kidney, the right ovary, salpinx, and uterine horn run against the roof of the abdominal cavity.

Figure 41A. Mouse, female. Abdominal topography (ventral aspect).

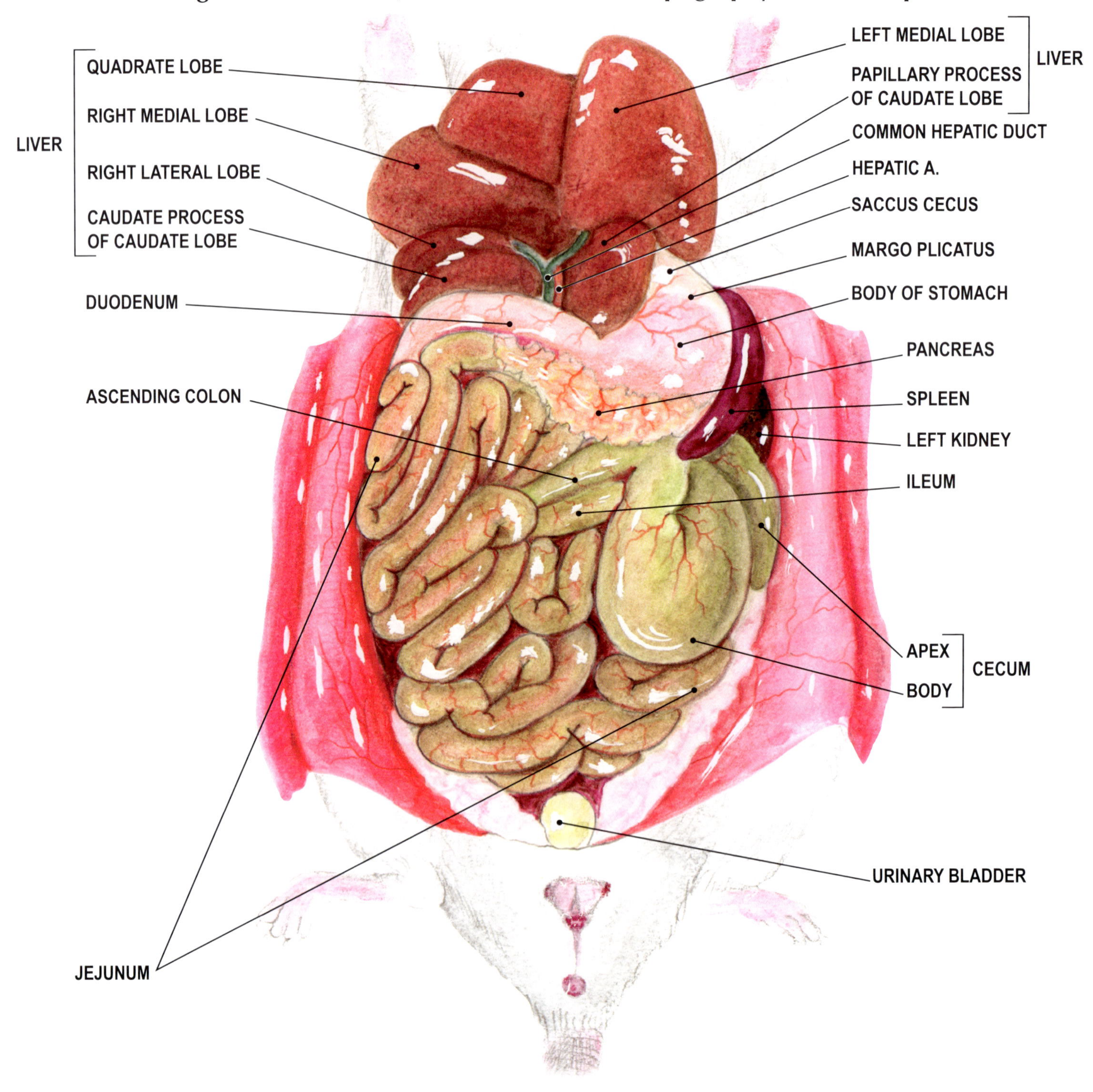

Figure 41B. Rat, female. Abdominal topography (ventral aspect).

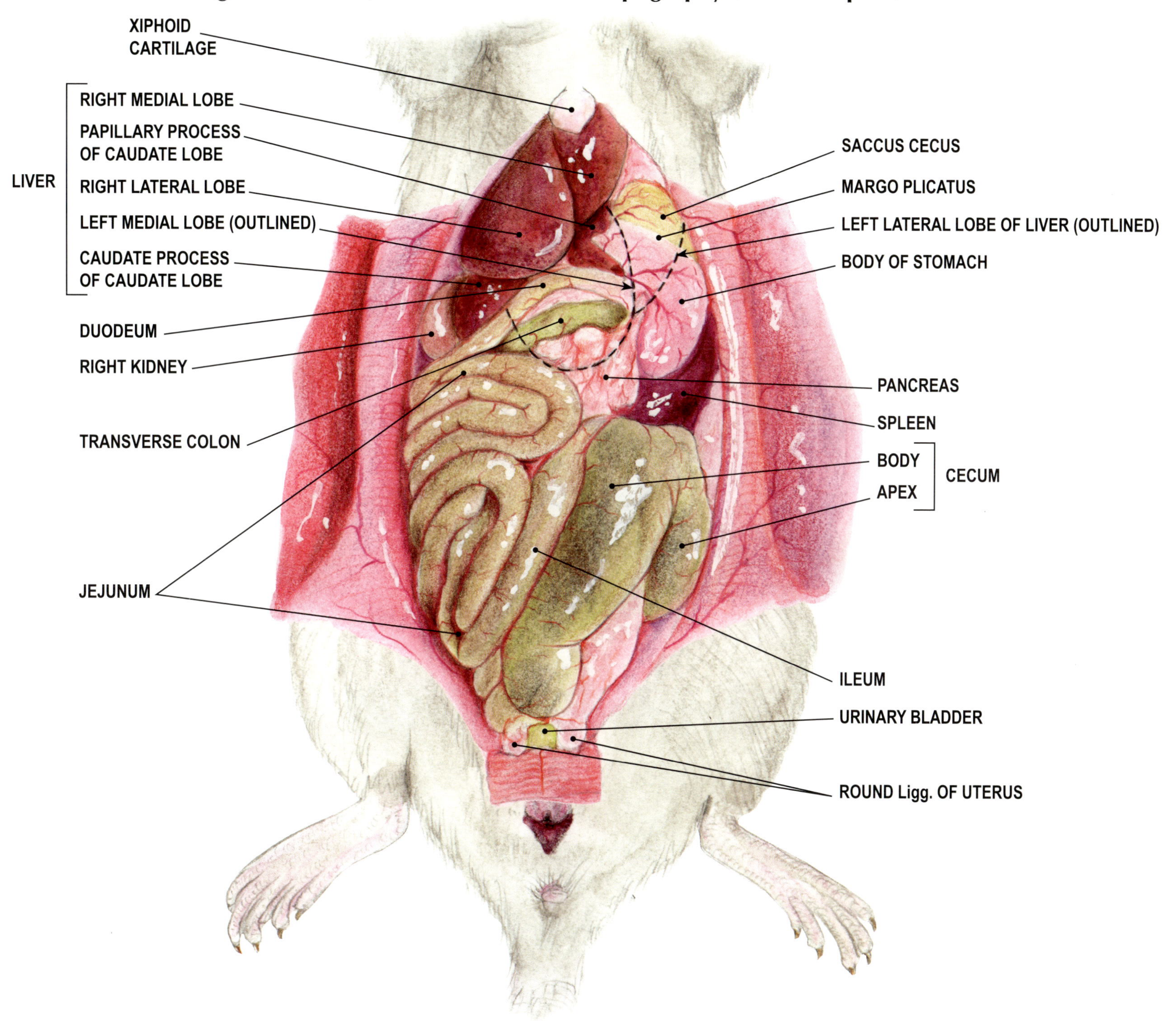

Figure 41A. Mouse, female. Abdominal topography (ventral aspect).

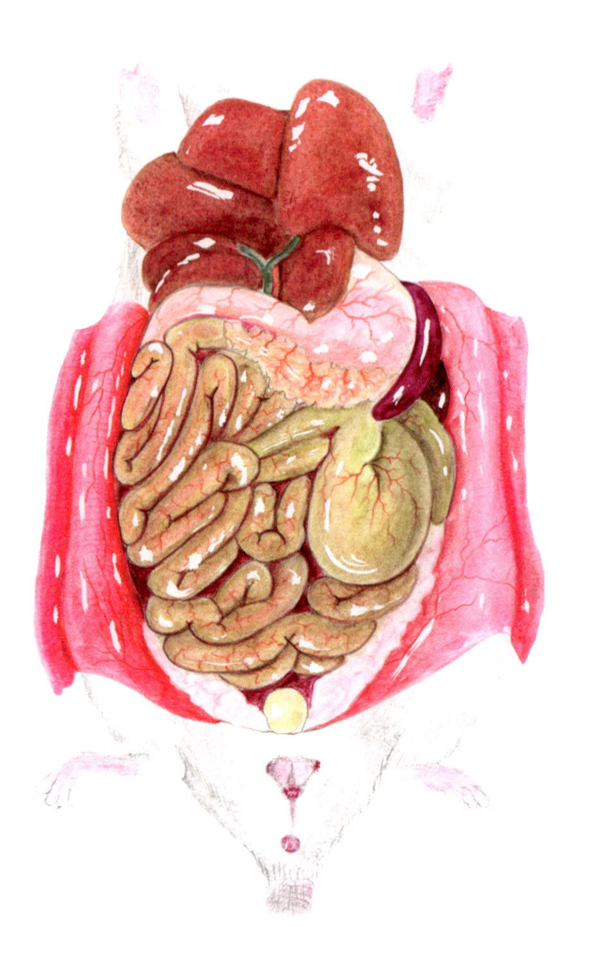

Figure 41B. Rat, female. Abdominal topography (ventral aspect).

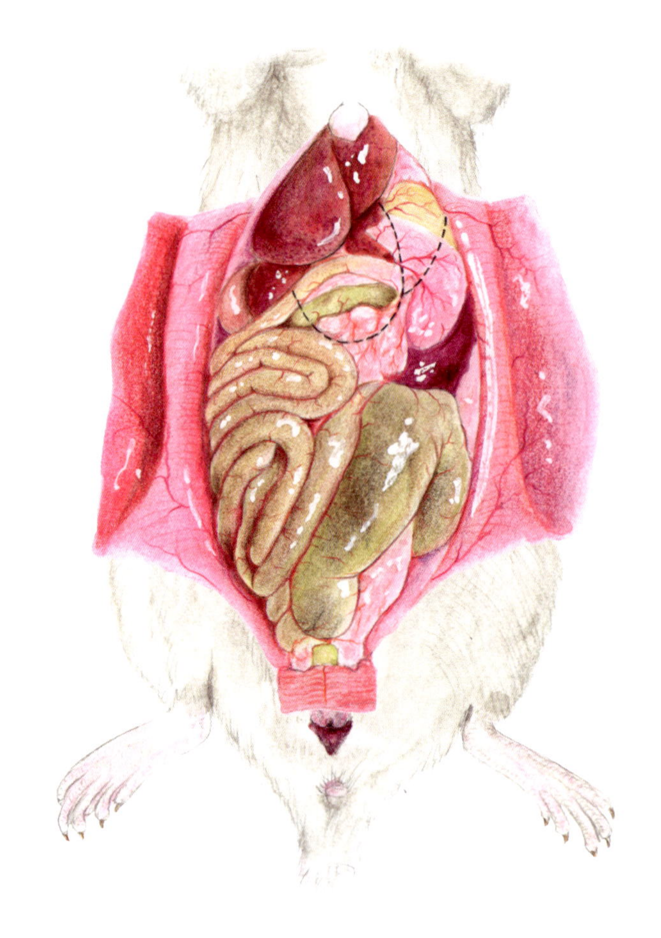

Figure 41. Abdominal topography (ventral aspect).

The cecum is movable in both species, but there are differences in shape and position between the species. This variation occurs between the two species due to diet and food consumption.

Mouse

The liver is reflected cranially to expose all lobes of the liver. Starting from the bottom left, on the figure, shown are the caudate process of the caudate lobe; the right lateral, right medial, quadrate, and left medial lobes; and the papillary process of the caudate lobe. The liver reflected cranially is also shown in Figure 43.

The right and left hepatic ducts and the common hepatic duct are accompanied by the hepatic artery.

The gallbladder is not shown in this figure (see Figure 43).

The pancreas is more compact and more closely associated with the stomach and duodenum in the mouse than in the rat.

Rat

From the right, shown are the caudate process of the caudate lobe, the right lateral and right medial lobes, and the papillary process of the caudate lobe. The left medial and left lateral lobes are outlined to allow the visualization of overlying structures (duodenum, stomach, and pancreas).

The liver reflected cranially is shown in Figure 43.

Lobules/acini of the pancreas are scattered in the greater omentum; these are not easy to distinguish from fat. Therefore, a total pancreatectomy is not possible in the rat.

Figure 42A. Mouse, female. Liver, stomach, and intestines (ventral aspect), intestines displaced.

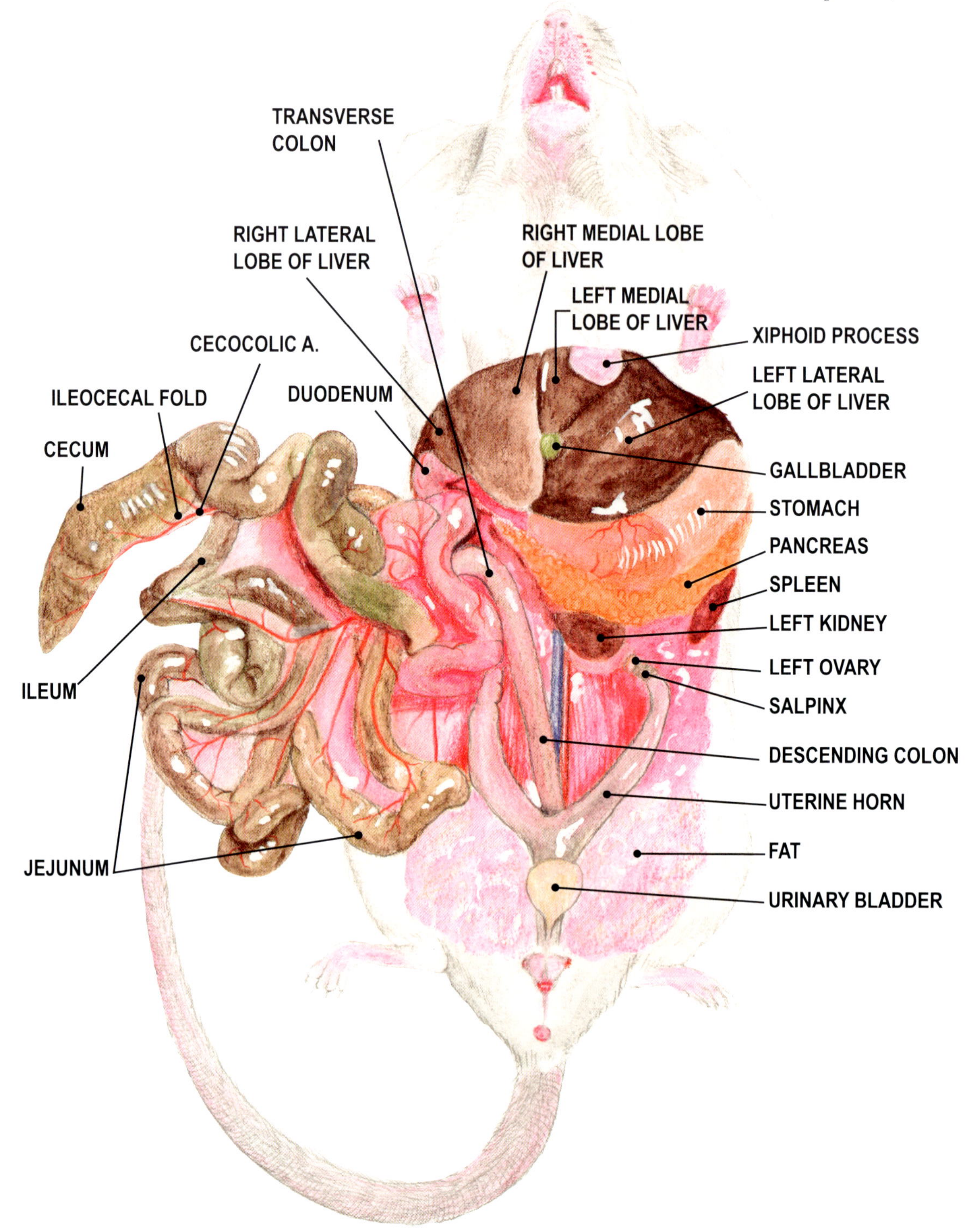

Figure 42B. Rat, male. Liver, stomach, and intestines (ventral aspect), intestines displaced.

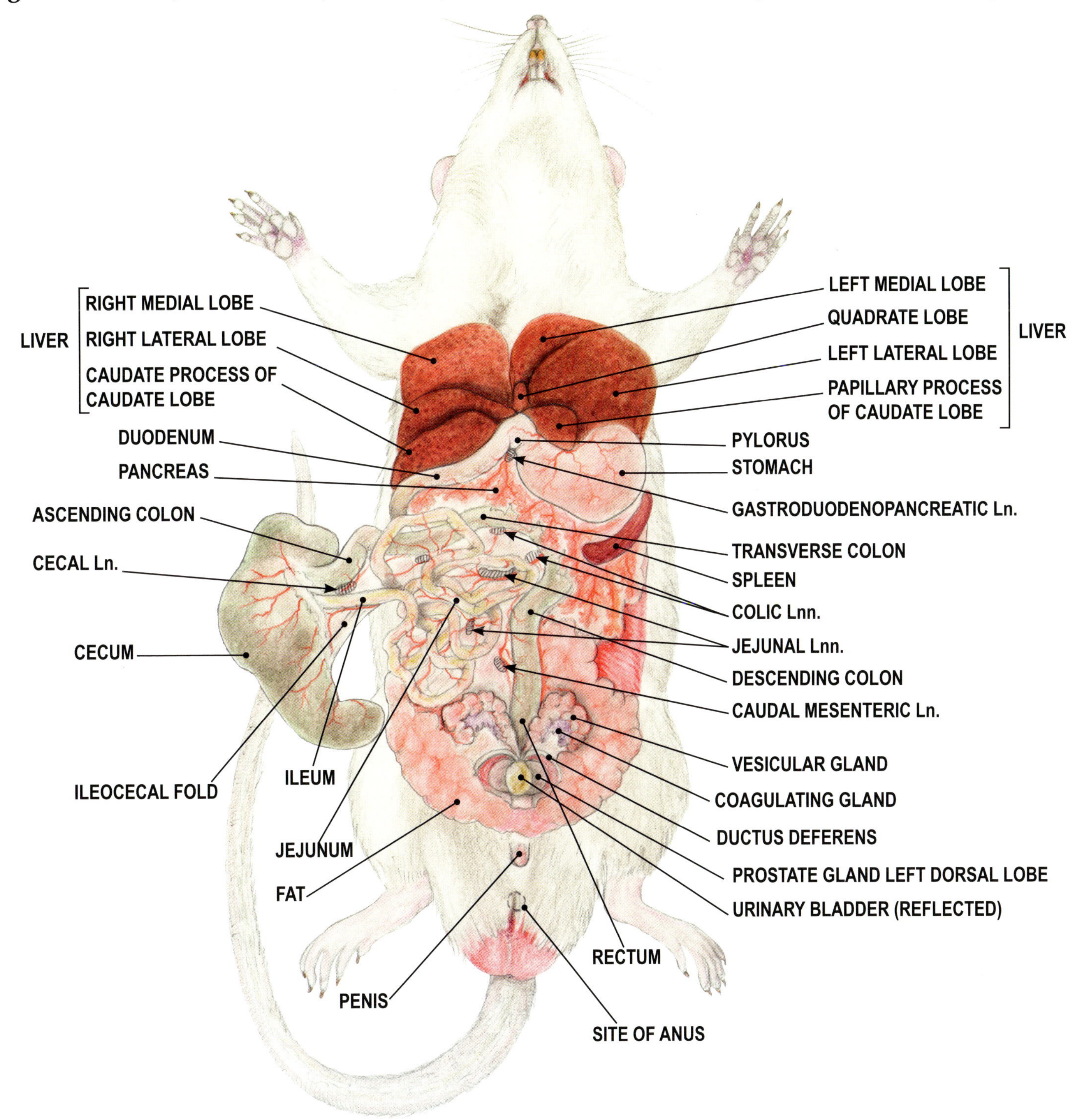

Figure 42A. Mouse, female. Liver, stomach, and intestines (ventral aspect), intestines displaced.

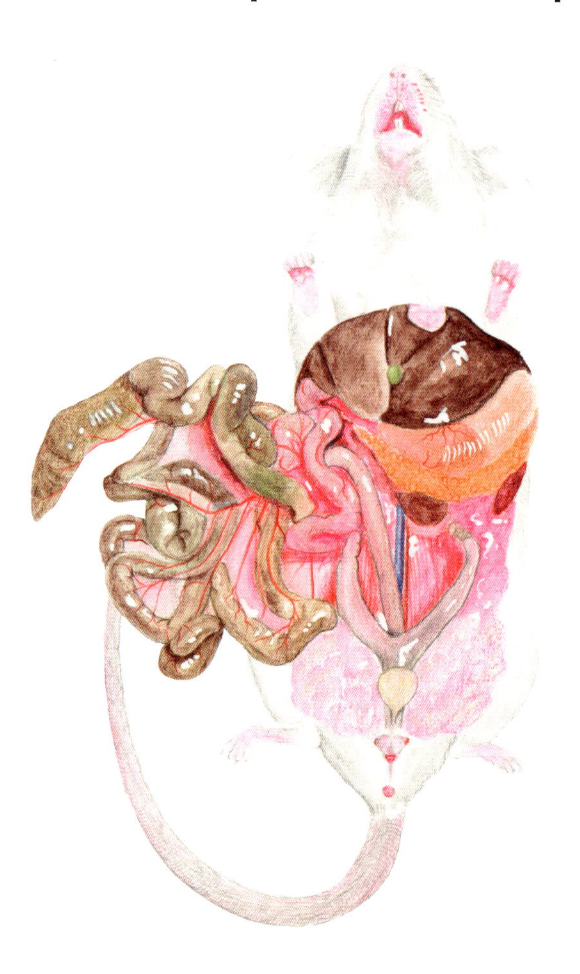

Figure 42B. Rat, male. Liver, stomach, and intestines (ventral aspect), intestines displaced.

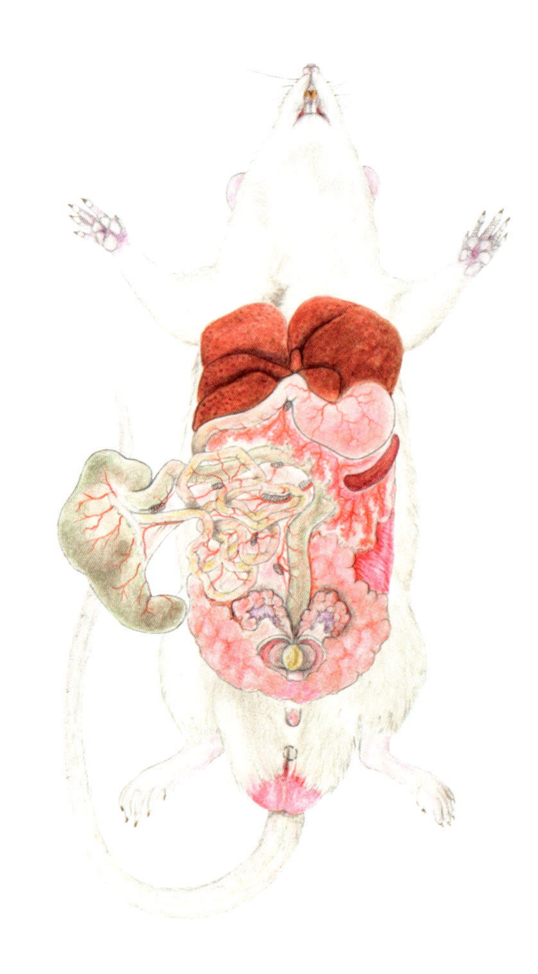

Figure 42. Liver, stomach, and intestines (ventral aspect), intestines displaced.

Mouse

In the figure, the xiphoid cartilage appears off midline because the weight of the stomach and liver press on the left abdominal wall, thus tilting the xiphoid cartilage.

The liver, in a natural position, shows the following lobes: right lateral, right medial, left medial, and left lateral. The caudate lobe, consisting of a caudate process and a papillary process, and the quadrate lobe cannot be seen in this figure.

Unlike the rat, the mouse has a gallbladder. Both rats and mice have bile ducts.

The pancreas extends along the greater curvature of the stomach and the cranial part of the duodenum. Portions of the pancreas lie in association with the spleen, liver, and right kidney.

The apex of the cecum is pointed in the mouse. Because the ileocecal fold is very short, the cecum is not held in a steady position and can move freely when mixing ingesta. Therefore, the cecum can be seen in the left, right, or ventral aspect of the abdomen.

Rat

The liver, reflected cranially, shows the following lobes: caudate process of the caudate lobe, right lateral, right medial, left medial, quadrate, left lateral, and papillary process of the caudate lobe.

The rat lacks a gallbladder. The bile duct is not shown (see Figure 43).

The pancreas of the rat has a wider spread, extending along a greater distance of the duodenum and caudally in the omentum. Clusters of pancreatic lobules/acini are usually scattered in the superficial leaf of the greater omentum, which makes a total pancreatectomy practically impossible. The pancreas lies in contact with the spleen, but has less association with the liver and kidneys than in the mouse.

The apex of the cecum is rounded, not pointed as in the mouse. The rat has a larger ileocecal fold and a cecocolic fold (between the cecum and the ascending colon), which hold the cecum in a more fixed position than in the mouse. The apex is highly movable, which is important for cecal mixing.

The following lymph nodes are exposed: gastroduodenopancreatic, jejunal, cecal, colic, and caudal mesenteric.

In the mouse and the rat, as in all species, the ileocecal fold joins the greater curvature of the ileum to the lesser curvature of the cecum. The end of the attachment of the ileocecal fold to the ileum marks the junction between the jejunum and the ileum. Also typical of all species, the mesentery (mesojejunum and mesoileum) attach to the lesser curvatures of the jejunum and the ileum, respectively, which are therefore called the mesenterial border of the intestines. The major curvature of all intestines, opposite to the mesenterial border is called the antimesenterial border of the intestines. The mesenterial arteries run parallel to the lesser (mesenterial) curvature, and the antimesenterial arteries run parallel to the greater (antimesenterial) curvature.

The mucosa of the jejunum and ileum is rich in aggregated nodules of lymphoid tissue called Peyer's patches, which are not covered by villi. The Peyer's patches cannot be shown in the figures because they are inside of the jejunum and ileum.

Figure 43A. Mouse. Upper abdominal structures (ventral aspect), liver reflected cranially.

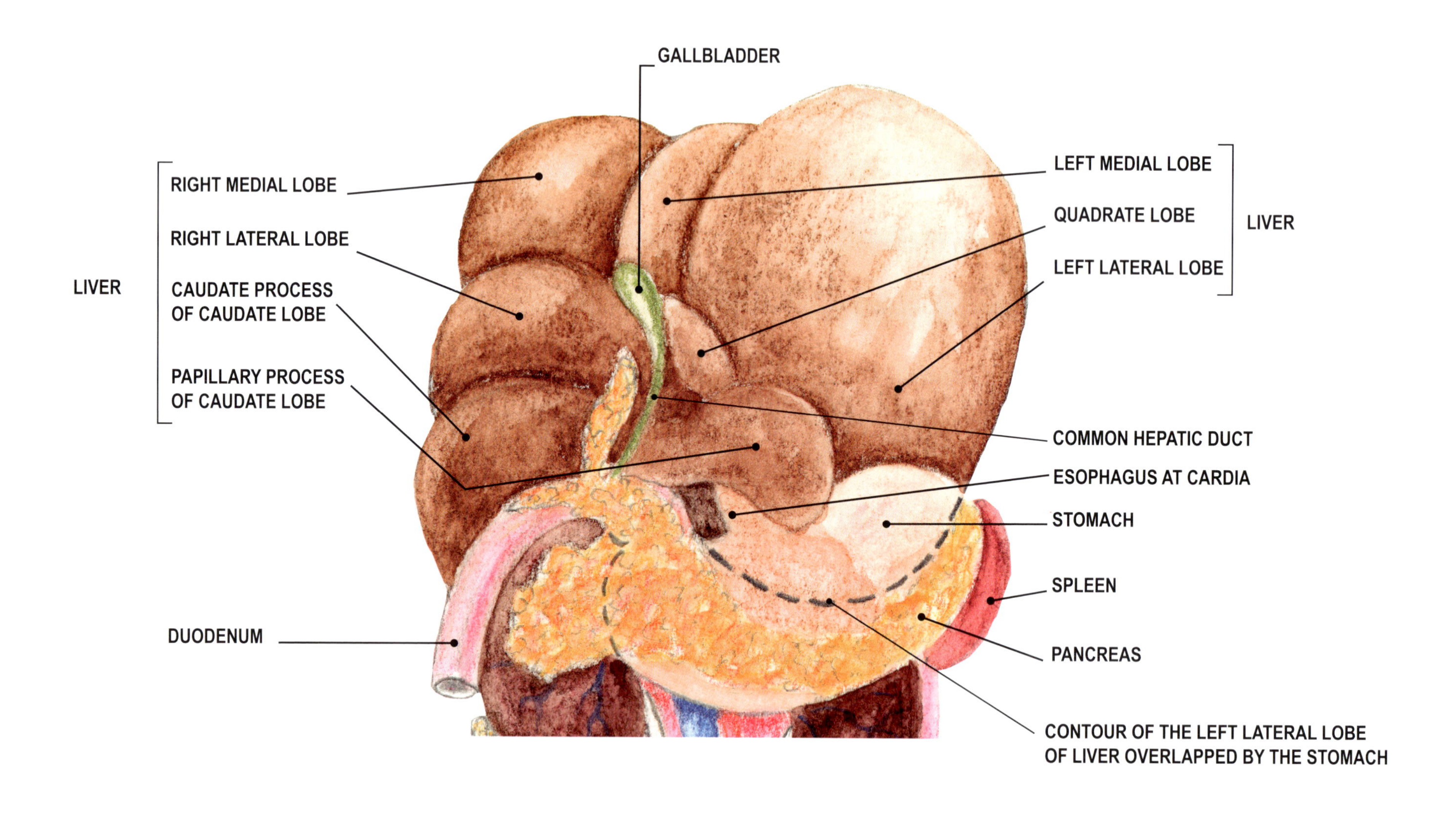

Figure 43B. Rat. Upper abdominal structures (ventral aspect), liver reflected cranially.

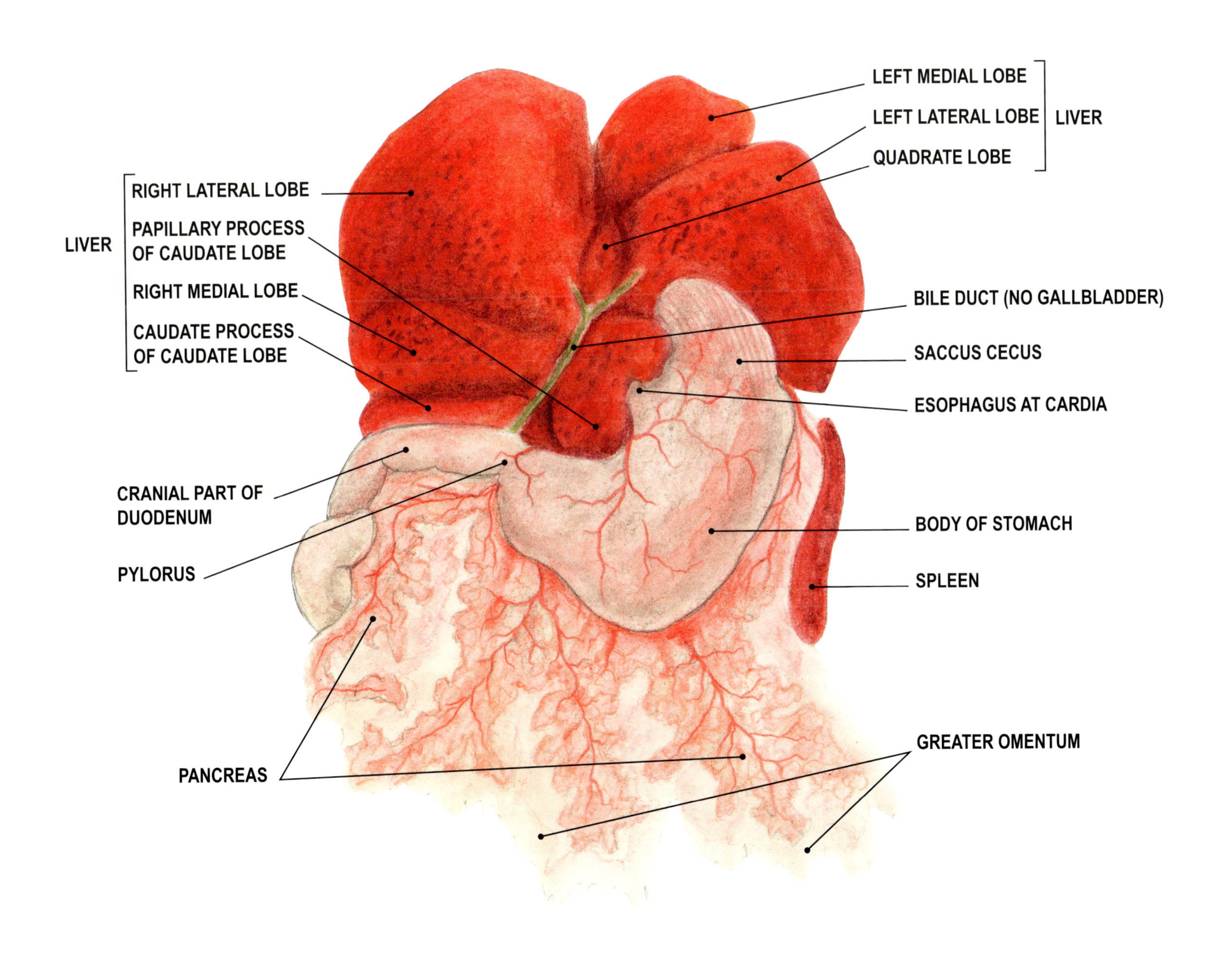

Figure 43A. Mouse. Upper abdominal structures (ventral aspect), liver reflected cranially.

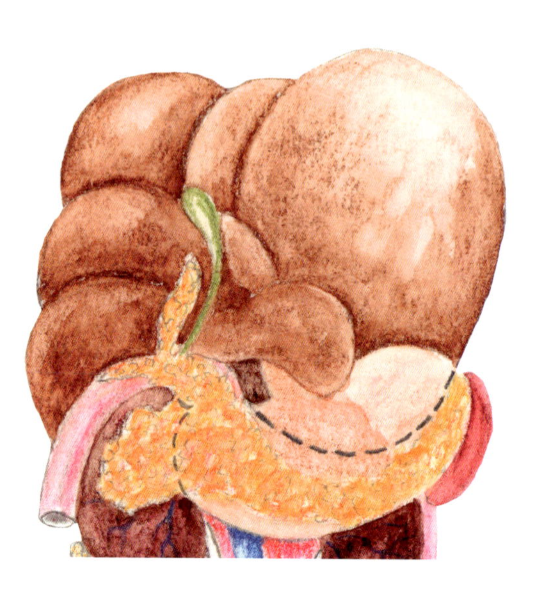

Figure 43B. Rat. Upper abdominal structures (ventral aspect), liver reflected cranially.

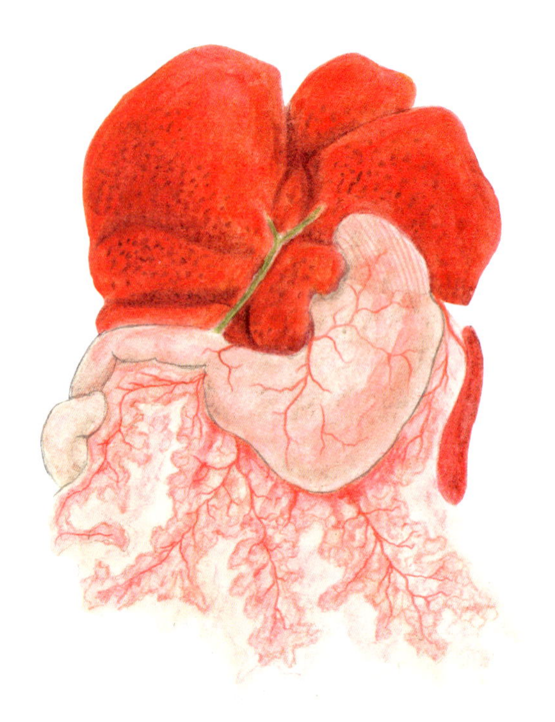

Figure 43. Upper abdominal structures (ventral aspect), liver reflected cranially.

In both species, the liver is reflected cranially to show all the lobes and the lack of a gallbladder in the rat. See Figure 42 for the position of the liver lobes in situ.

Figure 44A. Mouse. Stomach, distal esophagus, and proximal duodenum (internal aspect). Left, median section through the long axis. Right, section through the major curvature.

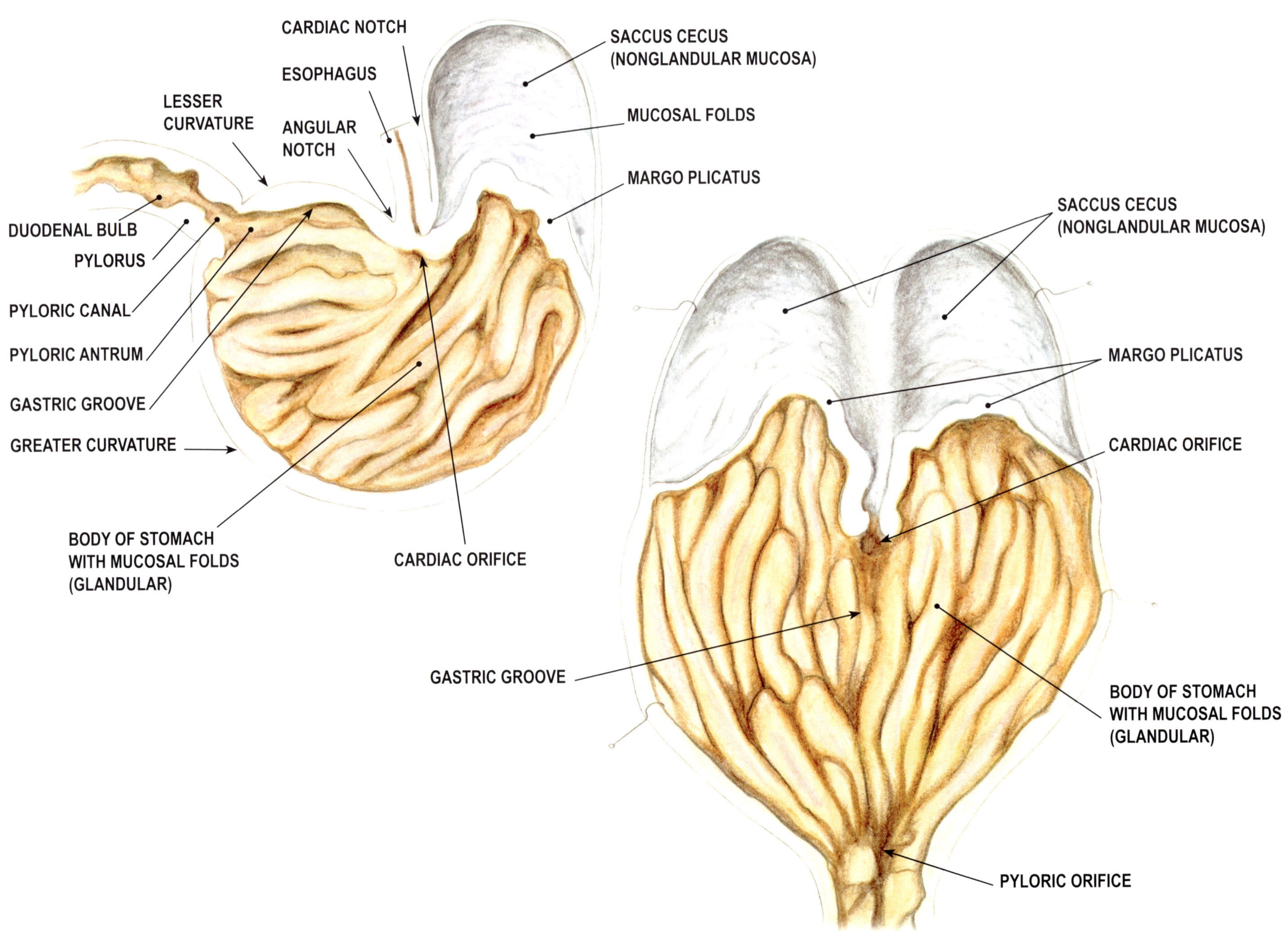

Figure 44B. Rat. Stomach, distal esophagus, and proximal duodenum (internal aspect). Left, median section through the long axis. Right, section through the major curvature.

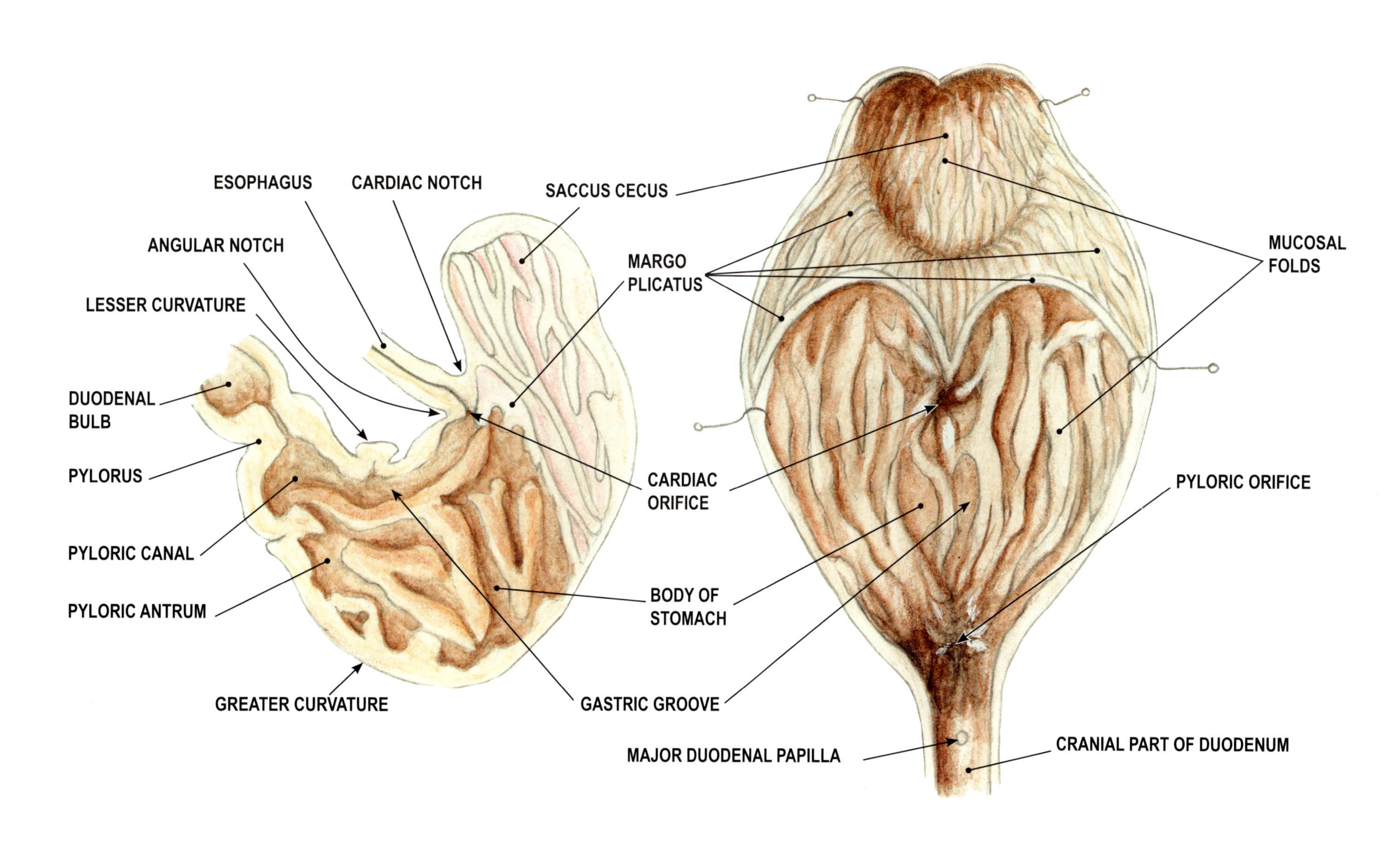

Figure 44A. Mouse. Stomach, distal esophagus, and proximal duodenum (internal aspect). Left, median section through the long axis. Right, section through the major curvature.

Figure 44B. Rat. Stomach, distal esophagus, and proximal duodenum (internal aspect). Left, median section through the long axis. Right, a section through the major curvature.

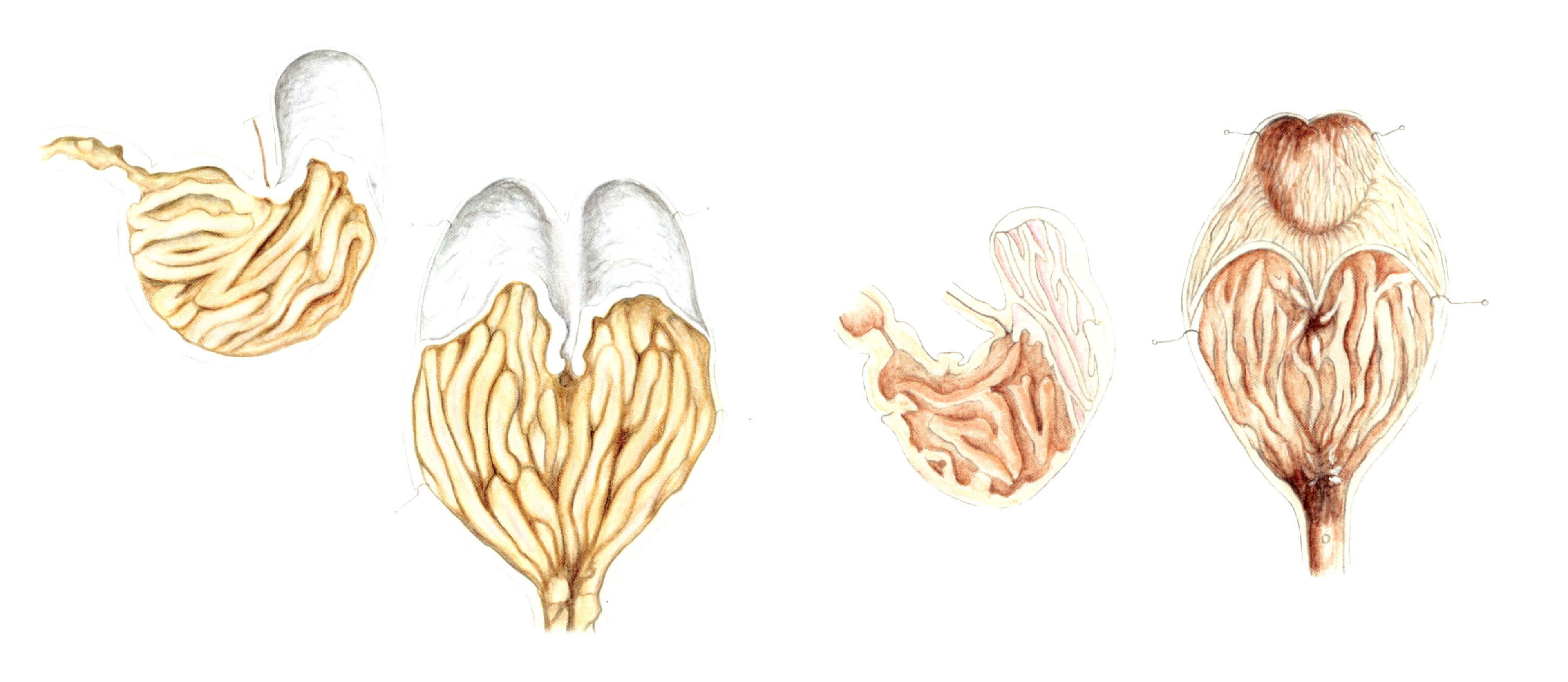

Figure 45B. Rat, female. Parietal lymph nodes of the roof of the abdominal cavity (ventral aspect).

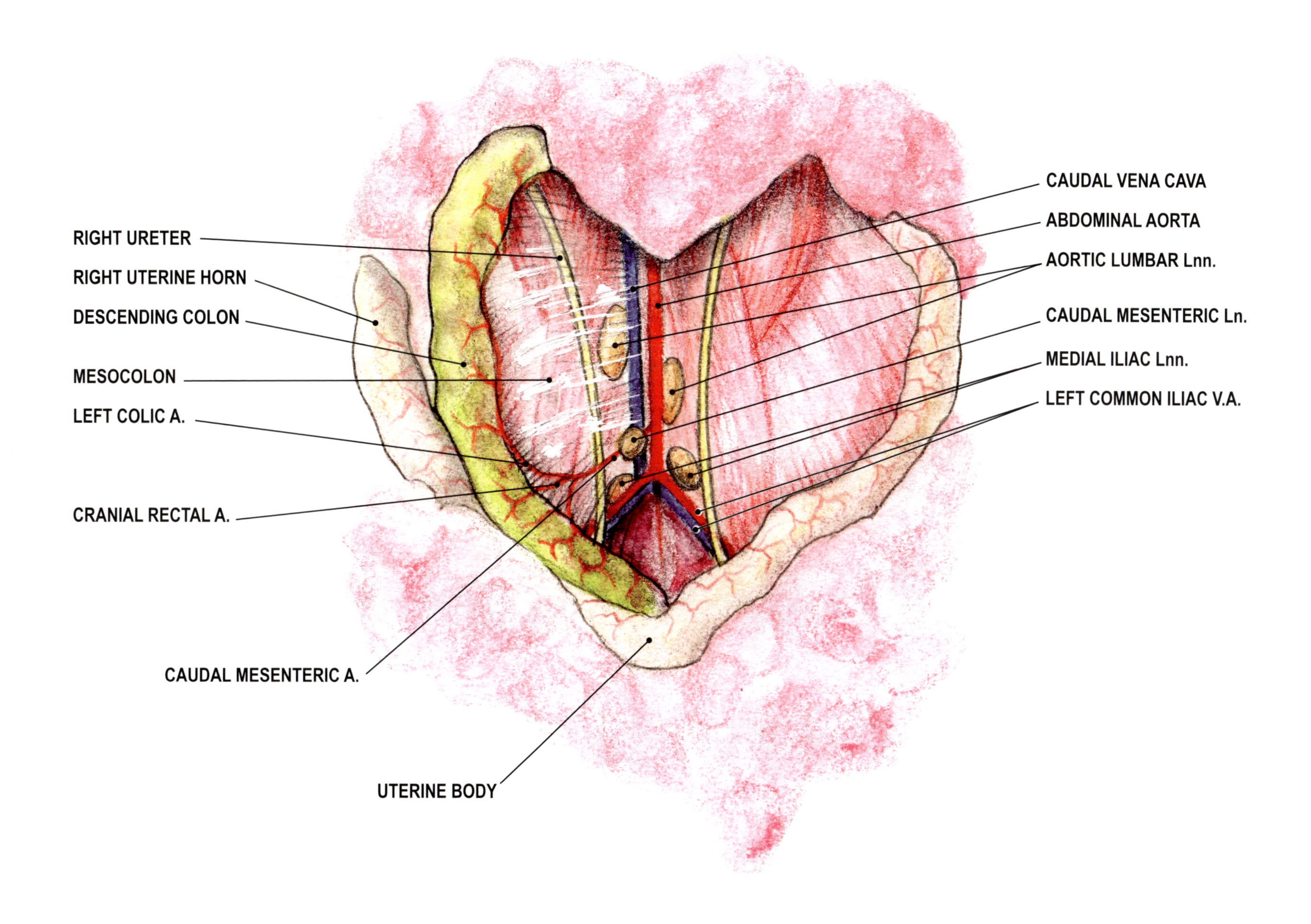

Figure 45A. Mouse, female. Parietal lymph nodes of the roof of the abdominal cavity (ventral aspect).

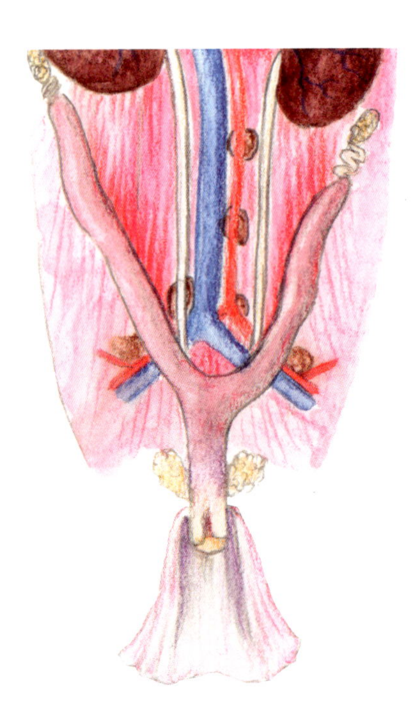

Figure 45B. Rat, female. Parietal lymph nodes of the roof of the abdominal cavity (ventral aspect).

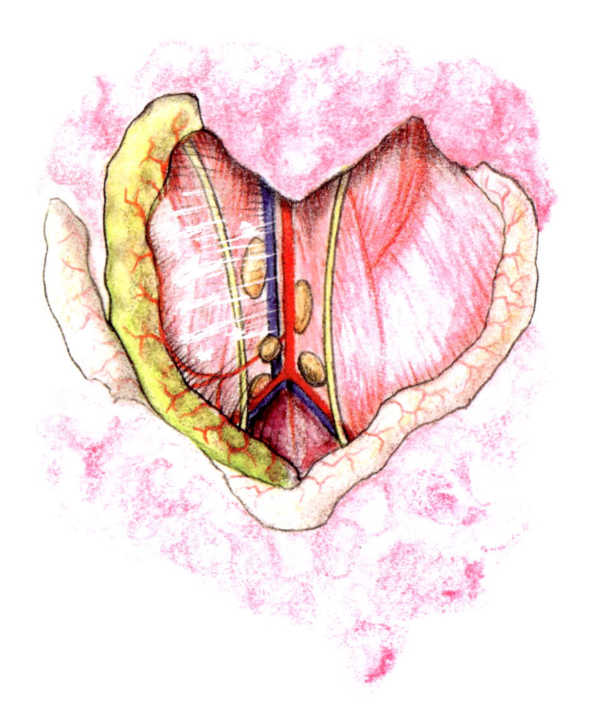

Figure 45. Parietal lymph nodes of the roof of the abdominal cavity (ventral aspect).

There are two major categories of lymph nodes in all species: muscular and cavitary. The cavitary lymph nodes are parietal and visceral. The parietal lymph nodes are located on the walls of the thoracic and abdominal cavities (dorsal or ventral), and are named according to the bones or arteries they are close to.

Particular attention should be given to the caudal mesenteric lymph node, which is often required for PCR confirmation of serologically diagnosed mouse parvovirus. However, this lymph node is not easy to find. According to Van den Broeck et al, it can be found consistently lying along the caudal mesenteric artery, between the two laminae of the descending mesocolon.[14]

MALE UROGENITAL APPARATUS

Figure 46A. Mouse. Male caudal abdominal and pelvic viscera including the testicle (left lateral aspect).

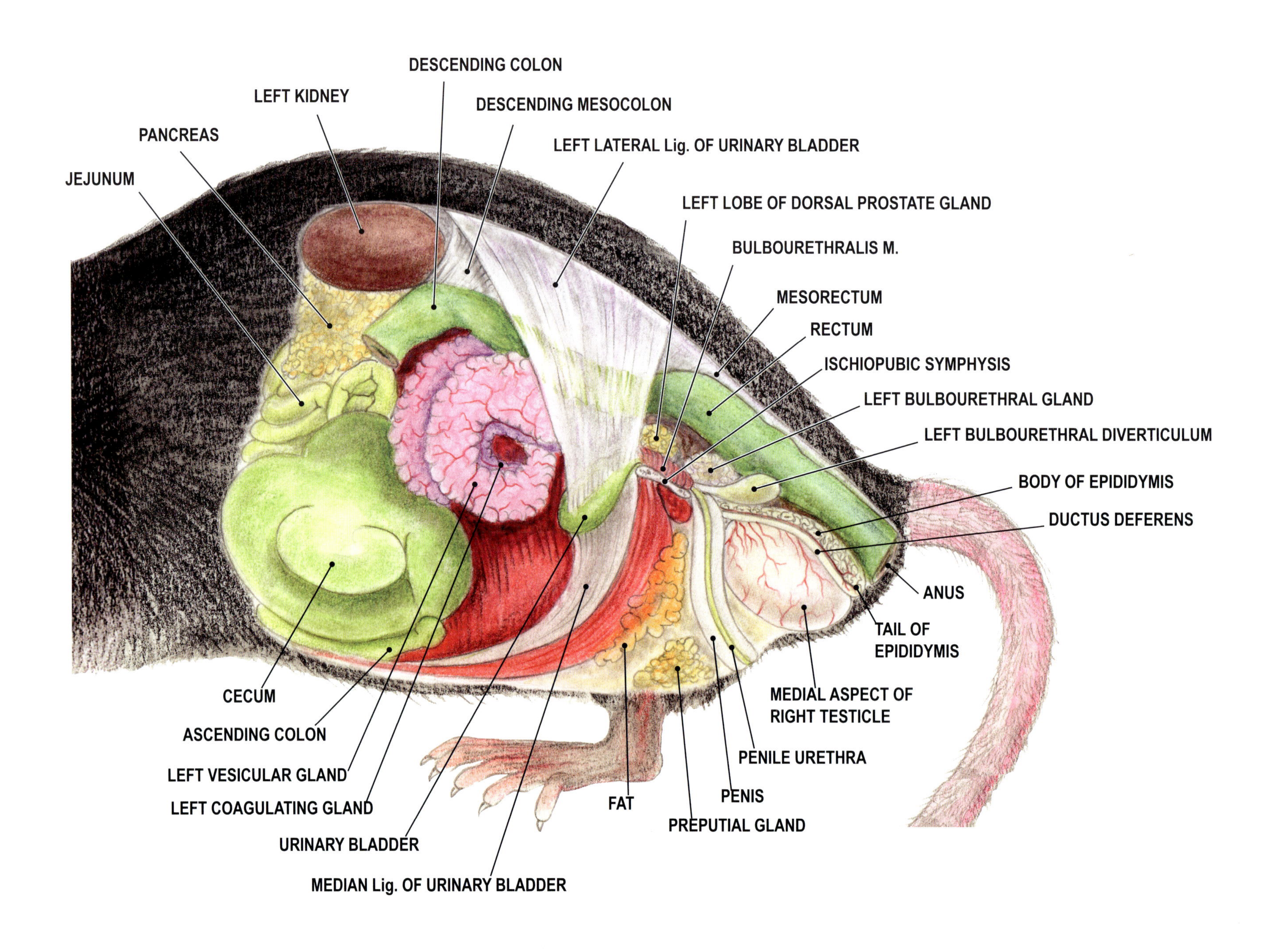

Figure 46B. Rat. Male caudal abdominal and pelvic viscera including the testicle (left lateral aspect).

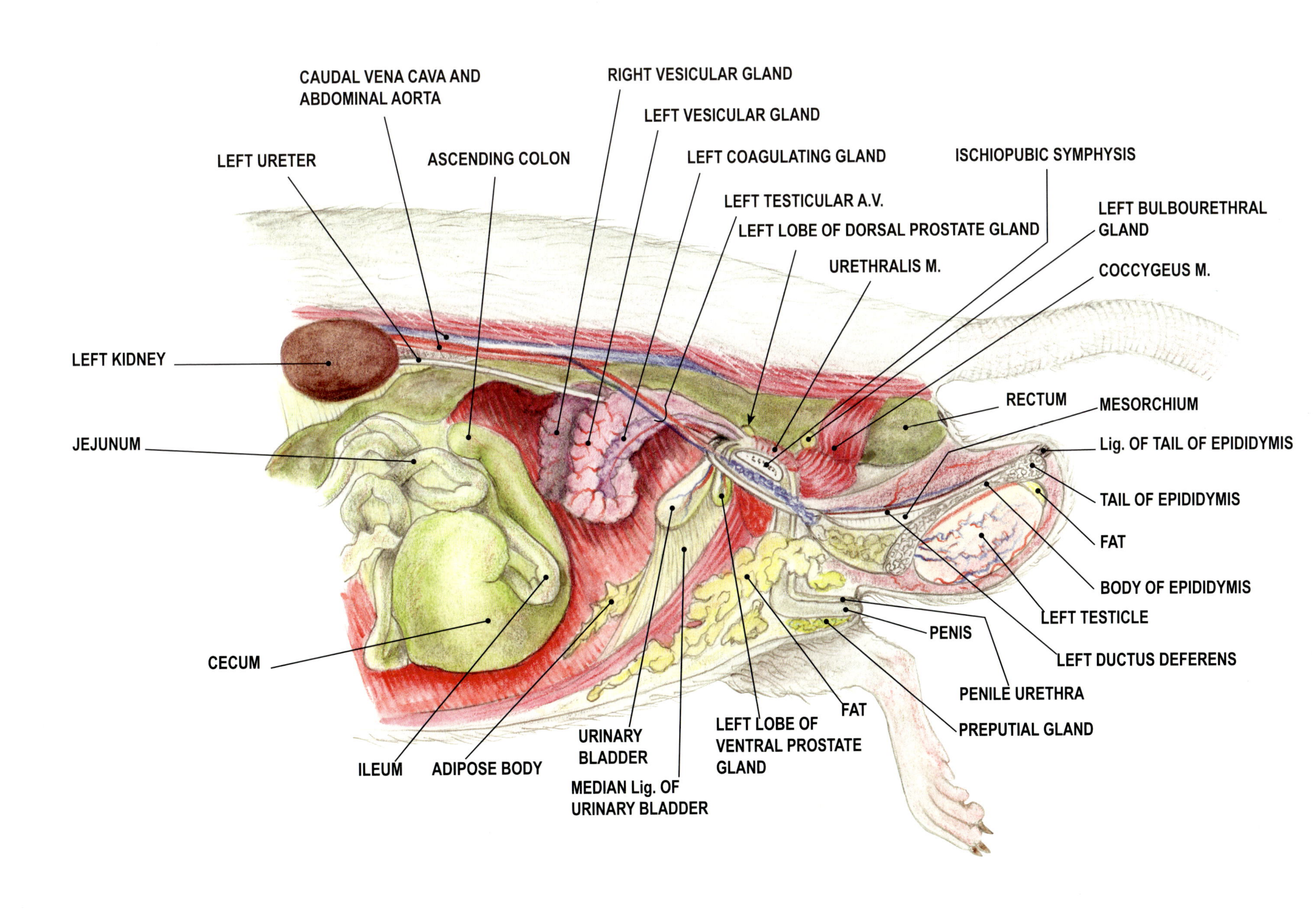

Figure 46A. Mouse. Male caudal abdominal and pelvic viscera including the testicle (left lateral aspect).

Figure 46B. Rat. Male caudal abdominal and pelvic viscera including the testicle (left lateral aspect).

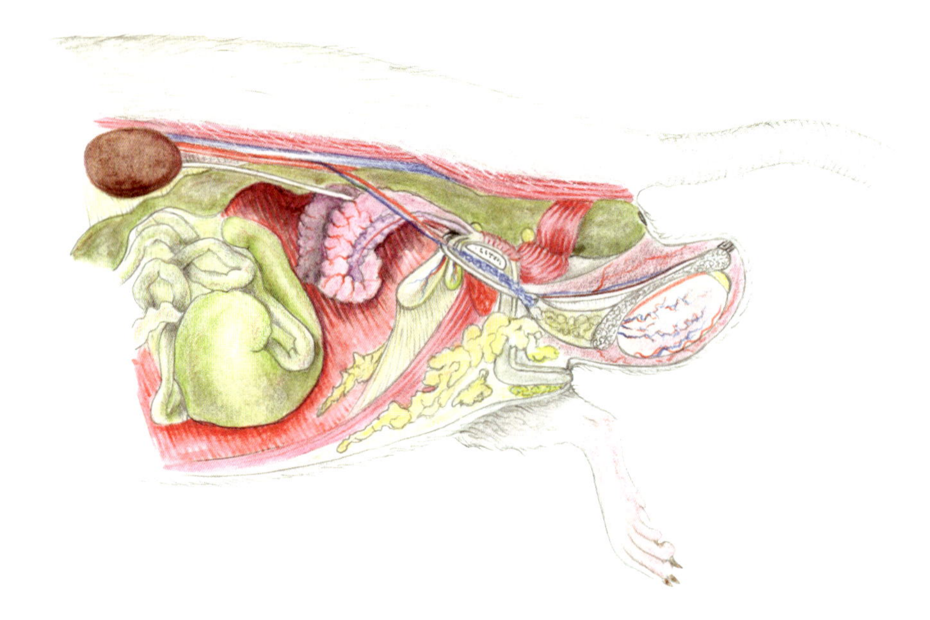

Figure 46. Male caudal abdominal and pelvic viscera including the testicle (left lateral aspect).

The inguinal canal is positioned similarly in all rodents (including mice and rats). Like in the rabbit, it remains open in adulthood. Contraction of the cremaster muscle (not shown) retracts the testicles into the abdomen. When males fight, they retract the testicles in defense against castration by biting attacks.

Mouse

Non-reproductive structures shown: jejunum, cecum, ascending colon, descending colon with the descending mesocolon, rectum, anus, pancreas, left kidney, and the urinary bladder with the median and left lateral vesical ligaments.

Annex glands of the genital apparatus are not fully exposed as they are in the previous figures; some are overlapped by the left lateral vesical ligament. The annex glands are shown in Figure 50 from the dorsal perspective.

The vesicular glands are large in a sexually mature mouse.

The coagulating glands follow the lesser curvature of the vesicular glands.

The epididymis (body and tail), the ductus deferens, and the testicle (medial aspect) with its blood supply are shown.

The longitudinal section of the penis, the penile urethra, and the preputial glands are shown.

Rat

Non-reproductive structures shown: jejunum, ileum, cecum, ascending colon, descending colon, rectum, anus, left kidney and ureter, urinary bladder with only the median ligament of the urinary bladder, abdominal aorta, and the caudal vena cava.

The size of the vesicular glands, the dorsal and ventral lobes of the prostate gland, and the bulbourethral gland are proportionately smaller in the rat than in the mouse. The coagulating glands are large in sexually mature individuals.

The left testicle and epididymis (with ligaments related to both), the ductus deferens, the testicular artery and vein, and the blood supply to the left testicle are shown on the lateral aspect of the body. Also shown are the penis and the preputial gland.

The lateral ligaments of the urinary bladder are not shown; these ligaments are similar to those of the mouse.

Figure 47A. Mouse. Male reproductive and urinary apparatus (ventral aspect).

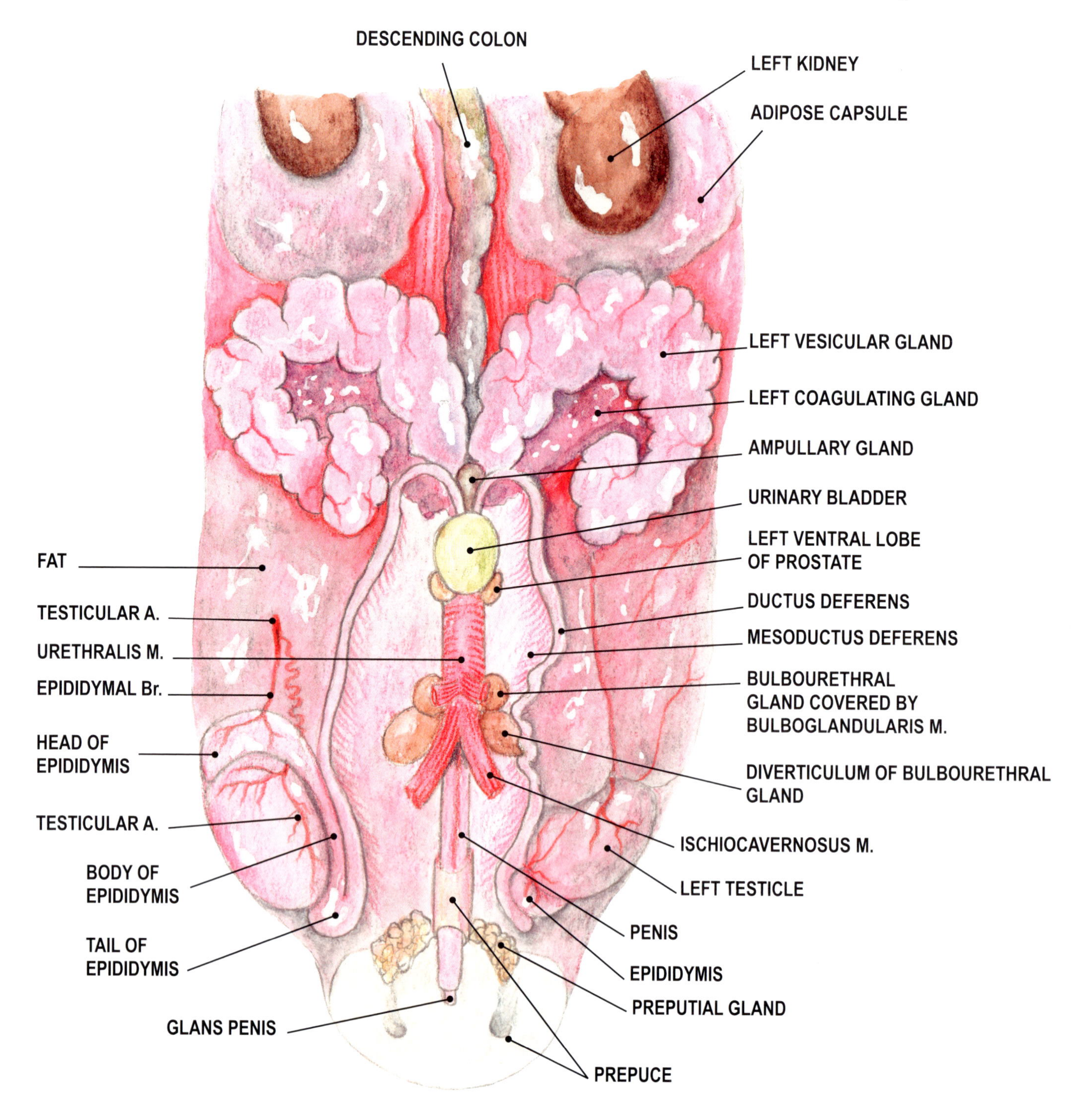

Figure 47B. Rat. Male reproductive and urinary apparatus (ventral aspect).

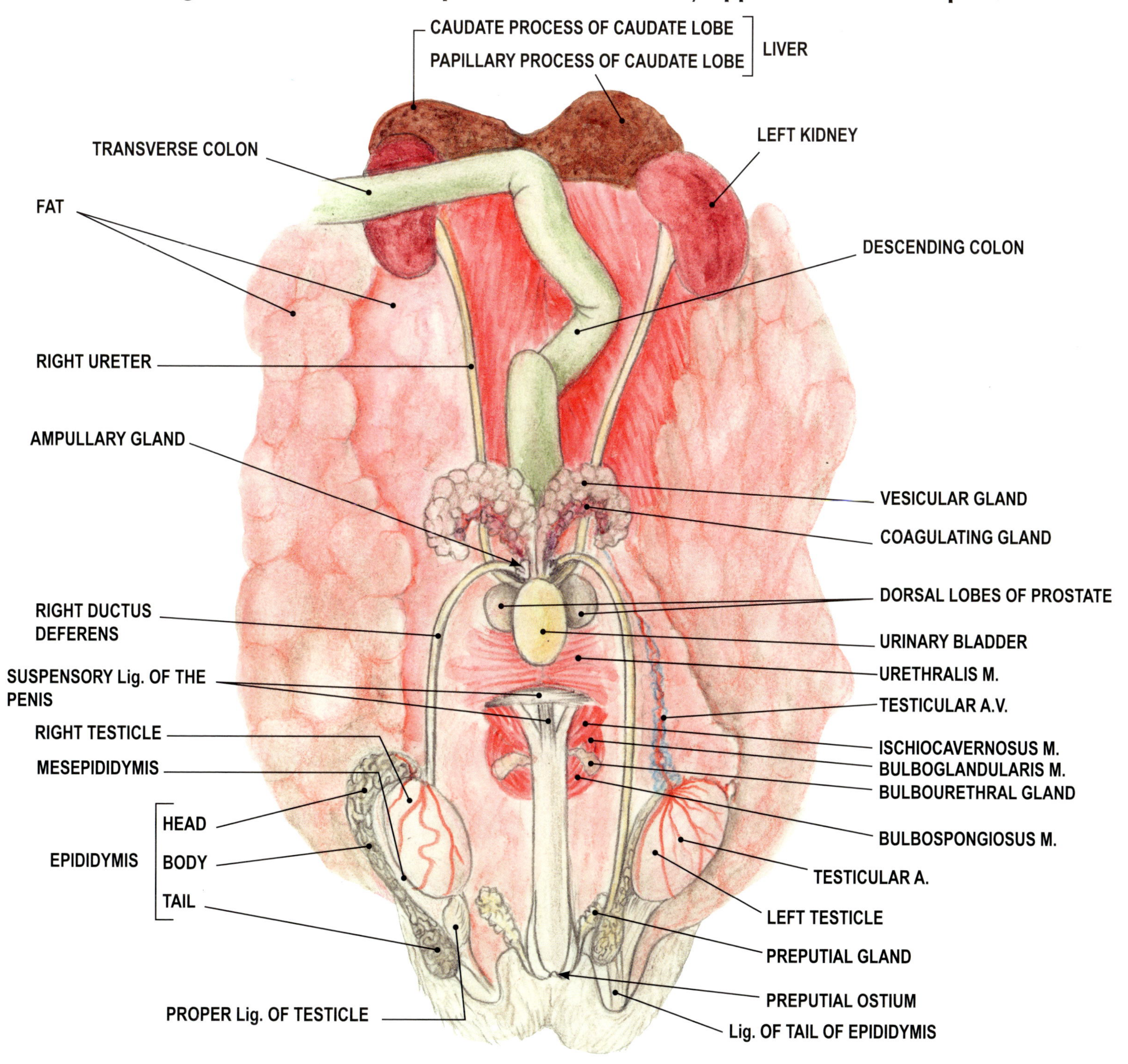

Figure 47A. Mouse. Male reproductive and urinary apparatus (ventral aspect).

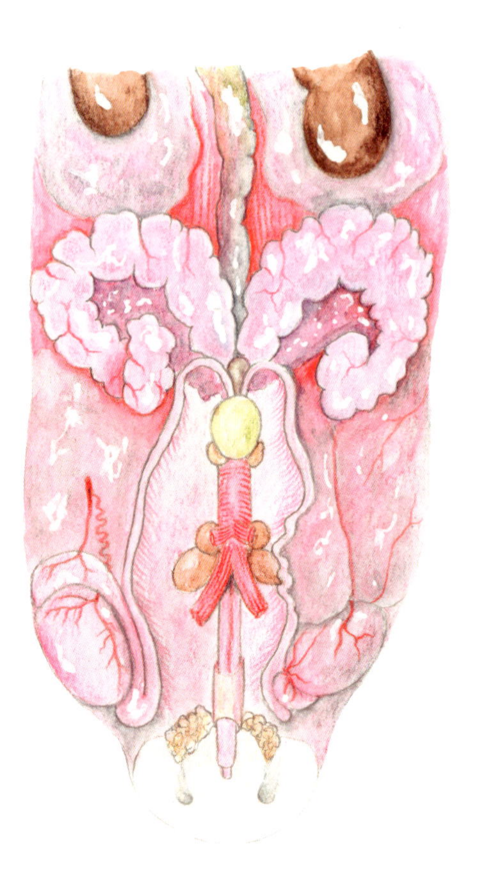

Figure 47B. Rat. Male reproductive and urinary apparatus (ventral aspect).

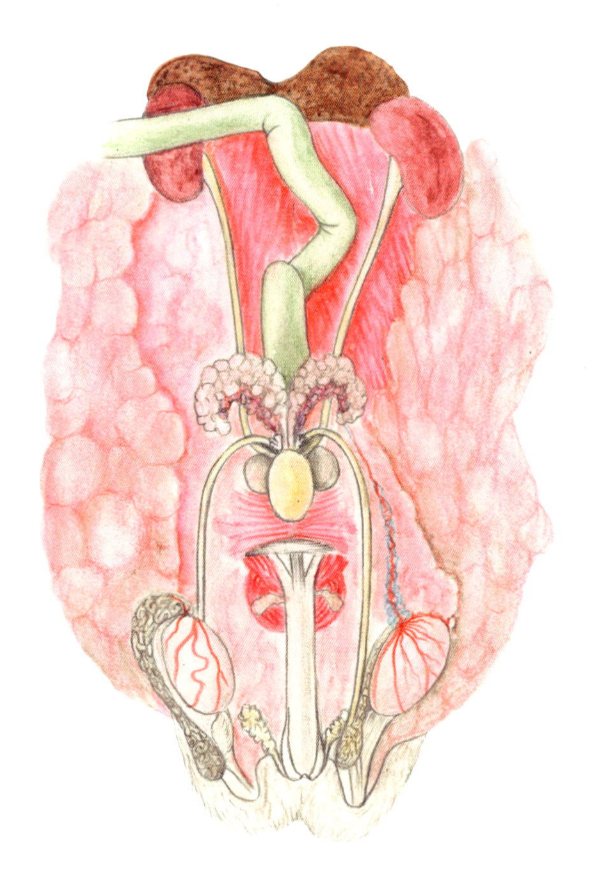

Figure 47. Male reproductive and urinary apparatus (ventral aspect).

In both species, the ureters are embedded in fat close to their insertion into the bladder at its dorsal aspect. The ureters are delicate structures that can be easily traumatized or inadvertently ligated in surgical manipulations.

In the mouse a bulbourethral diverticulum can be seen caudal to the corresponding bulbourethral gland. The rat lacks a bulbourethral diverticulum. See Figure 51 for details on the pelvic urethra and related structures.

Mouse

The ureters are covered in this specimen by an excessive amount of fat.

The urinary bladder is located between the ampullary gland of ductus deferens, the ductus deferentes, and the ventral lobes of the prostate.

Only the ventral lobes of the prostate can be exposed in dorsal recumbency.

The vesicular glands are proportionately larger and much more curved inwards than those of the rat.

The coagulating glands fill the space inside of the curvature of the vesicular glands.

Each ductus deferens is accompanied from its origin to the end by a wide mesoductus deferens.

The ductus deferens passes dorsally over the urinary bladder and opens on the roof of the prostatic urethra, leaving a small space for the ampullary gland. Before reaching the prostatic urethra, the wall of the ductus deferens is thicker and called the "ampulla of ductus deferens." As described in *Illustrated Veterinary Anatomical Nomenclature*, "The ampulla of ductus deferens is the enlargement of the terminal part of the ductus deferens, produced by a glandular thickening of its wall."[2] This segment of the ductus deferens is therefore not a true ampulla because the lumen is not wider than that of its other segments.

The pelvic urethra is covered by the urethralis muscle.

The bulbourethral gland, its diverticulum, and the annex muscles are shown from the ventral perspective.

The testicle (testis), epididymis head, body, and tail are visible.

The testicular artery and its epididymal branch are shown.

The penis, with the glans penis, the prepuce, and preputial glands are shown.

Penile vessels are shown in Figure 50.

Rat

The adipose capsules of the kidneys and the fat surrounding the ureters were removed to reveal the proximal portion of the ureters. Distally, the relationship between the ureters and the ductus deferentes is shown; the ductus deferentes cross the ureters ventrally.

The urinary bladder is associated with the ventral lobes of the prostate. The dorsal lobes of the prostate are shown partially.

The vesicular glands and the coagulating glands are proportionately smaller in the rat than in the mouse.

The ductus deferentes were dissected from the mesoductus deferentes and isolated for illustration.

Each epididymis shows a long tail, extending far past the caudal pole of the testicle. Therefore, the proper ligament of the testis and the ligament of the tail of the epididymis are comparatively longer in the rat than in the mouse.

The testicular artery and vein and the arterial supply of the testicle are shown in detail.

The bulbourethral gland, the bulboglandularis muscle, the base of the penis, and the suspensory ligaments of the penis are revealed because a ventral portion of the ischiocavernosus muscle was removed.

The penis is shown in relation to the preputial ostium and the preputial glands.

Figure 48A. Mouse. Testicle, epididymis, and spermatic cord (left lateral aspect).

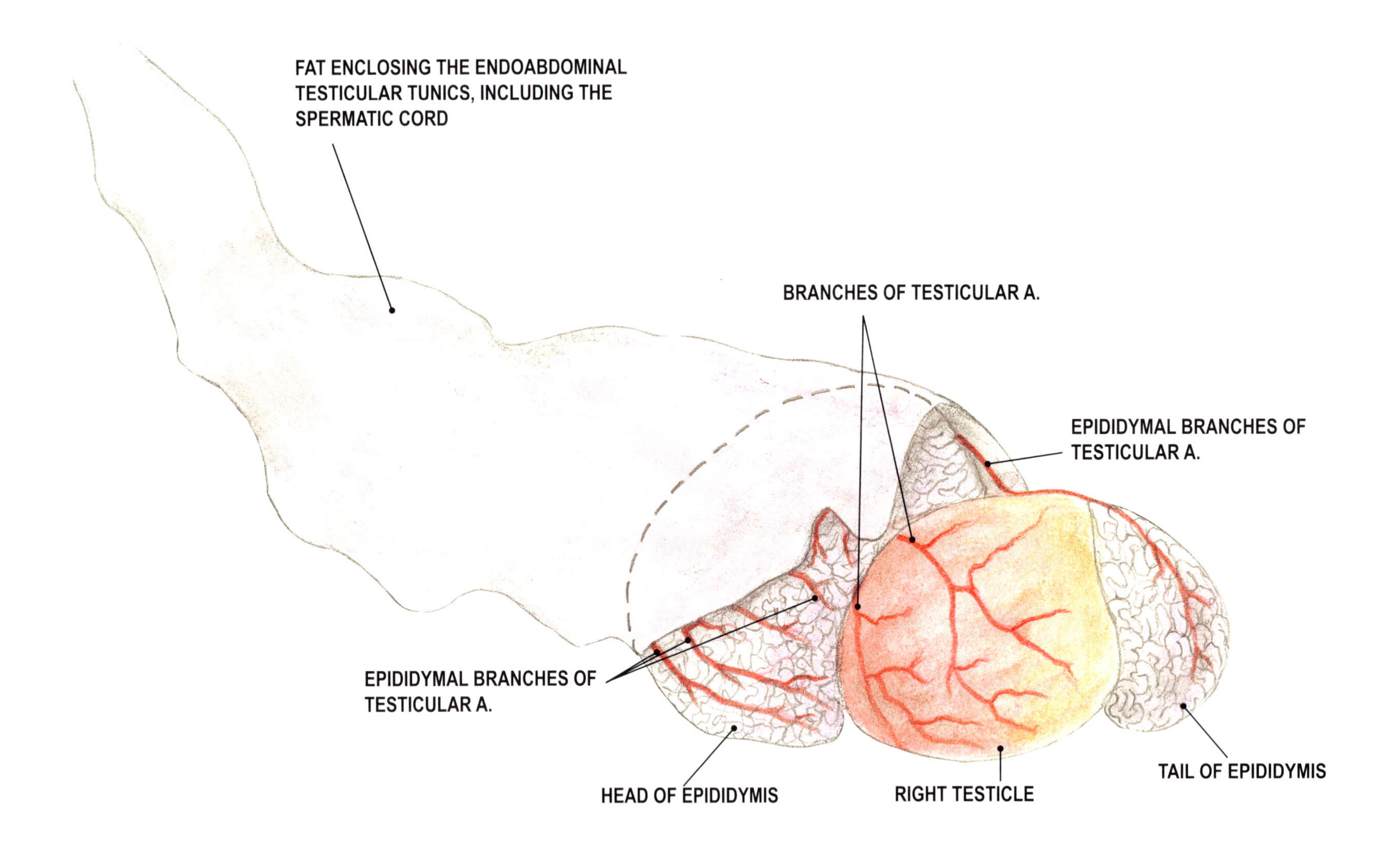

Figure 48B. Rat. Testicle, epididymis, and spermatic cord (left lateral aspect).

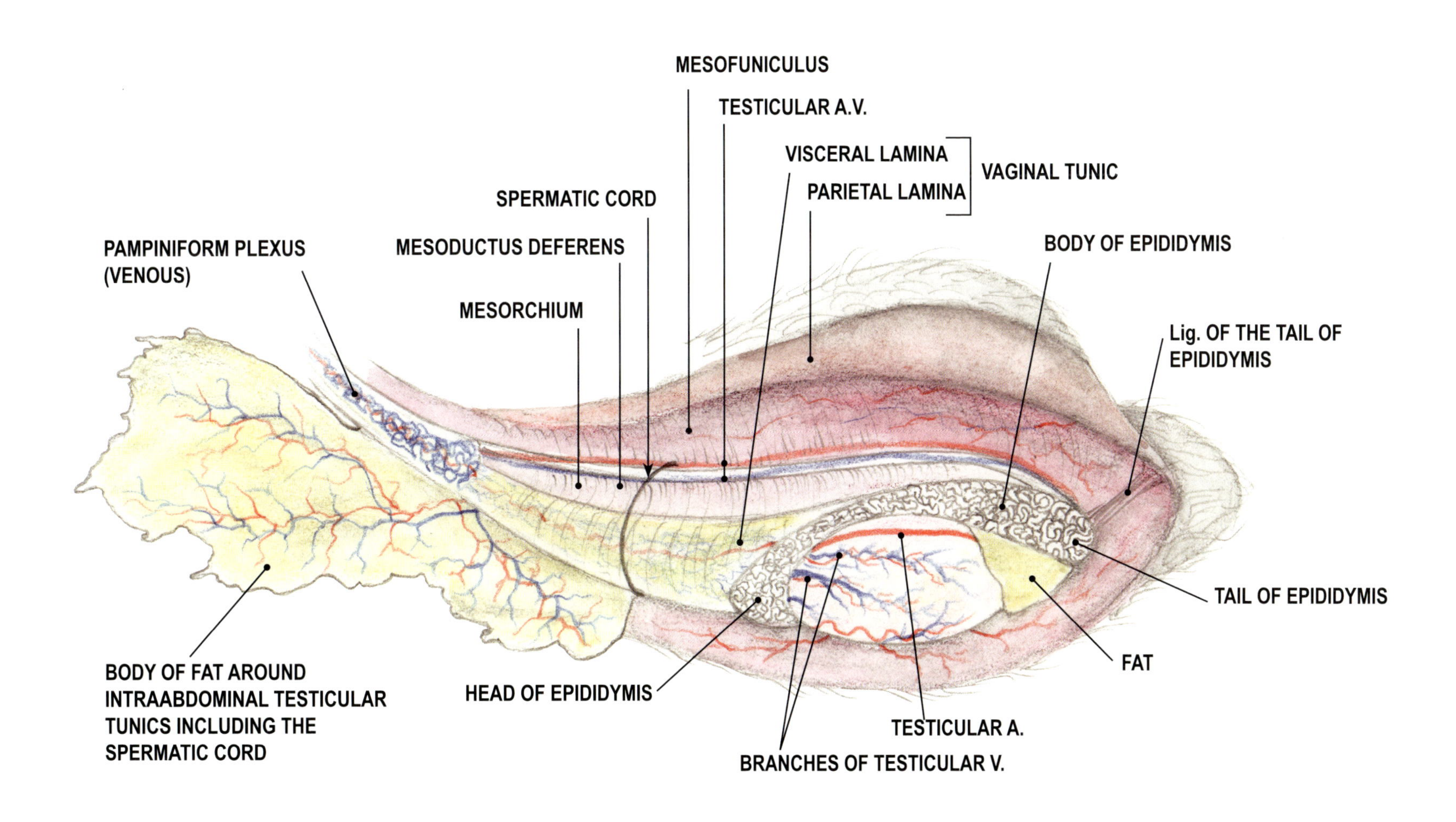

Figure 48A. Mouse. Testicle, epididymis, and spermatic cord (left lateral aspect).

Figure 48B. Rat. Testicle, epididymis, and spermatic cord (left lateral aspect).

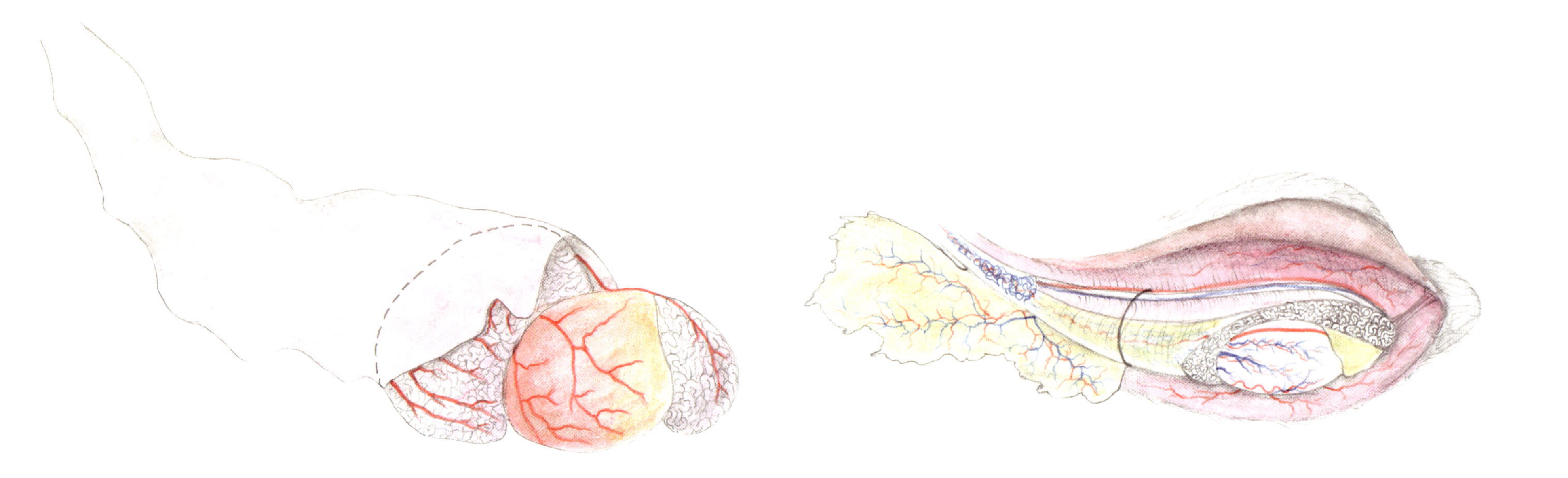

Figure 48. Testicle, epididymis, and spermatic cord (left lateral aspect).

Mouse

The testicle is ovoid in shape.

The mouse epididymis has a proportionately larger head and tail than that of the rat. These structures cover part of the caudal pole of the testicle. The body of the epididymis is hidden behind the testicle.

The testicular artery and its branches, as well as its epididymal branches, are conspicuous.

Rat

The inguinal and vaginal canals are opened to expose the structures within.

The testicle is ovoid in shape.

The testicular artery and vein lay parallel with each other on the lateral aspect of the testicle. The main artery runs close to the body of the epididymis, while branches of it run close to the dorsal border of the testicle.

The head, body and tail of the epididymis do not cover the testicle. A consistent amount of fat is interposed between the caudal pole of the testicle and the tail of the epididymis. The ligament of the tail of the epididymis is long. The proper ligament of the testicle (see Figure 49) cannot be seen from this perspective, being covered by fat.

The ductus deferens is accompanied by branches of the testicular artery and vein.

The approach for a closed castration is demonstrated by the curved line. The intact vaginal tunic, including the spermatic cord, is ligated and severed.* In an open castration, the vaginal tunic would be incised to expose the testicular vessels and spermatic cord for ligation and excision. The closed method of castration may prevent eventration of intestinal loops and minimize postsurgical infection because the vaginal canal leading into the peritoneal cavity is blocked by the ligature.

*The spermatic cord includes the ductus deferens and its artery and vein, the mesorchium, and the blood supply to the testicle and epididymis. The spermatic cord is surrounded by the visceral lamina of the vaginal tunic.

The inguinal canal extends between the superficial and the deep inguinal rings. The vaginal canal is the space between the parietal and visceral laminae of the vaginal tunic. The vaginal tunic is the corresponding name of the peritoneum inside of the inguinal canal.

Figure 49A. Mouse. Testicle, epididymis, and spermatic cord (right medial aspect).

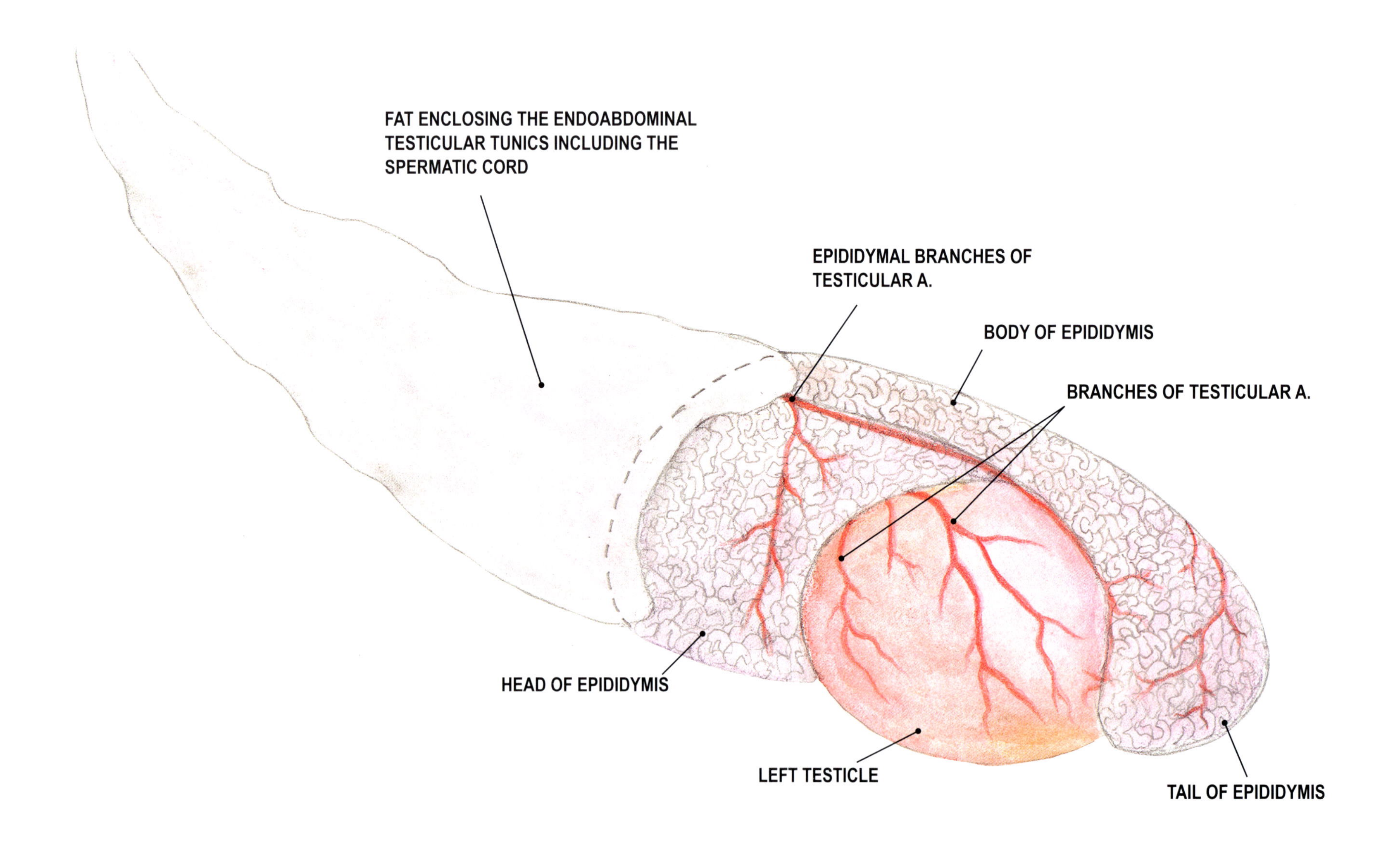

Figure 49B. Rat. Testicle, epididymis, and spermatic cord (right medial aspect).

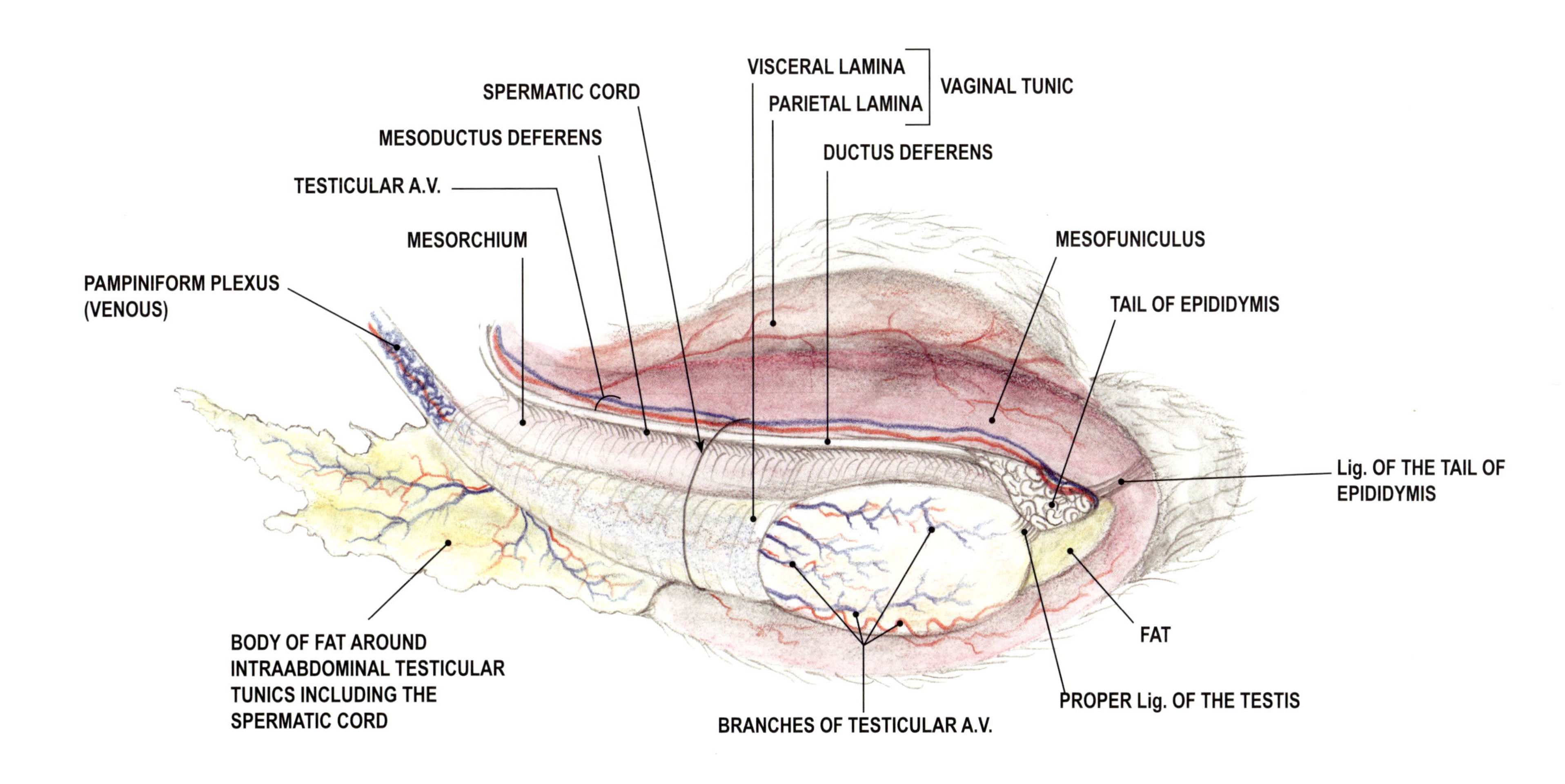

Figure 49A. Mouse. Testicle, epididymis, and spermatic cord (right medial aspect).

Figure 49B. Rat. Testicle, epididymis, and spermatic cord (right medial aspect).

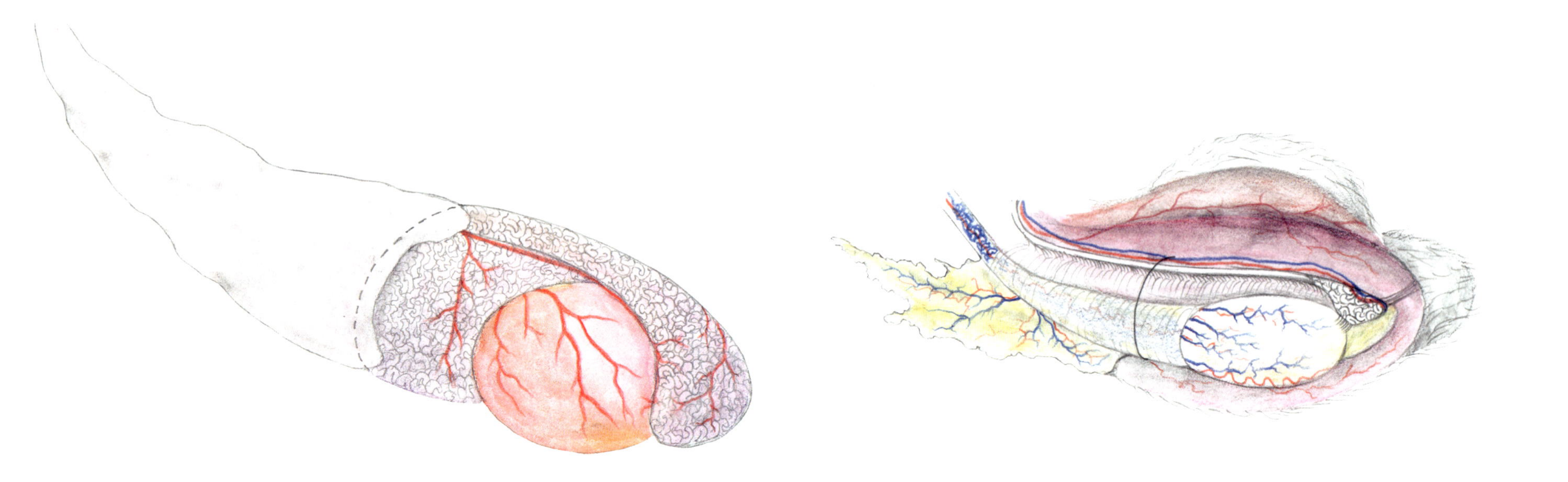

Figure 49. Testicle, epididymis, and spermatic cord (right medial aspect).

Mouse

The testicle is ovoid in shape, similar to the lateral aspect.

The mouse epididymis has a proportionately larger head, body, and tail than that of the rat. These structures cover part of the caudal pole of the testicle.

The ductus deferens is overlapped by the body of the epididymis.

The testicular artery and its branches, as well as its epididymal branches, are large, as on the lateral aspect.

Rat

The inguinal and vaginal canals are opened to expose the spermatic cord (see definition on page 193) and associated structures.

The testicle has an ovoid shape. A wide expanse of the testicle is visible medially because the epididymis is not attached to the epididymal border of the testicle.

Of the epididymis, only the tail is exposed, followed by the ductus deferens. Both the ligament of the epididymis tail and the proper ligament of the testis are shown.

The blood supply is conspicuous on the testicle and the ductus deferens. The arteries and veins run parallel with each other in the long axis of the testicle.

The parietal and visceral laminae of the vaginal tunic, connected by the mesofuniculus, and the spermatic cord are widely exposed. The curved line encircles the intact tunica vaginalis, as though to be ligated for a closed castration (described on page 193).

Figure 50A. Mouse. Penis, reflected caudally (dorsal aspect).

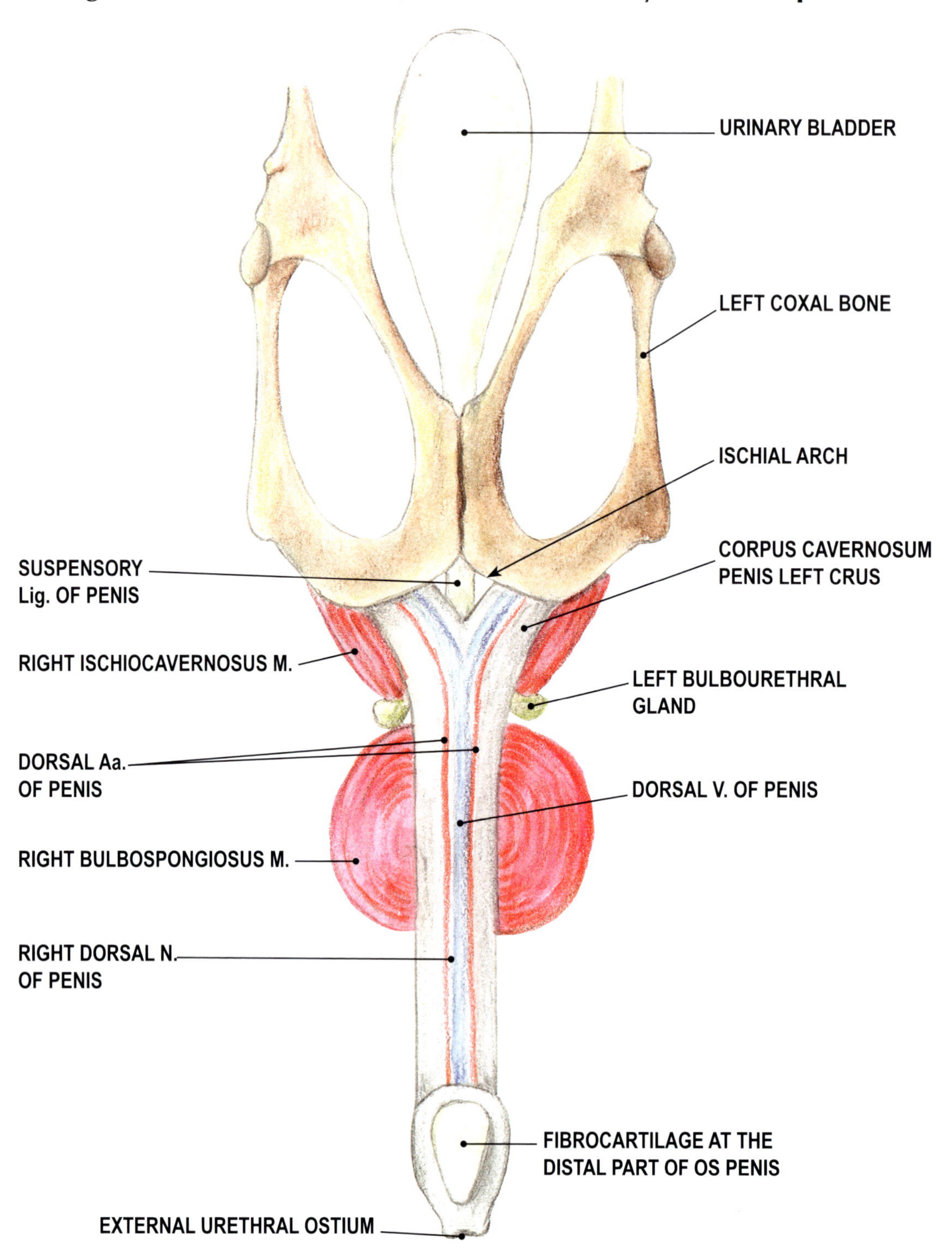

Figure 50B. Rat. Penis, reflected caudally (dorsal aspect).

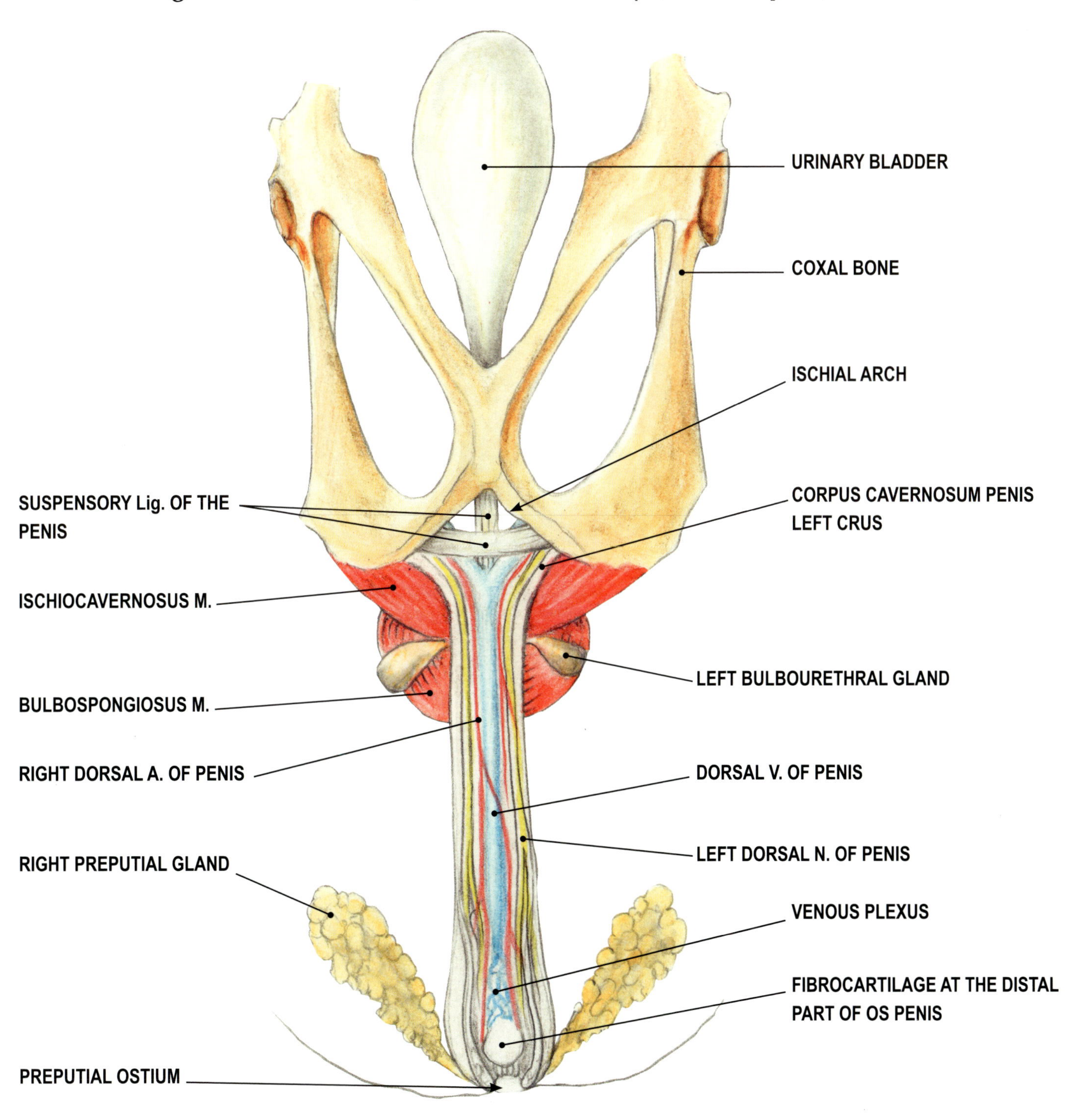

Figure 50A. Mouse. Penis, reflected caudally (dorsal aspect).

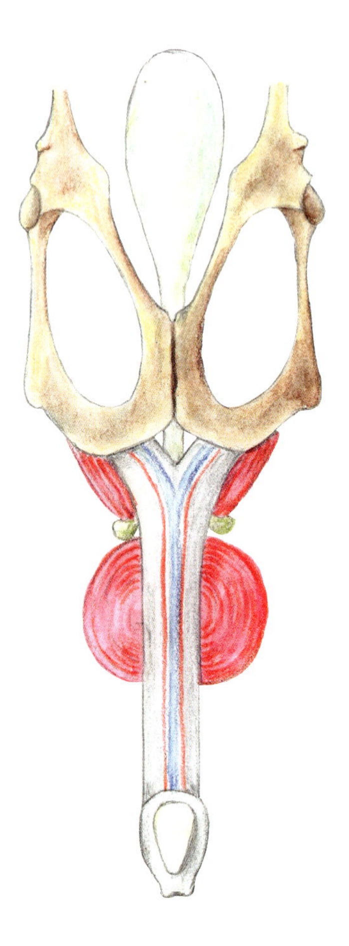

Figure 50B. Rat. Penis, reflected caudally (dorsal aspect).

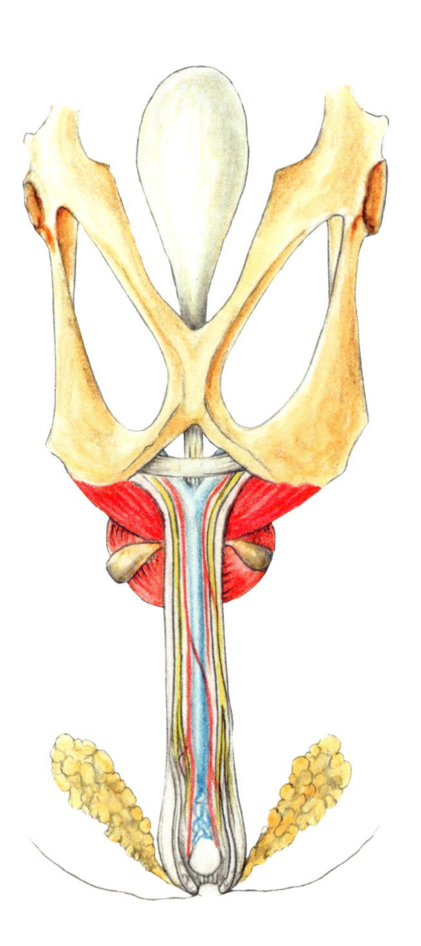

Figure 50. Penis, reflected caudally (dorsal aspect).

The figures show the penis in both species as straightened; however, in life it is bent about 90° caudally. Starting from the ischiatic tuberosities, the corpus cavernosum penis, with symmetrical roots, extends a little past the bent angle. From there the os penis continues up to the free end of the penis, as part of the glans penis.

Shown is the smallest component of the os penis (located at the distal end), which is a cartilage in immature animals. In the figures, the cartilage surrounds the distal part of the proximal segment of the os penis like a cap (see Figure 52). In rats, this cartilage is prone to ossify with age and becomes fused with the distal end of the os penis.

The preputial glands are located in the subcutaneous tissue in both species. The illustrations show these glands only in the rat (see Figure 46 for these glands in both species).

Mouse

The penis is extended for visualizing the vasculature. The bulbospongiosus muscle is therefore extended caudal to the bulbourethral gland. In the natural state, the relationship between these two structures is similar to that in the rat.

The dorsal vein of the penis does not originate from a venous plexus. The vein, which is the most important structure clinically, is prominently located on the midline and is easy to identify with magnification.

The two symmetrical dorsal nerves of the penis accompany the vein on both sides, followed by the dorsal arteries of the penis. Their relationship may vary with individuals; one nerve may be closer to the vein or the corresponding artery.

Rat

The penis is extended less than in the mouse figure.

The dorsal vein of the penis originates from a venous plexus, unlike the mouse.

Symmetrical dorsal arteries and nerves of the penis parallel the dorsal vein of the penis. Their relationship may vary with individuals.

Figure 51A. Mouse. Proximal urethra and urethral recess (dorsal aspect, positioned with head up), median section through urethral recess.

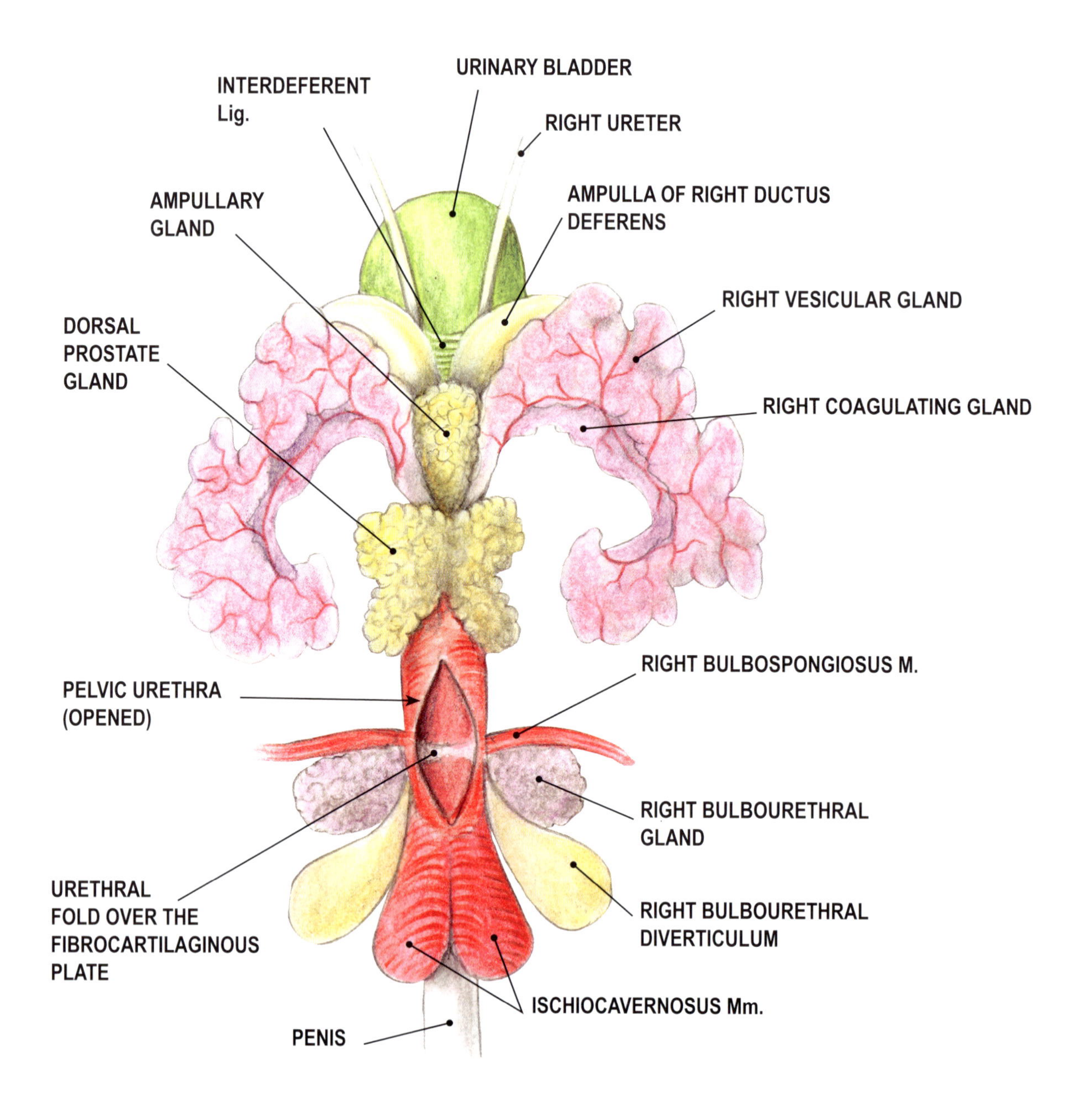

Figure 51B. Mouse. Latex cast of pelvic urethra (lateral aspect, positioned with head to the left).

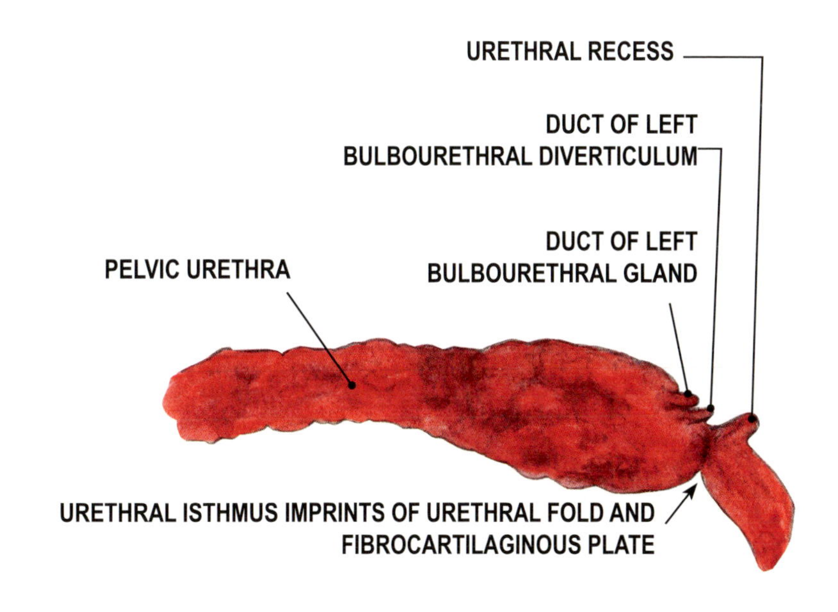

Figure 51C. Mouse. Latex cast of pelvic urethra (dorsal aspect, positioned with head to the left).

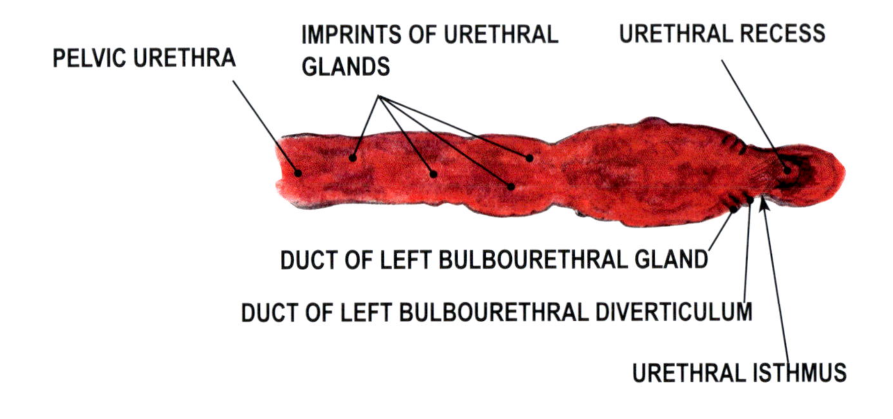

Figure 51D. Rat. Proximal urethra and related structures (median section, positioned with head to the left).

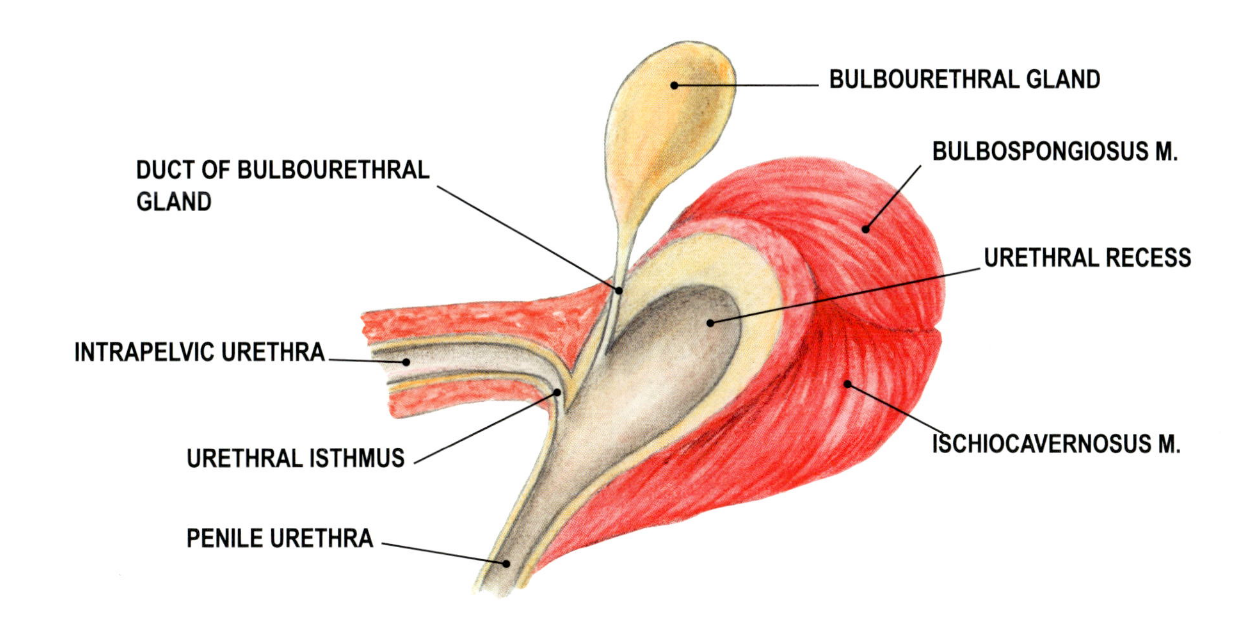

Figure 51. Male urethra and urethral recess.

A. Mouse. Proximal urethra and urethral recess (dorsal aspect, positioned with head up), median section through urethral recess.

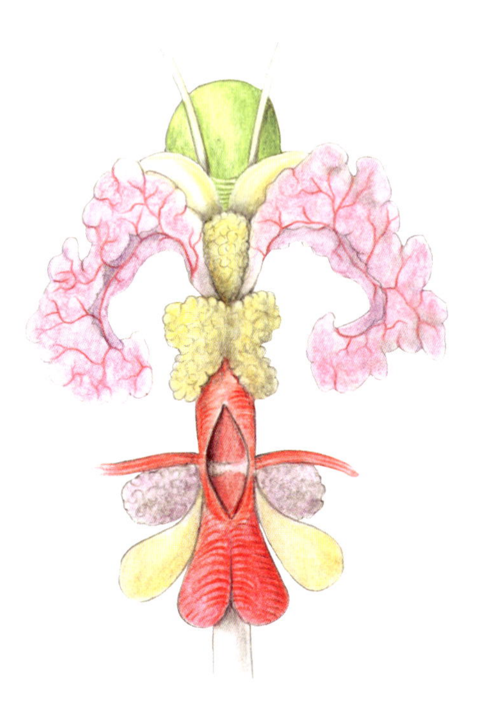

B. Mouse. Latex cast of pelvic urethra (lateral aspect, positioned with head to the left).

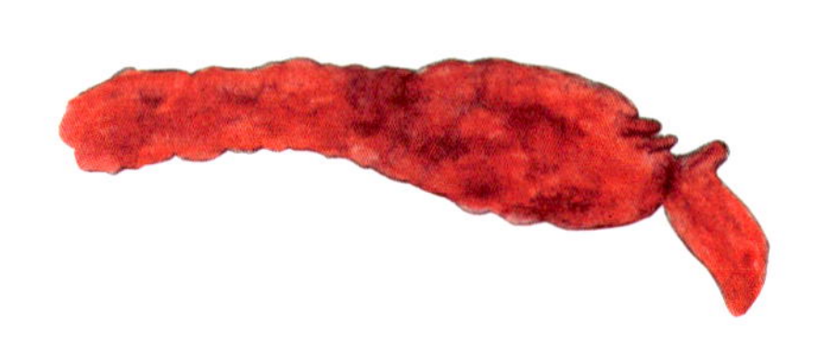

C. Mouse. Latex cast of pelvic urethra (dorsal aspect, positioned with head to the left).

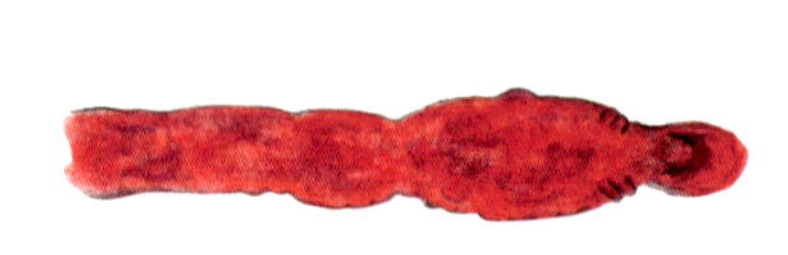

D. Rat. Proximal urethra and related structures (median section, positioned with head to the left).

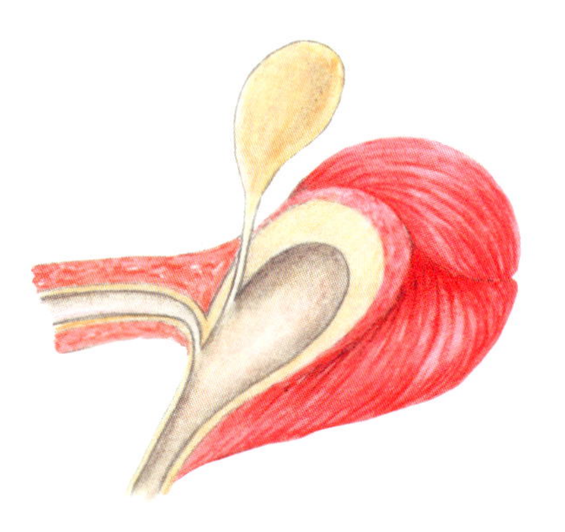

Figure 51. Male urethra and urethral recess.

In both species, the ureters are embedded in fat close to their insertion into the bladder at its dorsal aspect. The ureters are delicate structures that can be easily traumatized or inadvertently ligated in surgical manipulations.

The rat lacks a bulbourethral diverticulum, but the urethral recess is present. In the mouse, in addition to the urethral recess, a bulbourethral diverticulum caudal to the corresponding bulbourethral gland can be seen.

Mouse (Figures A-C)

The ductus deferentes pass dorsally over the urinary bladder, crossing the ureters, and they widen considerably, becoming the ampullae of ductus deferentes, which open on the roof of the prostatic urethra. The ampullae are connected by the interdeferent ligament and leave a small space for the unique gland of the ampullae of the ductus deferentes (the ampullary gland); see page 189 for details. In the mouse, the ampullary gland appears very different from that of the rat and the domestic mammals.

The butterfly-shaped dorsal prostate gland is fully exposed in this view. It is in contact with the caudal apex of the ampullary gland.

The urethral recess, though not mentioned in the literature, was found consistently in the mouse.

Through a median section of the roof of the pelvic urethra, a mucosal fold covering a fibrocartilaginous plate was exposed on the urethral floor. This plate, also not mentioned in the literature, is located at the transition of the pelvic urethra and the urethral isthmus. The plate extends laterally and attaches to the ischium.

To expose the urethral recess and highlight additional information about the pelvic urethra, a cast of the pelvic urethra and the initial part of the penile urethra was prepared.* By examining the cast from the lateral and dorsal perspectives, the spatial relationships among the urethra, annex glands, and ducts can be better appreciated from the interior aspect of the pelvic urethra. In the lateral perspective, positions are clearly identified for the ducts of the bulbourethral gland and diverticulum; the urethral recess; and the imprints of the urethral glands, the urethral fold, and the fibrocartilaginous plate at the urethral isthmus. The dorsal perspective of the urethra (not illustrated in the literature, leaving the reader to imagine the relationship among these structures) offers a complementary view of the symmetrical ducts, the unique urethral recess, the imprints of the urethral glands, and the enlargement of the caudal half of the pelvic urethra.

An enlargement of the urethral lumen can be seen in the caudal half of the pelvic urethra, abruptly narrowed at the isthmus. Note that the bulbourethral diverticulum is for temporary storage of the secretion of the corresponding bulbourethral gland.

Rat (Figure D)

A median section of the urethra shows the pelvic urethra starting, on the left, from the urinary bladder (not shown in the figure). The pelvic urethra continues to the right with the urethral isthmus (the bent narrow lumen) and the penile urethra. The bulbourethral gland and duct and the urethral recess are also shown. The urethral recess extends dorsally of the penile urethra; here the duct of the bulbourethral gland opens. Note that here an artistic license was used to show in the median section a bulbourethral gland and duct, which are symmetrical structures. The pelvic urethra is surrounded by the urethralis muscle (not labeled), whereas the urethral recess is in contact with the bulbospongiosus and the ischiocavernosus muscles.

The urethral recess has been described elsewhere in the rat.[6]

The topography and relationships of the bulbourethral gland duct are slightly different from those of the mouse:

- The bulbourethral gland duct opens in the urethral recess, close to the transition between the urethral isthmus and the penile urethra.
- The caudal half of the pelvic urethra is not wide in the rat as in the mouse.

*The latex cast was obtained by an injection of latex into the urinary bladder via cystocentesis of a mouse euthanized by an intraperitoneal injection of a euthanasia solution. (Euthanizing mice with carbon dioxide may result in a urethral plug formed by the excretion of the vesicular and coagulating glands. Information courtesy of Dr. Cynthia Besch-Williford, University of Missouri-Columbia.) A needle, attached to a syringe, was inserted into the bladder for infusing the latex. The urinary bladder was ligated at the insertion point of the needle. The needle and syringe were left in place for 24 h, to prevent the reflux of latex, while the latex hardened. Then the needle and syringe were removed. The roof of the pelvic urethra was opened longitudinally to remove the cast before the illustration was created.

Figure 52A. Mouse. Penis (median section), stained by hematoxylin and eosin.

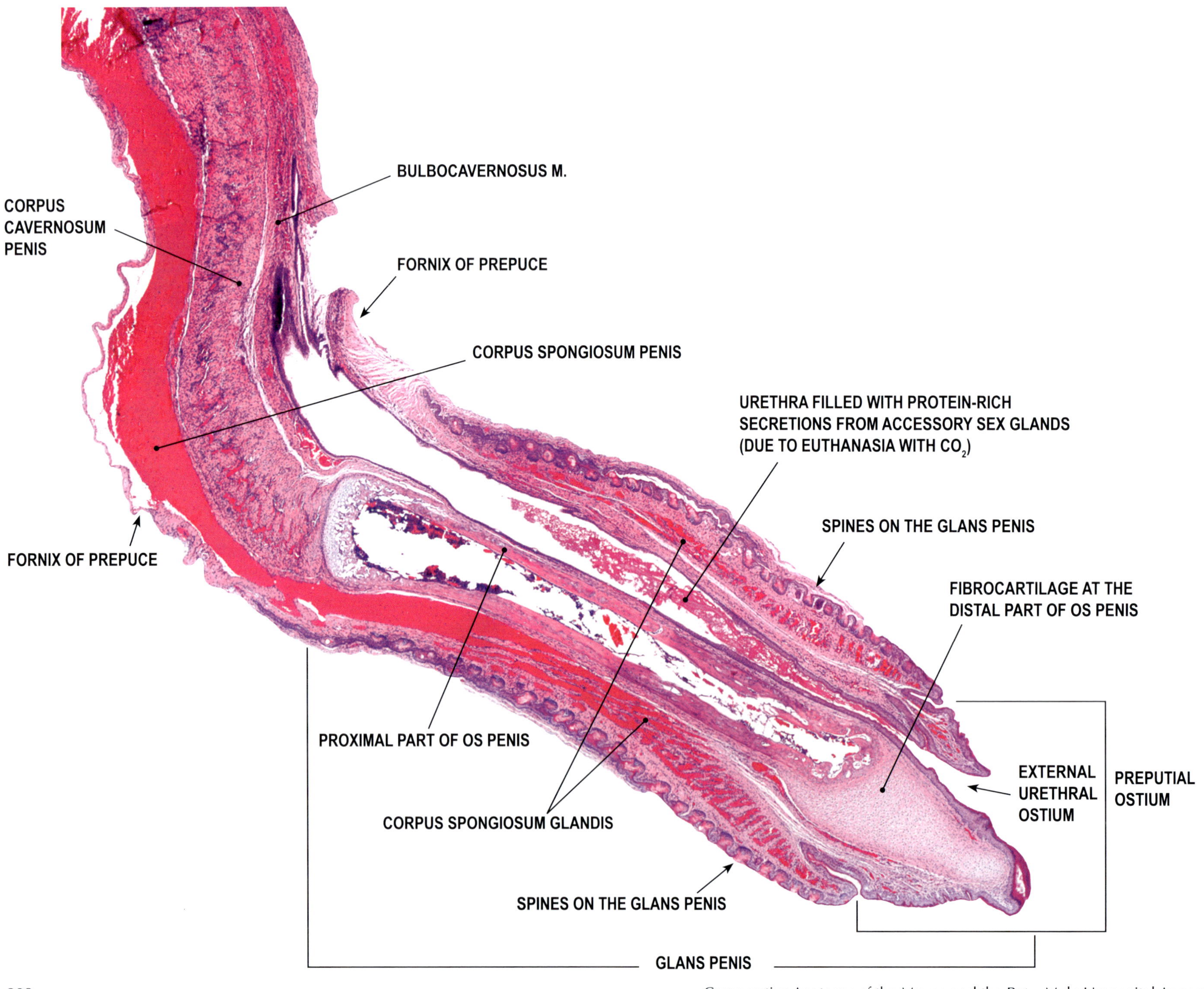

Figure 52B. Rat. Penis (median section), stained by hematoxylin and eosin.

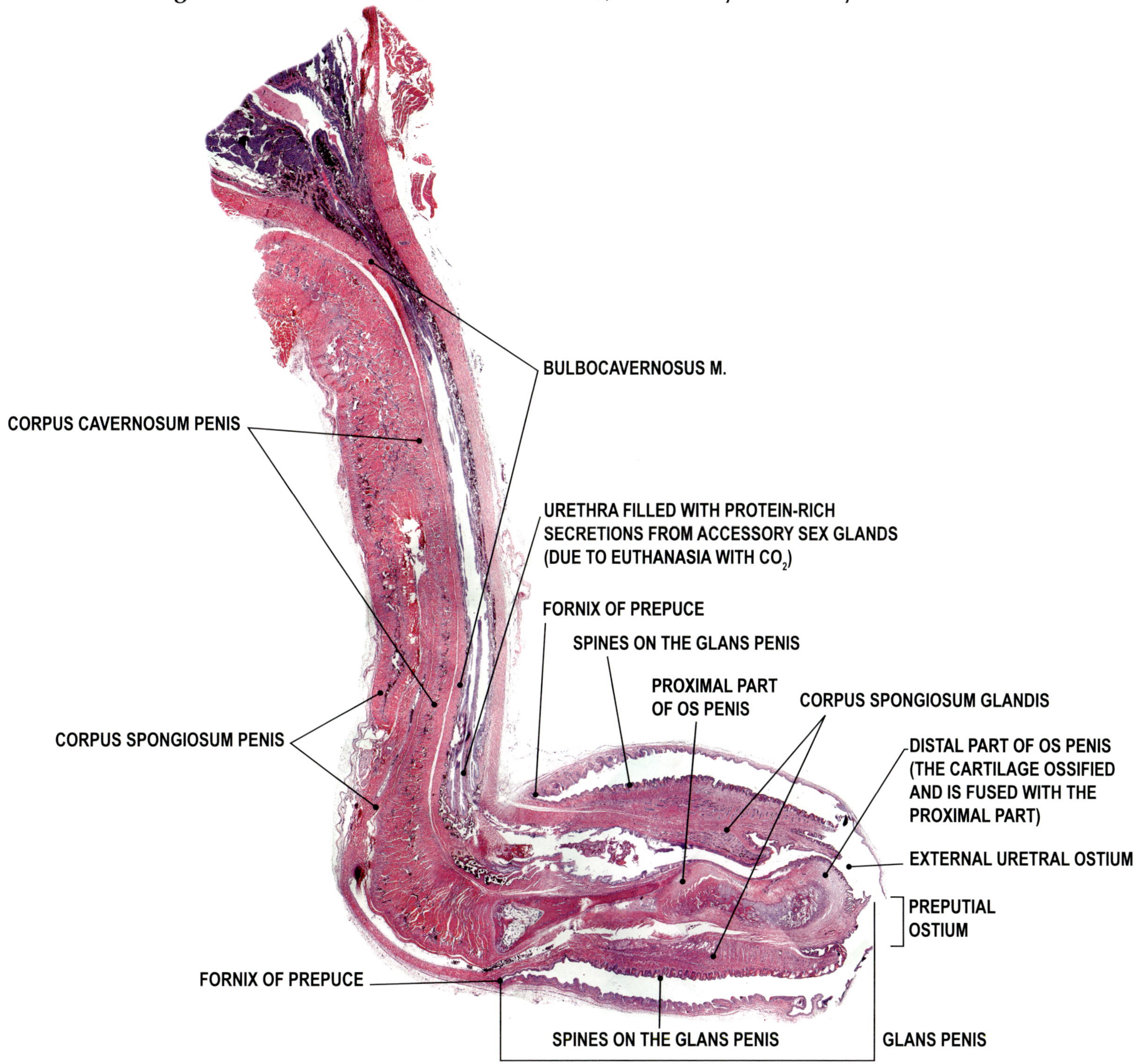

BULBOCAVERNOSUS M.

CORPUS CAVERNOSUM PENIS

URETHRA FILLED WITH PROTEIN-RICH SECRETIONS FROM ACCESSORY SEX GLANDS (DUE TO EUTHANASIA WITH CO_2)

FORNIX OF PREPUCE

SPINES ON THE GLANS PENIS

PROXIMAL PART OF OS PENIS

CORPUS SPONGIOSUM GLANDIS

CORPUS SPONGIOSUM PENIS

DISTAL PART OF OS PENIS (THE CARTILAGE OSSIFIED AND IS FUSED WITH THE PROXIMAL PART)

EXTERNAL URETRAL OSTIUM

PREPUTIAL OSTIUM

FORNIX OF PREPUCE

SPINES ON THE GLANS PENIS

GLANS PENIS

Figure 52A. Mouse. Penis (median section), stained by hematoxylin and eosin.

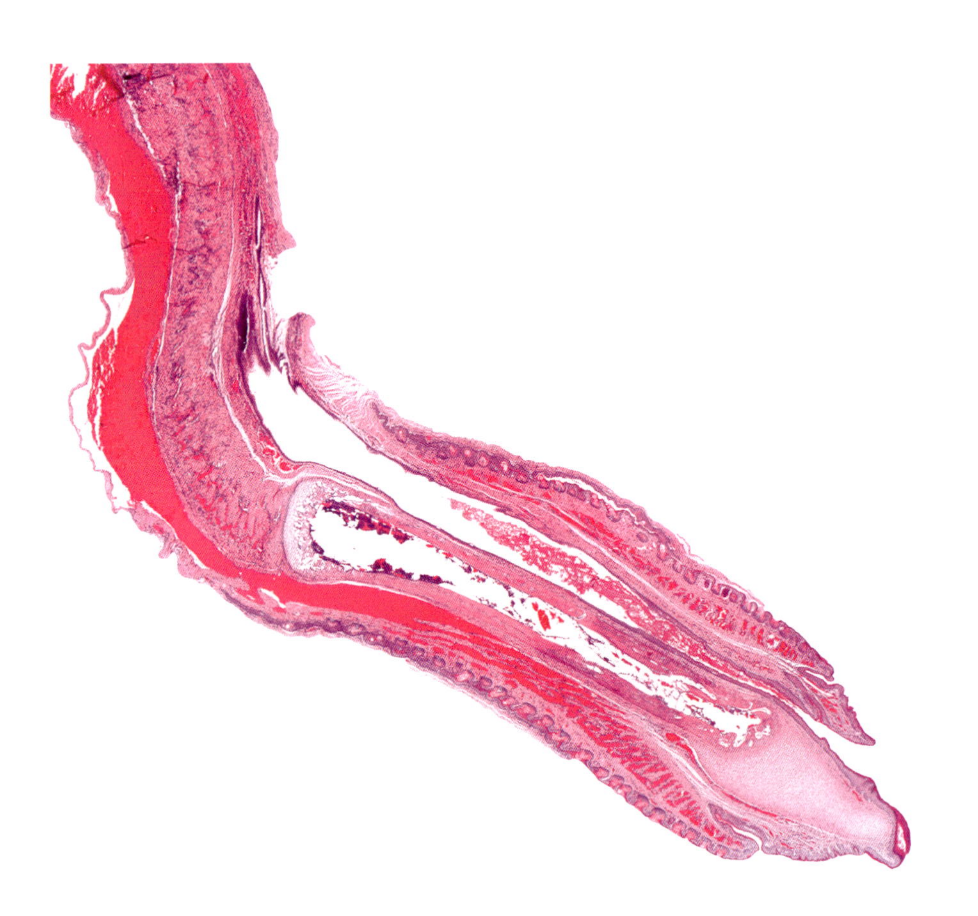

Figure 52B. Rat. Penis (median section), stained by hematoxylin and eosin.

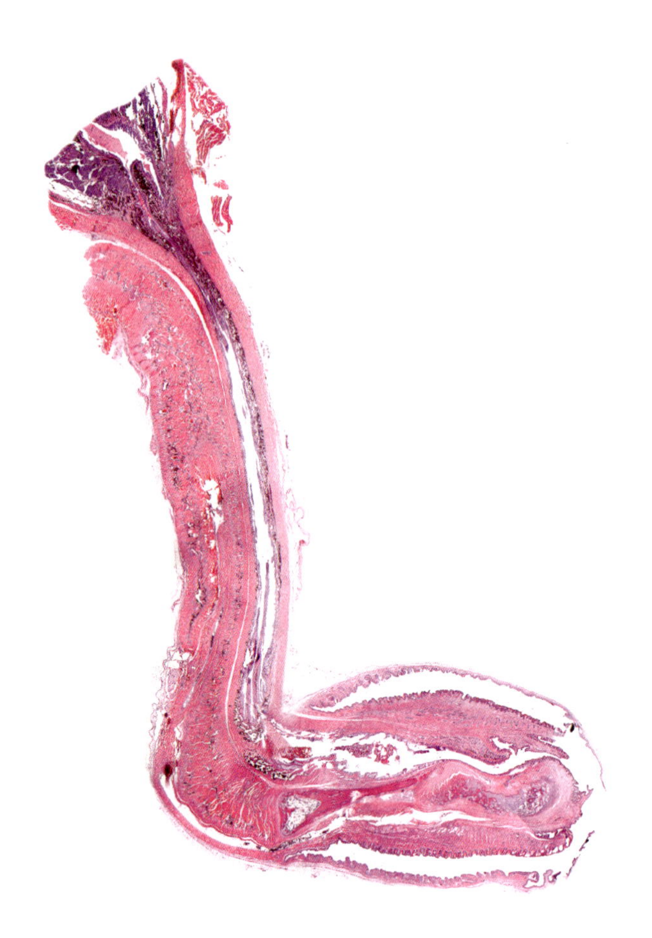

Figure 52. Penis (median section), stained by hematoxylin and eosin.

In both species, the penis is shown in the resting position, vertical and bent caudally, with the glans penis horizontal, as in the median sections of the mouse and rat body (see Figure 46). Each figure is a composite of images photographed at 5× magnification.

An important feature in both species is the large size of the corpus spongiosum glandis, which surrounds the os penis and the penile urethra all around.

In the resting position, the penile urethra in the glans penis is located dorsal, not ventral, to the os penis. During erection and copulation, the penis exits the prepuce and the glans penis turns cranially. Therefore, the urethra takes its place ventral to the os penis, as in any other species with an os penis.

The os penis is located in continuation of the corpus cavernosum penis. It consists of two bony segments: one proximal (the longest) and one distal (the smallest). The proximal segment develops by endochondral ossification in the proximal half, while the distal half develops by membranous ossification. At 1–3 days of age, the proximal segment has a hyaline cartilage formed in the proximal half and a membrane bone formed in the distal half. These two components fuse immediately, and the process of endochondral ossification continues in the proximal segment. At 2–3 weeks of age, the distal segment develops into a fibrocartilage, which is prone to ossification in rats.[9]

Female Urogenital Apparatus

Figure 53A. Mouse. Female caudal abdominal and pelvic viscera (left lateral aspect).

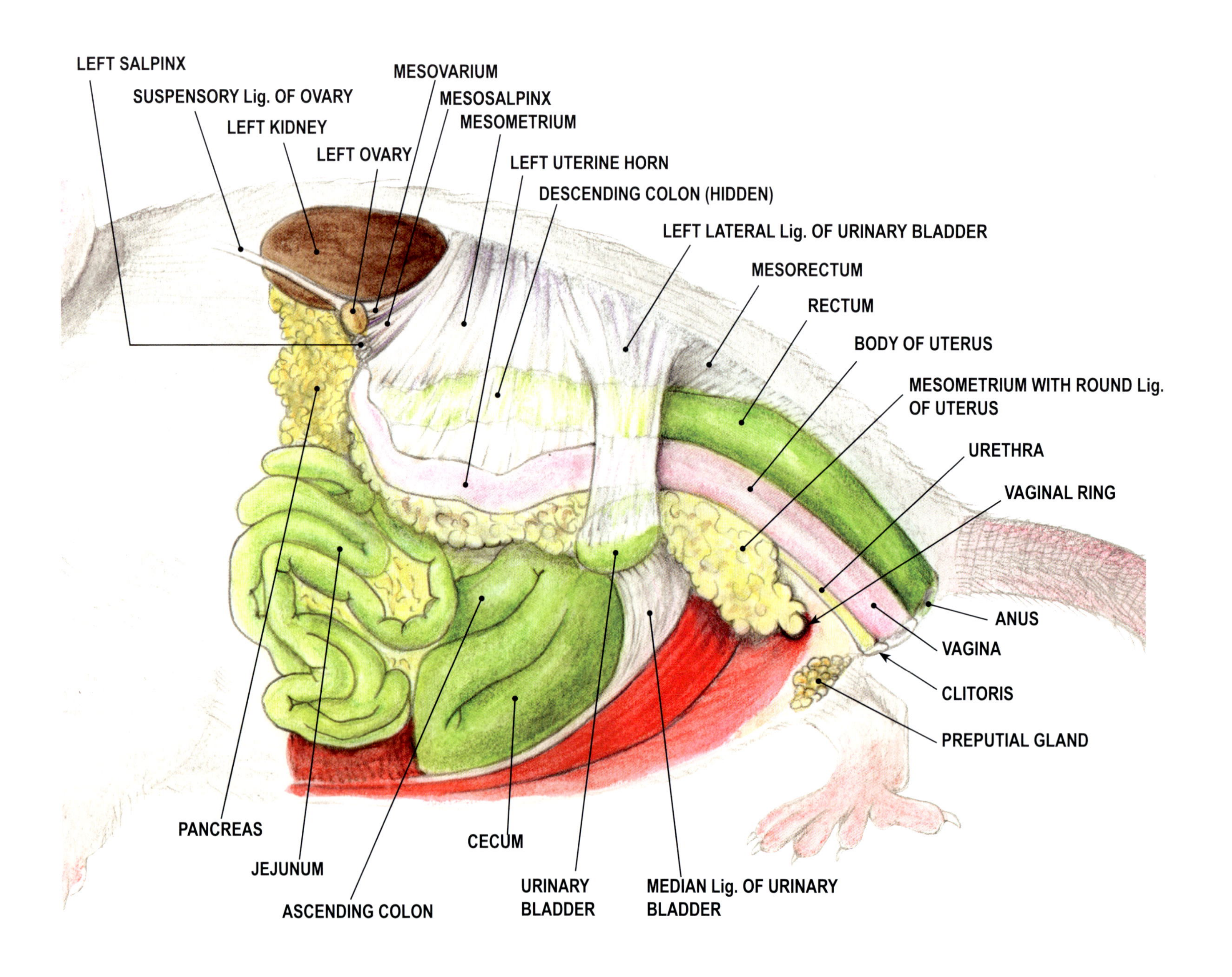

Figure 53B. Rat. Female caudal abdominal and pelvic viscera (left lateral aspect).

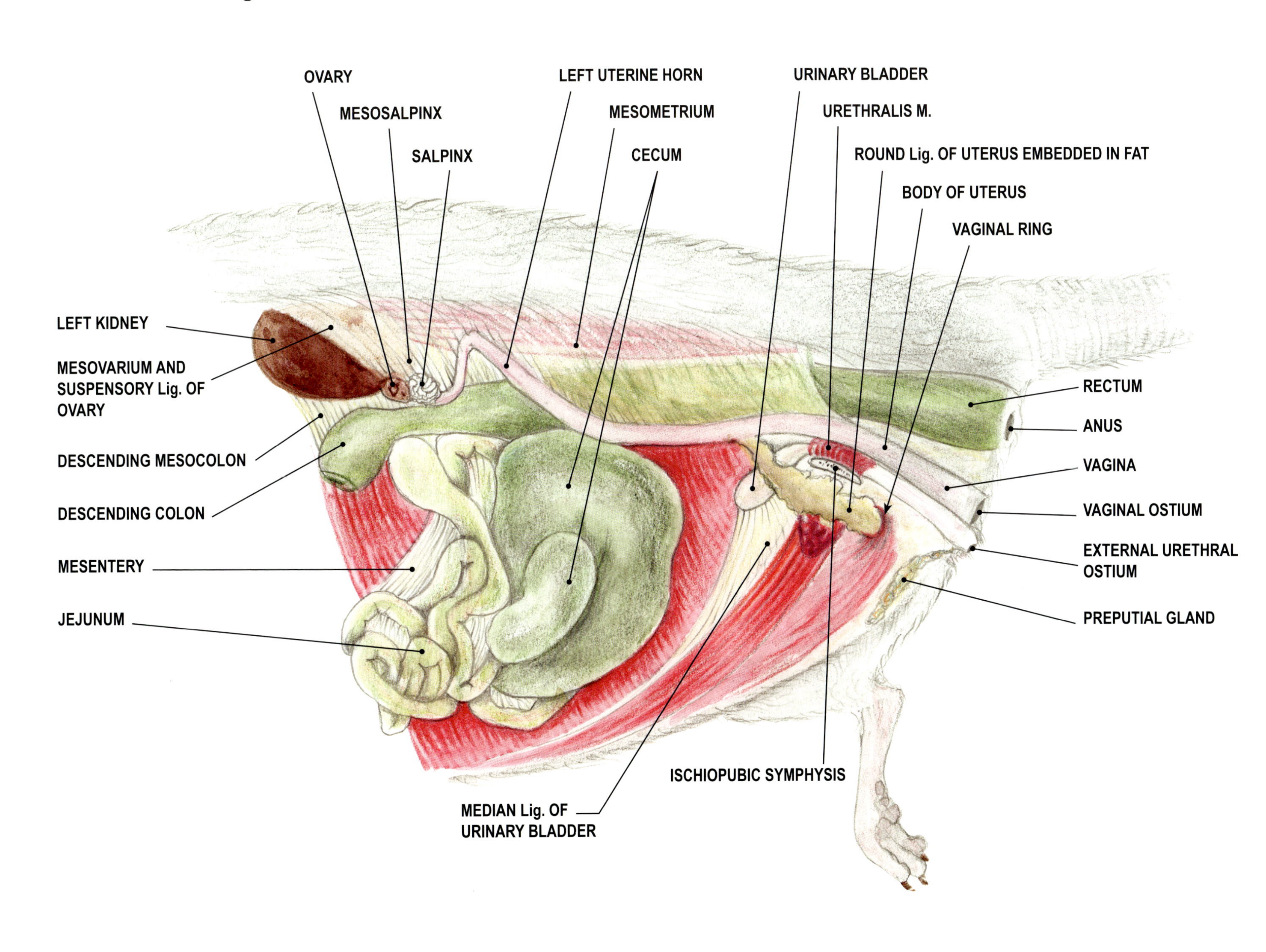

Figure 53A. Mouse. Female caudal abdominal and pelvic viscera (left lateral aspect).

Figure 53B. Rat. Female caudal abdominal and pelvic viscera (left lateral aspect).

Figure 53. Female caudal abdominal and pelvic viscera (left lateral aspect).

The amount of fat in the broad ligament depends on the individual, not on the species.

Neonates and young juveniles of both species have a thin layer of epithelium (not shown) overlying the vaginal ostium. This layer is known as the vaginal plate; it degenerates in puberty to open the vagina. Magnification is needed to visualize the vaginal plate. The urethra opens separately from the vagina in both the mouse and rat; no vaginal vestibule is present. Therefore, the vaginal plate does not obstruct the urethra.

Mouse

Non-reproductive structures shown: jejunum, cecum, ascending colon, descending colon (shown through transparency behind the mesometrium), rectum suspended by the mesorectum, anus, pancreas, left kidney, urinary bladder with the left lateral and median ligaments, and urethra with the external urethral ostium (covered by the clitoris in this figure).

Female genital apparatus shown: Left ovary, salpinx (fallopian or uterine tube), uterine horn, body of the uterus, vagina, vaginal ostium, clitoris, and the preputial gland. The preputial glands (left and right) are associated with the urethral ostium.

The visceral peritoneum has several interconnected folds connecting the different major parts of the internal genitals: the ovary (mesovarium), salpinx (mesosalpinx), and uterus (mesometrium). The generic name of these folds is the broad ligament; it is embedded in fat.

The suspensory ligament of the ovary originates from the free border of the mesovarium. It extends up to the diaphragm. The mesometrium extends to the deep inguinal ring (not shown), surrounding the round ligament of the uterus. The round ligament is accompanied by the external pudendal artery and vein. See Figure 55 for more details.

Rat

Non-reproductive structures shown: jejunum (suspended by the mesentery), cecum, descending colon, rectum and anus, left kidney, urinary bladder and median ligament, pelvic urethra (surrounded by the urethralis muscle), the extrapelvic urethra (not labeled), and the external urethral ostium.

The (left) ovary, salpinx, uterine horn, the body of uterus, the vagina and the vaginal ostium, and the preputial gland are components of the female genital apparatus.

The round ligament of the uterus is embedded in the fat of the broad ligament entering the vaginal ring, accompanied by the external pudendal artery and vein which enter the deep inguinal ring. (Details in Figure 55.)

As in the mouse, the suspensory ligament of the ovary, the mesovarium, mesosalpinx, and mesometrium are in continuation with one another.

Figure 54A. Mouse. Female reproductive and urinary apparatus (ventral aspect).

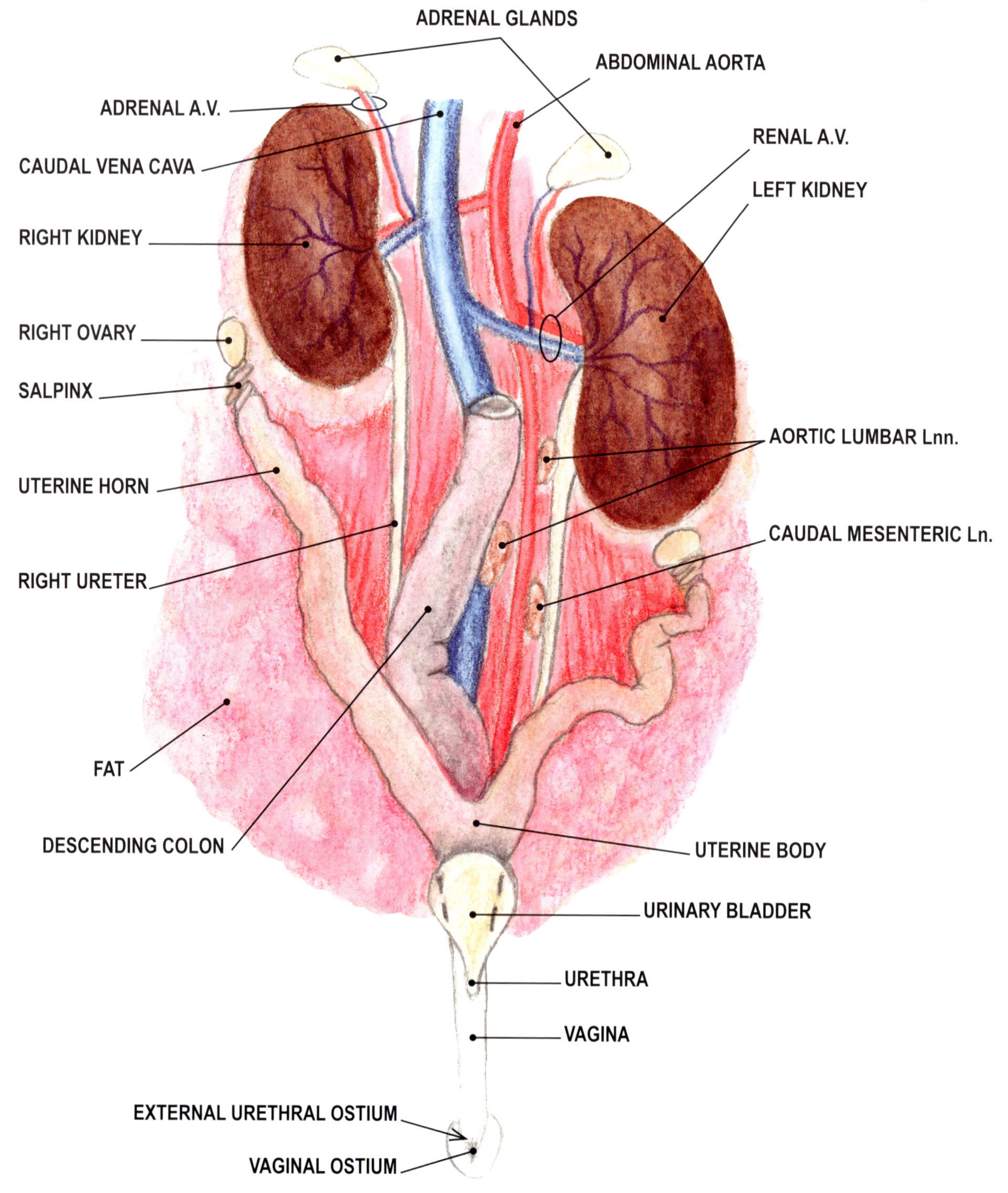

Figure 54B. Rat. Female reproductive and urinary apparatus (ventral aspect).

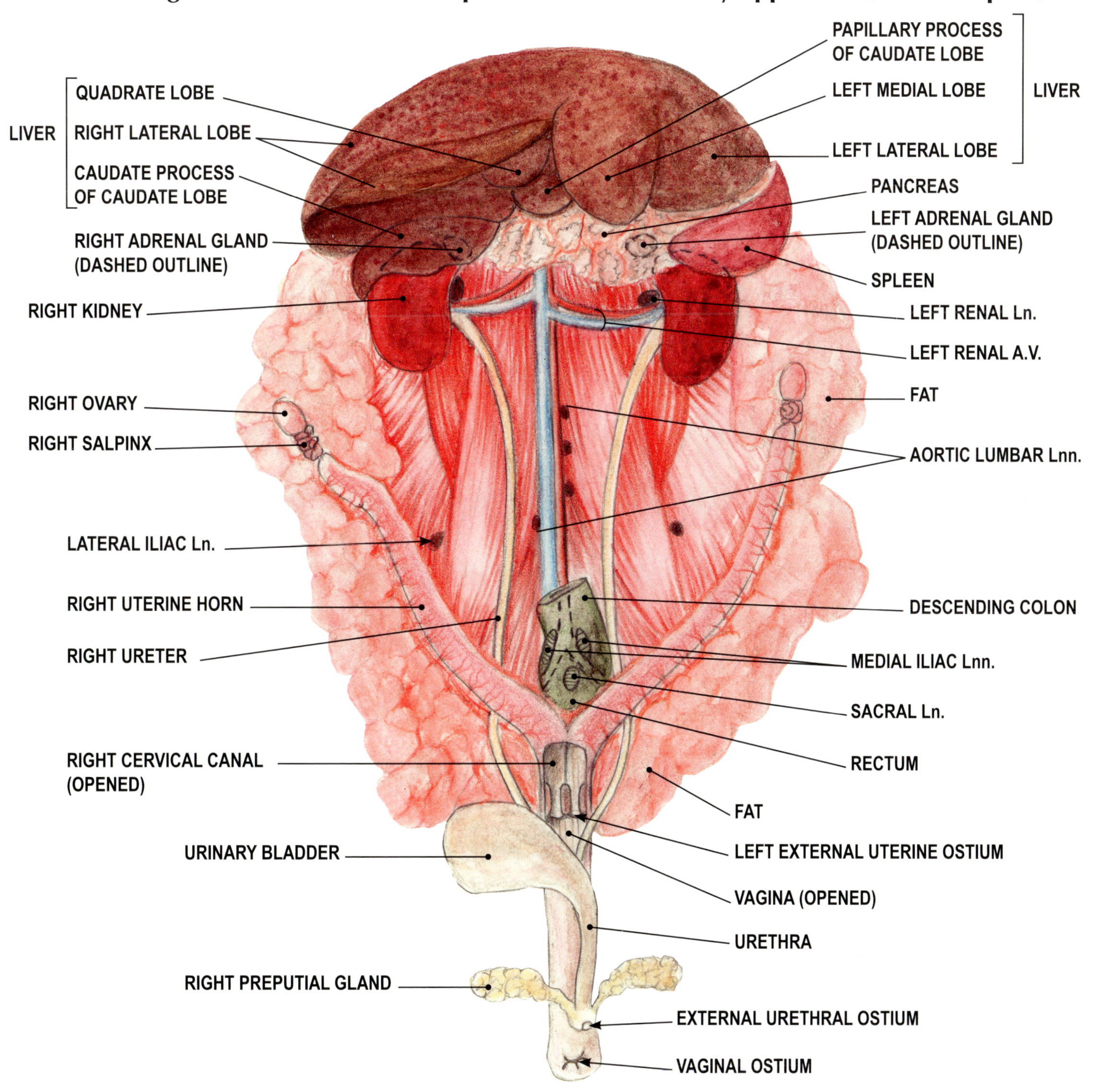

Figure 54A. Mouse. Female reproductive and urinary apparatus (ventral aspect).

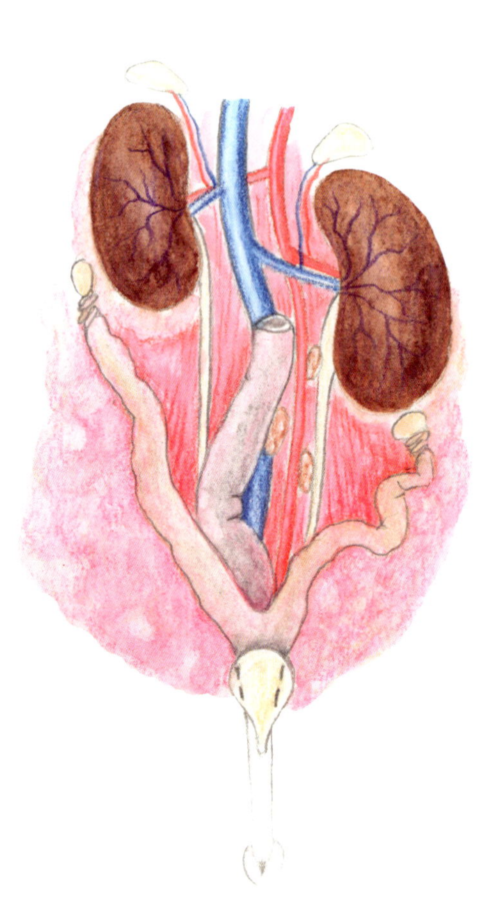

Figure 54B. Rat. Female reproductive and urinary apparatus (ventral aspect).

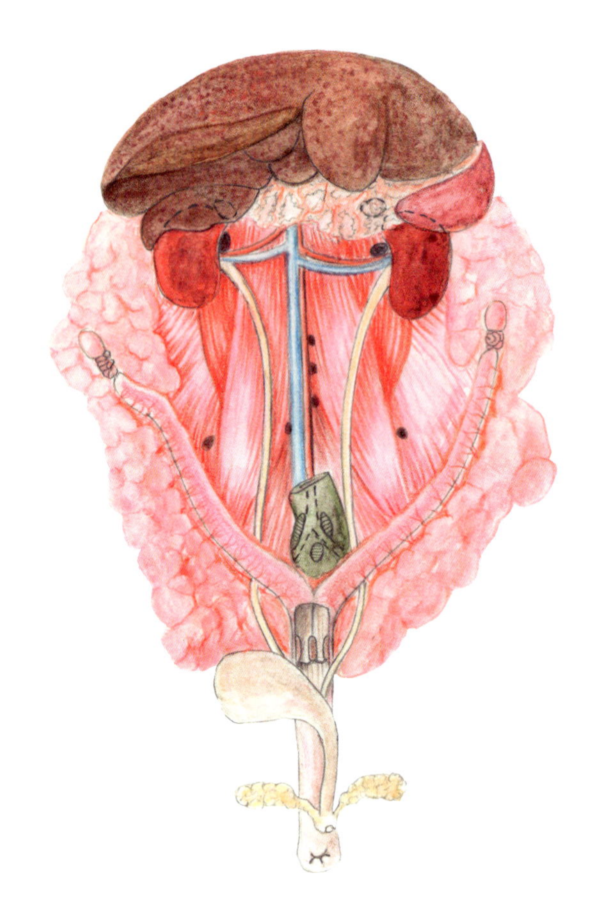

Figure 54. Female reproductive and urinary apparatus (ventral aspect).

The rat uterus, unlike that of the mouse, is classified as uterus duplex and vagina simplex, which means that there are two external uterine ostia opening into the vagina. The mouse has a uterus bicornis because the uterine horns join together in a single body, with one cervix and one uterine ostium.

Neonates and young juveniles of both species have a thin layer of epithelium (not shown) overlying the vaginal ostium. This layer is known as the vaginal plate; it degenerates in puberty to open the vagina. Magnification is needed to visualize the vaginal plate. The urethra opens separately from the vagina in both the mouse and rat; no vaginal vestibule is present. Therefore, the vaginal plate does not obstruct the urethra.

Mouse

The kidneys, with their respective blood supply, the adrenal glands, the ureters, the urinary bladder, and the descending colon are shown.

The ovaries, salpinges (fallopian or uterine tubes), uterine horns, and the uterine body are outlined laterally by an abundance of fat. The fat was removed from their medial aspects to expose the parietal lymph nodes on the roof of the abdominal cavity, the abdominal aorta, the caudal vena cava, and the descending colon.

The mouse has one cervical canal and one uterine ostium.

The preputial glands are shown in Figure 53.

The uterine vasculature is shown in Figure 55.

Rat

The liver (all lobes except the right medial lobe covered by the right lateral lobe), the pancreas, the spleen, and the descending colon are shown.

The kidneys (partially overlapped by cranial abdominal structures), the ureters, the urinary bladder, the urethra, the external urethral ostium, and the preputial glands are shown. The adrenal glands are concealed by other abdominal structures; their position is outlined.

The rat has two cervical canals and two uterine ostia.

The ovaries, the salpinges, the uterine horns with their blood supply, the uterine body, cervical canals, and the vagina are shown. The ventral floor of the cervical canals and vagina are opened to reveal the uterine ostia and a median wall separating the two uteri from each other.

The parietal lymph nodes on the roof of the abdominal cavity, the abdominal aorta, and the caudal vena cava are illustrated between the uterine horns.

Figure 55A. Mouse. Round ligament of the uterus (ventral aspect).

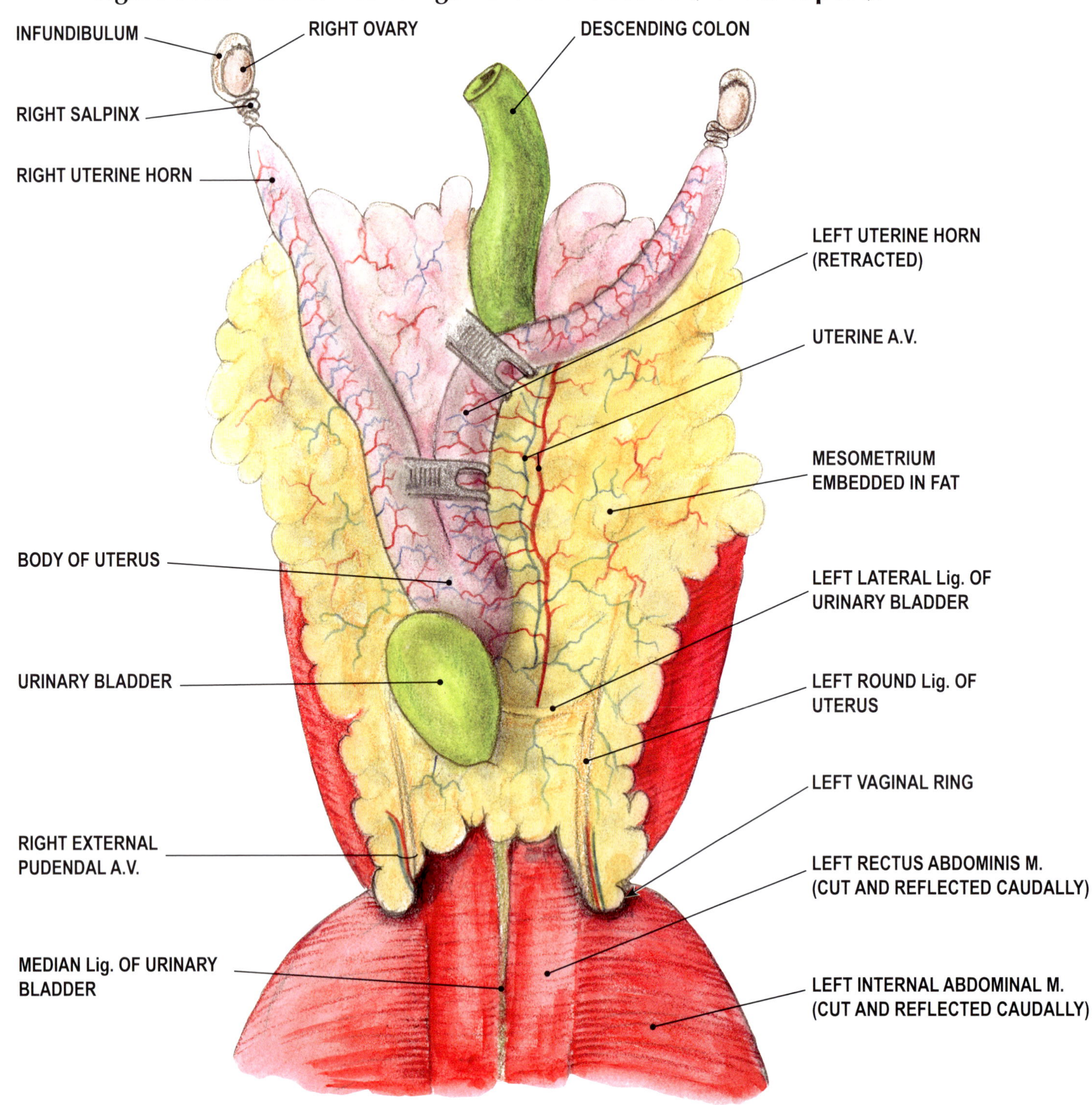

Figure 55B. Rat. Round ligament of the uterus (ventral aspect).*

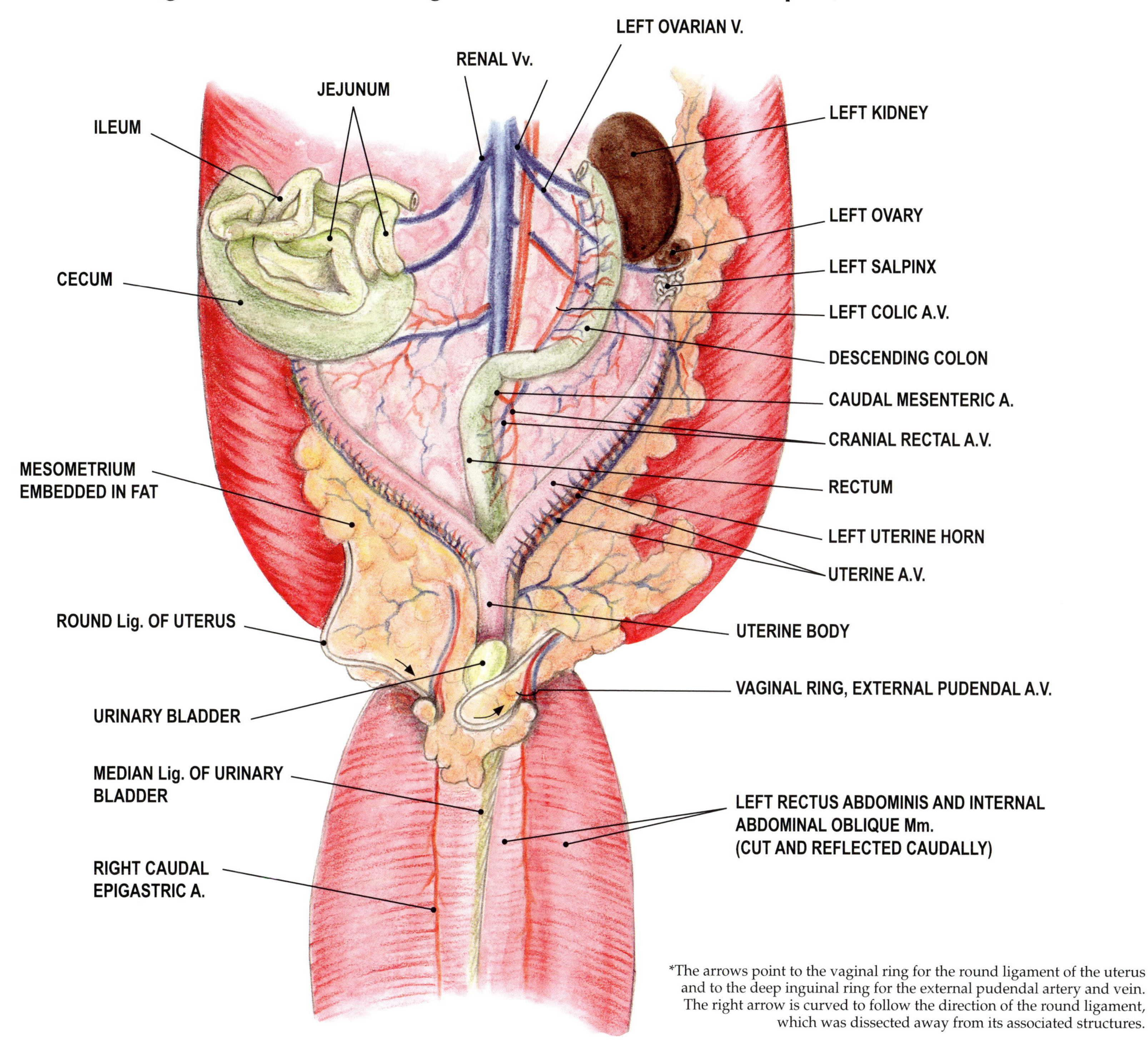

*The arrows point to the vaginal ring for the round ligament of the uterus and to the deep inguinal ring for the external pudendal artery and vein. The right arrow is curved to follow the direction of the round ligament, which was dissected away from its associated structures.

Figure 55A. Mouse. Round ligament of the uterus (ventral aspect).

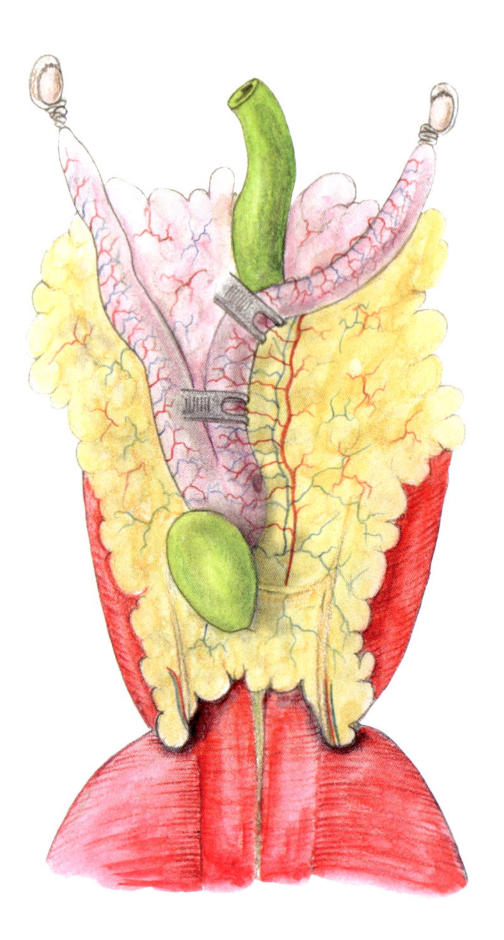

Figure 55B. Rat. Round ligament of the uterus (ventral aspect).*

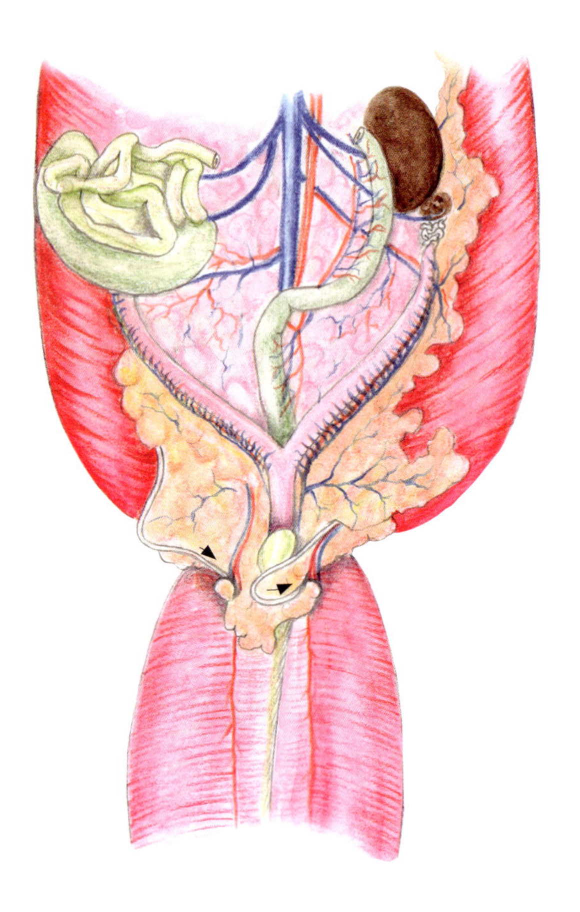

*The arrows point to the vaginal ring for the round ligament of the uterus and to the deep inguinal ring for the external pudendal artery and vein. The right arrow is curved to follow the direction of the round ligament, which was dissected away from its associated structures.

Figure 55. Round ligament of the uterus (ventral aspect).

In both species, the ventral aspect of the female genital apparatus is shown in relationship to the surrounding structures. In preparing these specimens, the ventral wall of the abdominal cavity was cut and reflected caudally to expose the round ligaments of the uterus embedded in the mesometrial fat and entering the vaginal rings.

The round ligament of the uterus, which is not commonly described in the mouse or rat, extends from the tip of the uterine horn to the vaginal ring enclosed in a lateral fold of the mesometrium (part of the broad ligament) attached to the uterus. (The mesometrium connects the parietal peritoneum to the visceral peritoneum of the uterus.) In each figure, the round ligament is embedded in the fat of the broad ligament. The round ligament enters the vaginal canal through the vaginal ring, and is paralleled up to the vaginal ring by the external pudendal artery and vein and the genital branch of the genitofemoral nerve.* These vessels and nerve pass together through the deep inguinal ring into the inguinal canal.

*The inguinal canal extends between the superficial and the deep inguinal rings. The vaginal canal is the space between the parietal and visceral laminae of the vaginal tunic. The vaginal tunic is the corresponding name of the peritoneum inside of the inguinal canal. The deep inguinal ring is the space outlined by the arcus inguinalis, caudally, and the caudal border of the internal abdominal oblique muscle, cranially. The vaginal ring is the ring-like parietal peritoneum which intimately lines the deep inguinal ring.

Pelvic Limb Vessels and Nerves

Figure 56A. Mouse. Pelvic limb, left (lateral aspect).

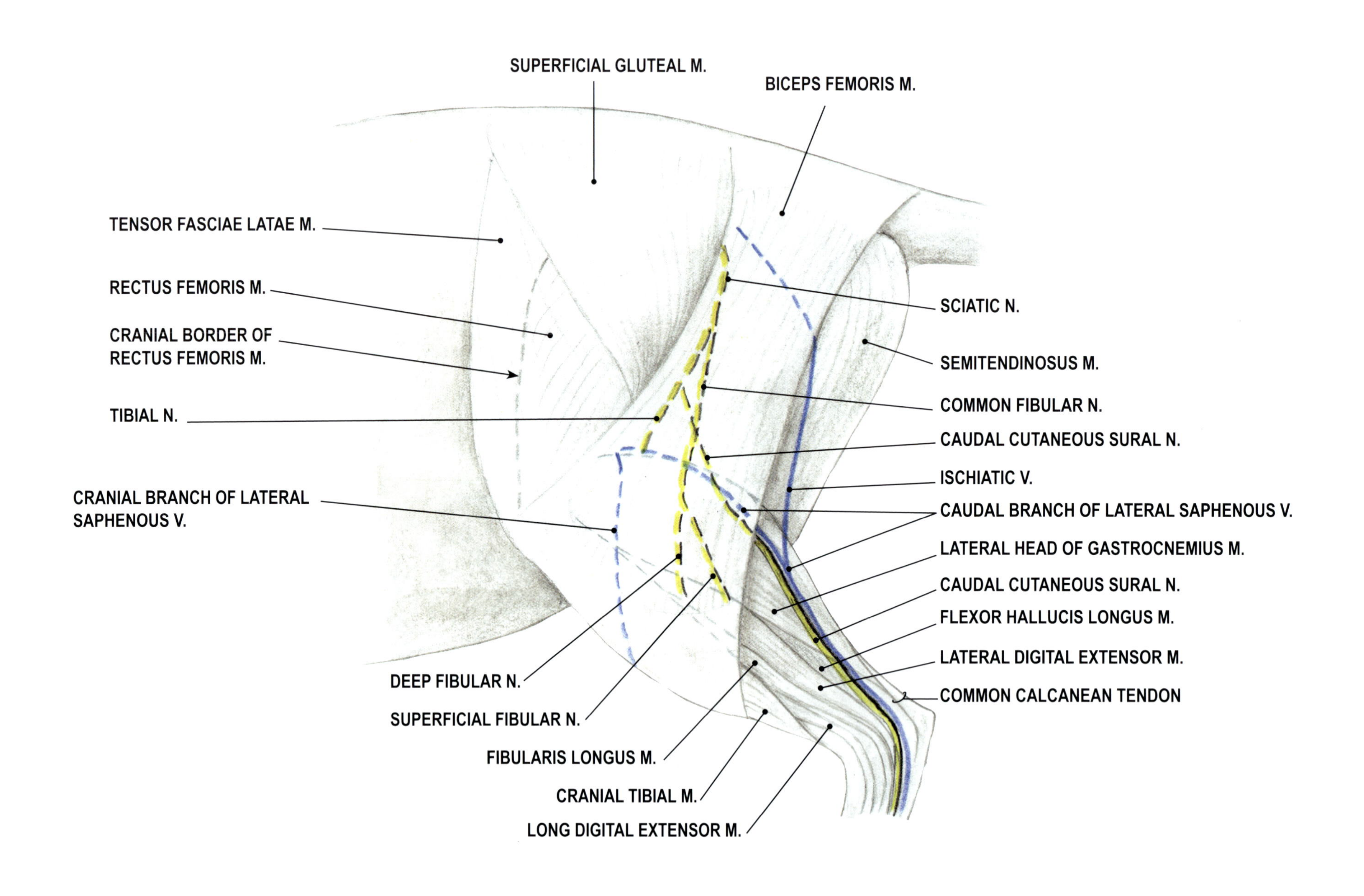

Figure 56B. Rat. Pelvic limb, left (lateral aspect).

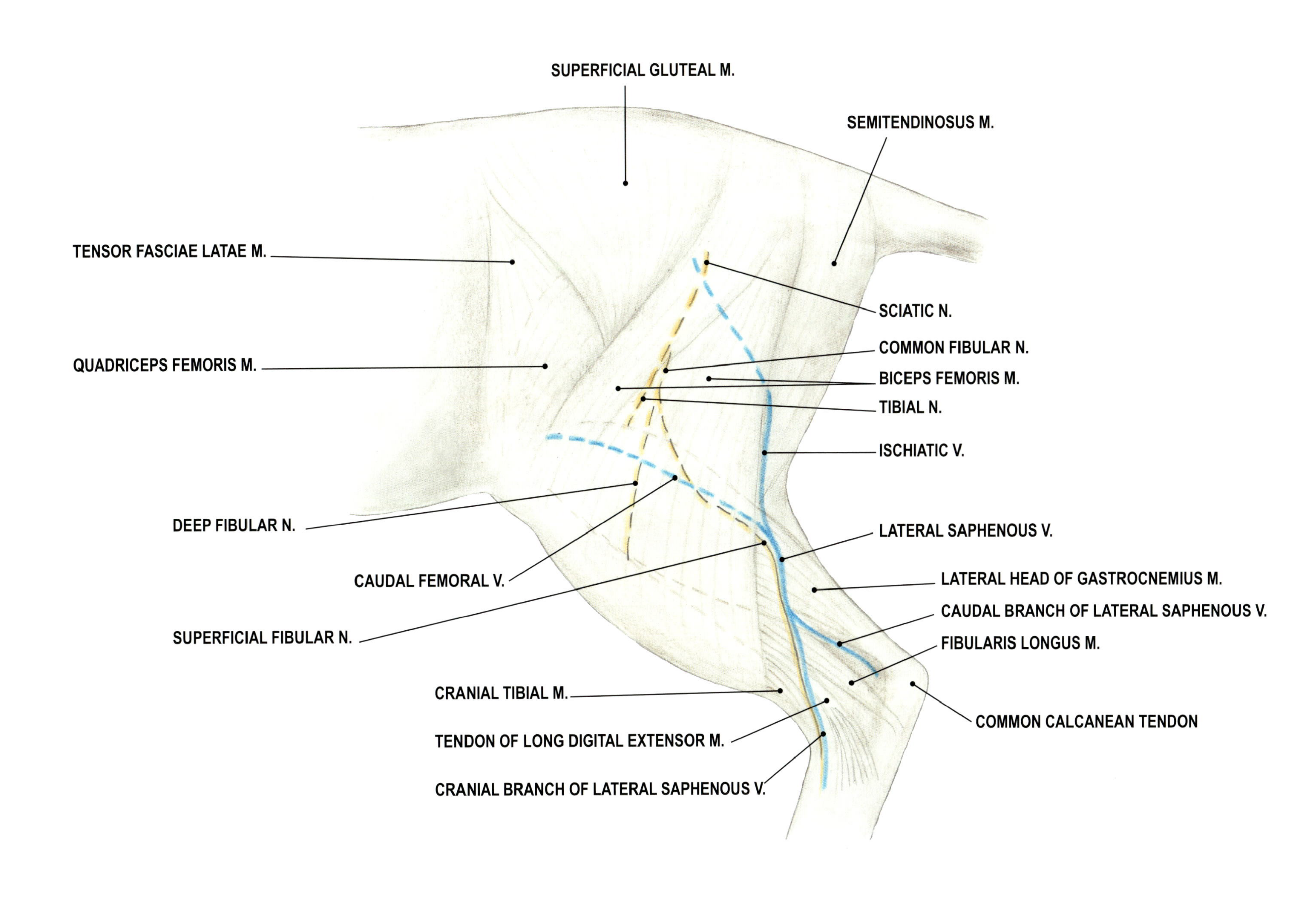

Figure 56A. Mouse. Pelvic limb, left (lateral aspect).

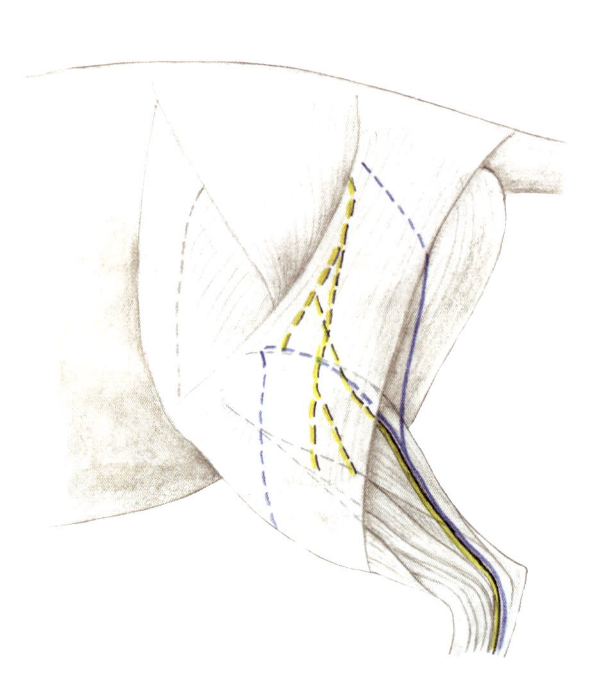

Figure 56B. Rat. Pelvic limb, left (lateral aspect).

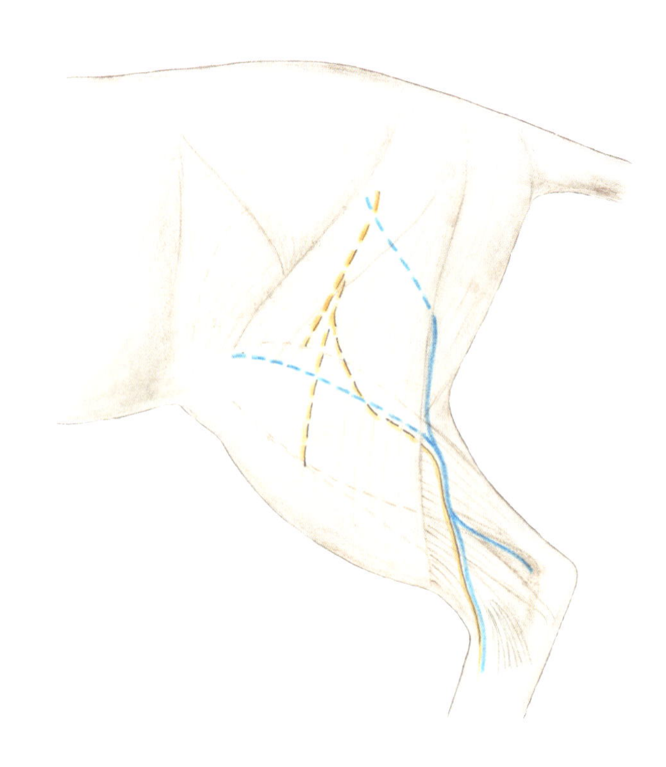

Figure 56. Pelvic limb, left (lateral aspect).

The major veins and nerves are shown for purposes of vascular procedures. The subcutaneous veins and nerves are shown in continuous lines, whereas those which run deep to muscles are shown in dashed lines. The muscles are outlined lightly and labeled in both species.

Mouse

The cranial and caudal branches of the lateral saphenous vein merge deep, proximally.

The ischiatic vein merges with the caudal branch of the lateral saphenous vein, which runs parallel to the caudal border of the crus* over the lateral head of the gastrocnemius muscle, and continues over the tarsus. The caudal branch of the lateral saphenous vein is superficial and can be seen through the skin.

The caudal branch of the lateral saphenous vein is accompanied by the caudal cutaneous sural nerve. The cranial branch (labeled) of this vein runs deep.

The sciatic nerve bifurcates to form the tibial nerve and the common fibular nerve.

- The tibial nerve delivers a branch, which is the caudal cutaneous sural nerve.
- The common fibular nerve bifurcates to form the deep and superficial fibular nerves.

Rat

As in the mouse, the ischiatic vein merges with the lateral saphenous vein. The lateral saphenous vein runs parallel to the caudal border of the biceps femoris muscle and crosses the lateral crus obliquely over the lateral head of the gastrocnemius muscle. In the rat, the pattern of the merger and bifurcation of the lateral saphenous vein is different from the mouse. The caudal femoral vein is considered the origin of the lateral saphenous vein, which arises at the site where the ischiatic vein joins with the caudal femoral vein.

At the limit between the lateral head of the gastrocnemius and the fibularis longus muscles, the lateral saphenous vein splits into a caudal branch and a cranial branch. The caudal branch runs in a caudal direction parallel to the ventral border of the lateral head of the gastrocnemius muscle. The cranial branch continues distally, and crosses obliquely the fibularis longus, the long digital extensor, and the cranial tibial muscles. The cranial branch is paralleled by the superficial fibular nerve and continues over the dorsolateral aspect of the tarsus.

The nerves are similar to those of the mouse.

*The crus is the anatomical region of the pelvic limb between the knee and tarsal joints, which in the common language is called "leg".

Figure 57A. Mouse. Pelvic limb, left (medial aspect).

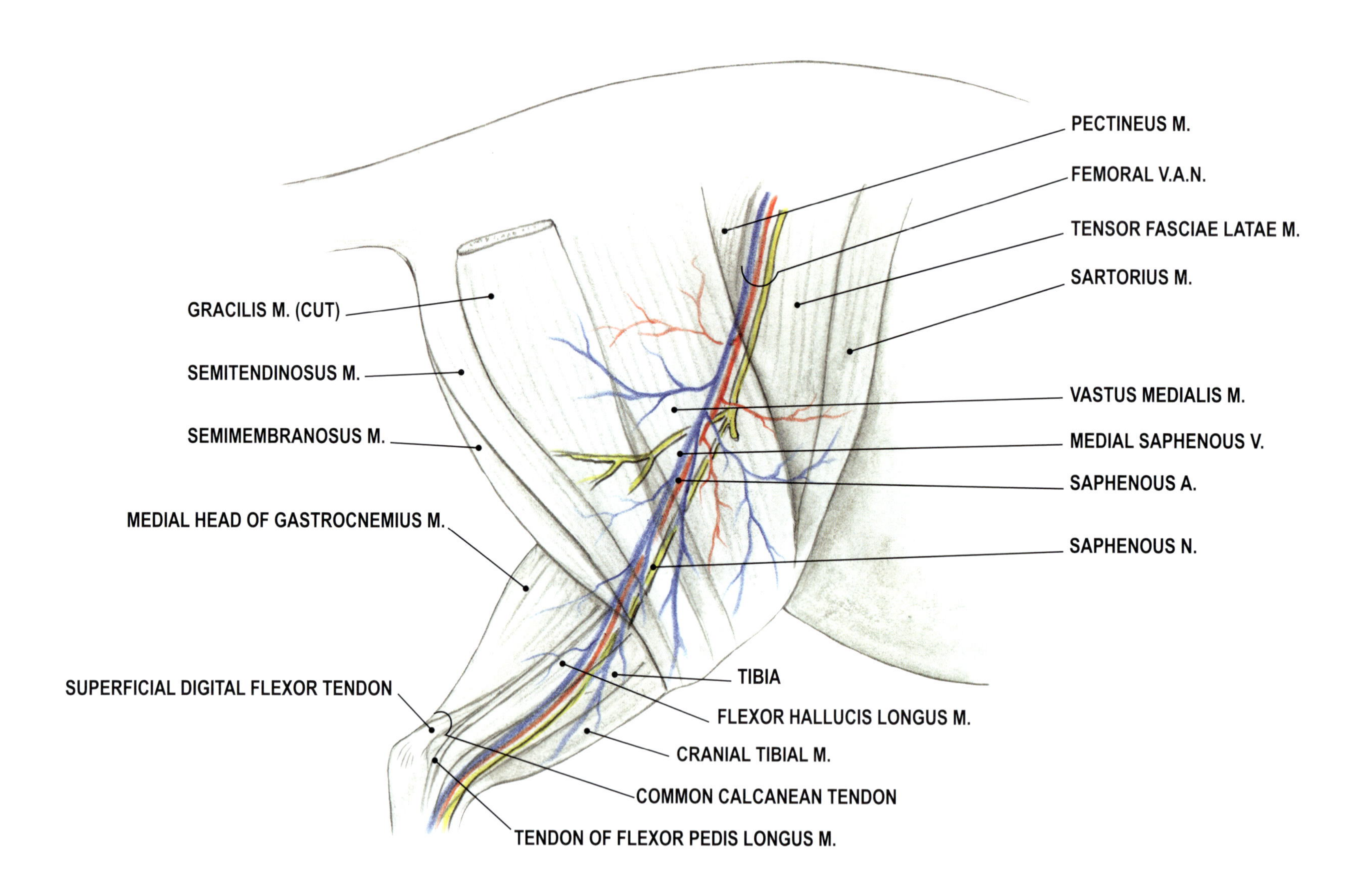

Figure 57B. Rat. Pelvic limb, left (medial aspect).

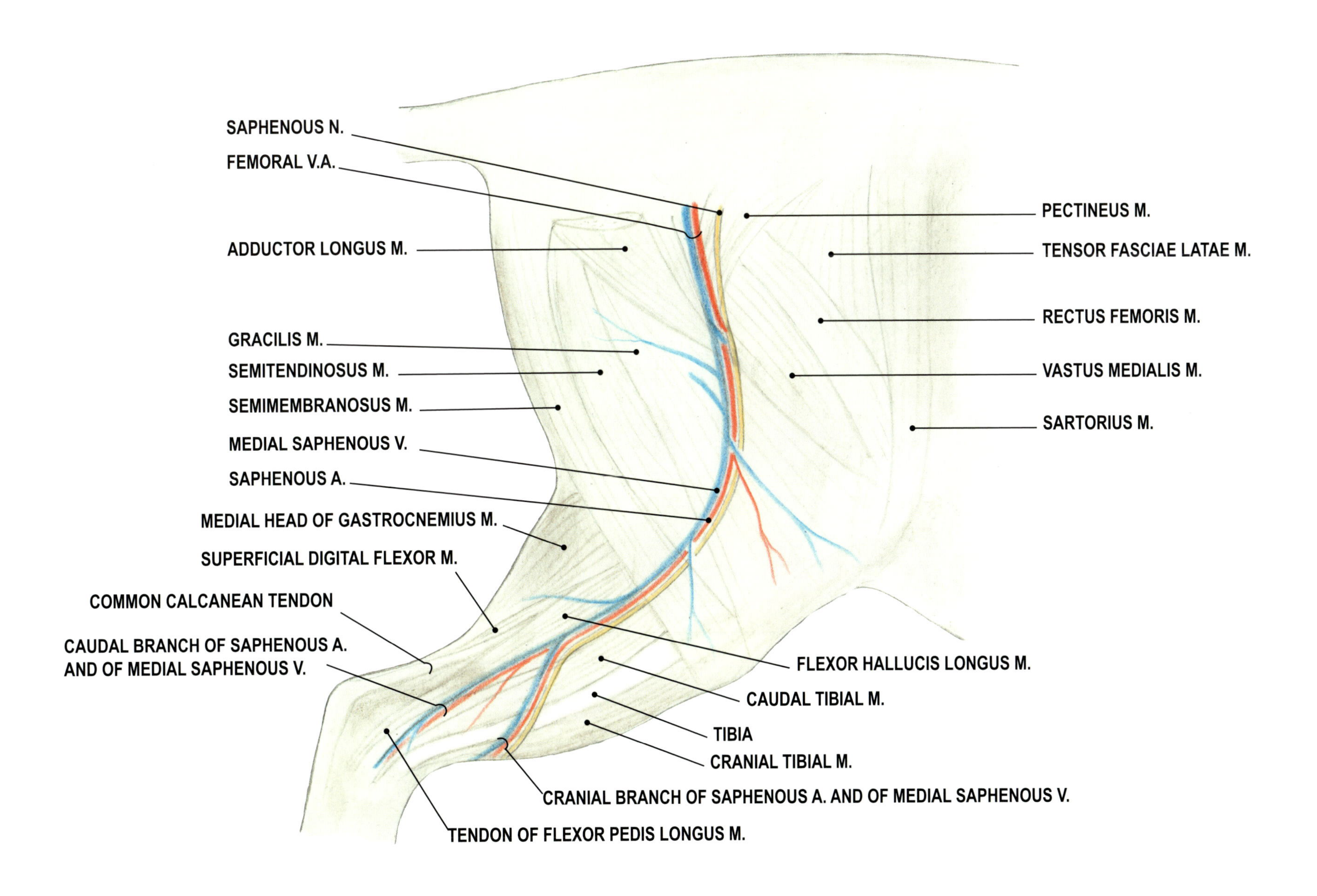

Figure 57A. Mouse. Pelvic limb, left (medial aspect).

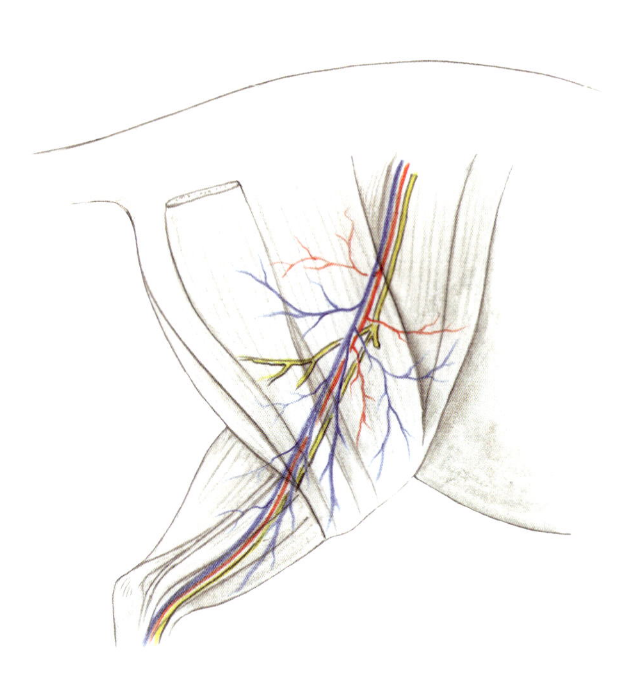

Figure 57B. Rat. Pelvic limb, left (medial aspect).

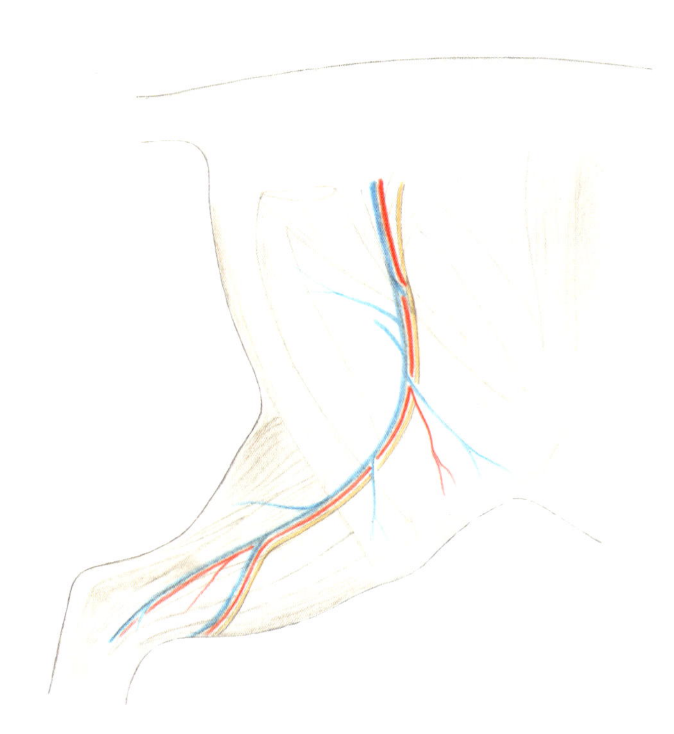

Figure 57. Pelvic limb, left (medial aspect).

In both species, the main vessels and nerves are shown, and the muscles are lightly outlined and labeled.

Mouse

Together, the femoral vessels and nerve cross obliquely over the tensor fasciae latae and vastus medialis muscles.

The saphenous vessels and nerve arise as a continuation of the femoral vessels and nerve, and obliquely cross the vastus medialis, gracilis, semitendinosus, and semimembranosus muscles.

The saphenous vessels and nerve course parallel to, and between, the flexor hallucis longus and cranial tibial muscles.

Rat

Together, the femoral vessels and the saphenous nerve bend from an initial vertical position and run ventrocaudally as the saphenous vessels and nerve. These vessels and nerve obliquely cross the pectineus, adductor longus, vastus medialis, gracilis, semitendinosus, and semimembranosus muscles.

The saphenous vessels and nerve course parallel to, and between, the flexor hallucis longus and cranial tibial muscles.

The saphenous artery and medial saphenous vein branch and form, respectively, the:

- Caudal branch of the saphenous artery and the medial saphenous vein.
- Cranial branch of the saphenous artery and the medial saphenous vein.

The saphenous nerve accompanies the cranial branch of the saphenous artery and the cranial branch of the medial saphenous vein.

Structures of the Tail

Figure 58A. Mouse. Base of the tail (cross section), stained by hematoxylin and eosin.

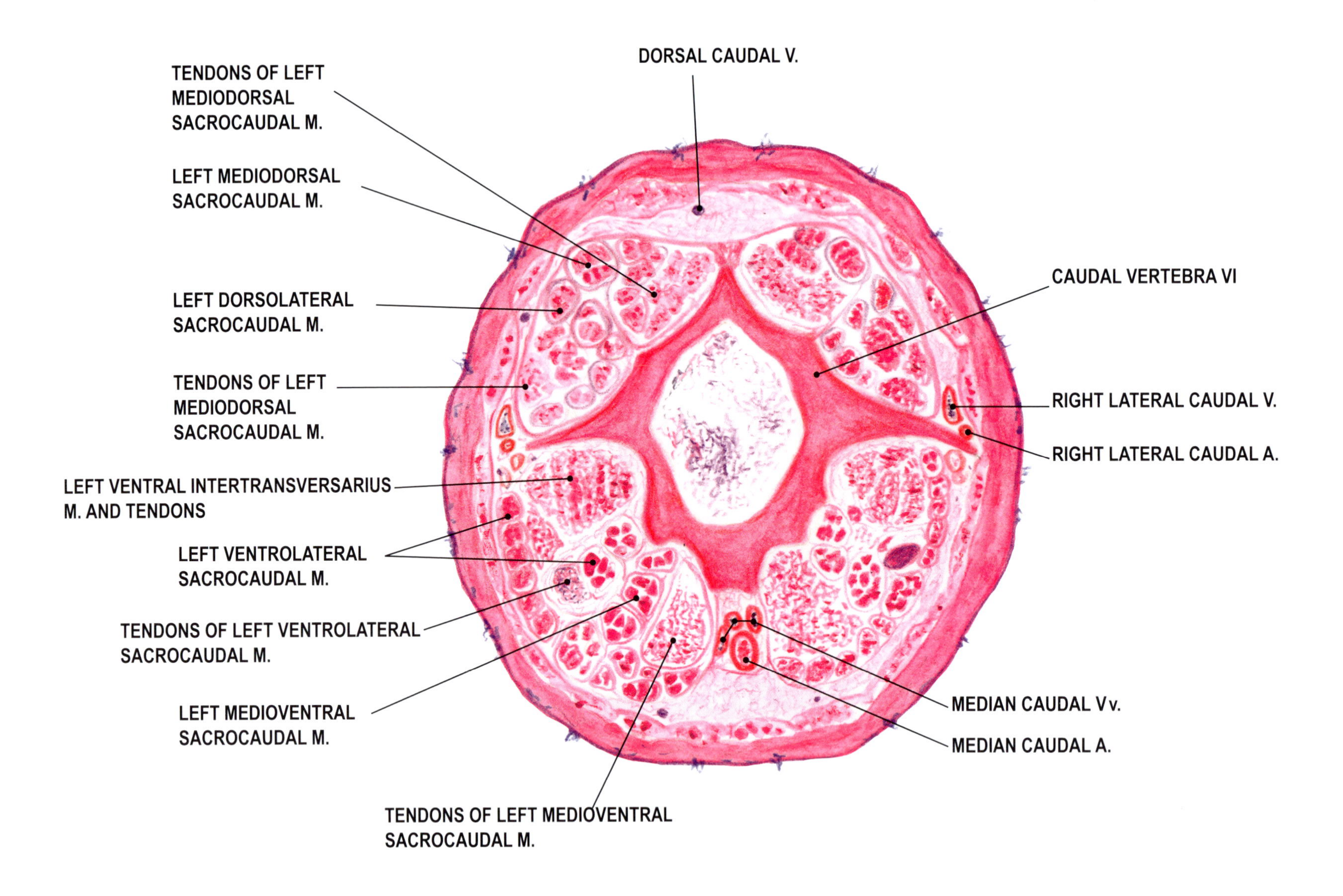

Figure 58B. Rat. Base of the tail (cross section), stained by hematoxylin and eosin.

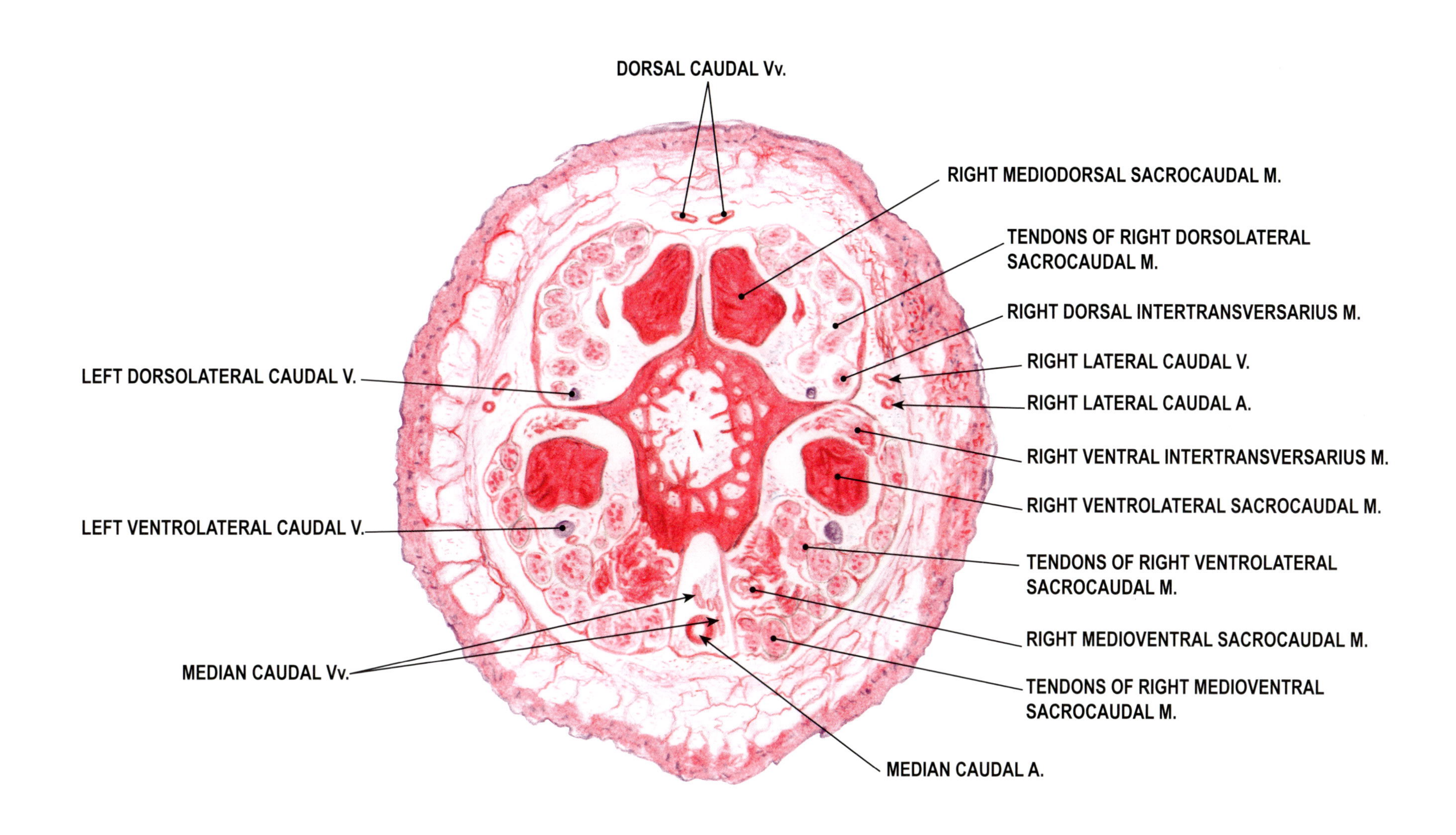

Figure 58A. Mouse. Base of the tail (cross section), stained by hematoxylin and eosin.

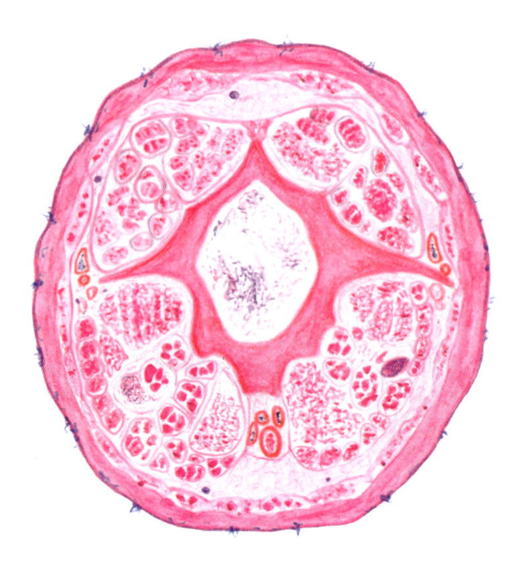

Figure 58B. Rat. Base of the tail (cross section), stained by hematoxylin and eosin.

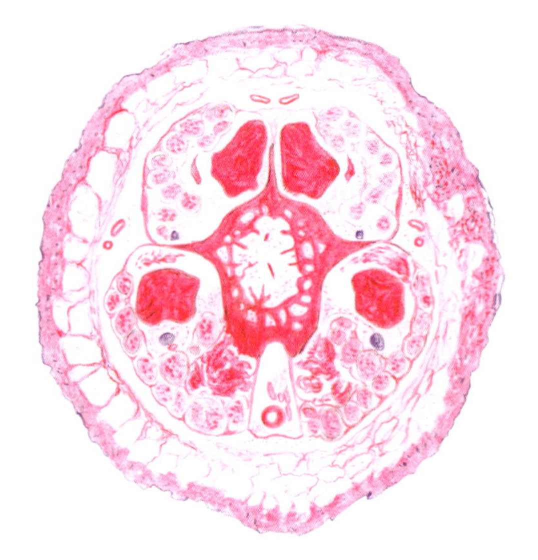

Figure 58. Base of the tail (cross section), stained by hematoxylin and eosin.

The specimens are illustrated at the base of the tail, approximately at the 6th caudal vertebra in the mouse and the 3rd caudal vertebra in the rat. Each figure is a composite of images photographed at 20× magnification. Nerves are not labeled because special procedures such as a Giemsa Gömöri stain would be required to reveal their location.

Mouse

This illustration shows only one dorsal caudal vein, compared with two veins shown in the rat (see right). This cross section was performed at the level of the 6th caudal vertebra; at this location, the two dorsal caudal veins merge into a single vein in both the mouse and the rat.

Mainly tendons can be found, due to the vertebral level of the cross section; they are shown as small light red dots. Of the tail muscles, few muscle fibers extend beyond the 4th to 5th caudal vertebra; these are shown as large dark red dots. The septa of the caudal fascia are thicker (stronger) on the dorsal aspect to bind down the extensor muscles of the tail, which are more powerful than the flexors.

The dermis is comparatively thinner than in the rat, and therefore the tail vessels are more superficial. As mice are occasionally lifted by the tail via tongs, it is important to be aware that tongs can damage the vessels and affect vascular access.

The inner surface of the vertebral canal is smooth.

Rat

The cross section was performed at the approximate level of the 3rd caudal vertebra to show the two dorsal caudal veins, which join with each other at the level of the 6th caudal vertebra and run as a single dorsal caudal vein to the end of the tail (see mouse, above).

In the specimen shown, the transverse processes are smaller than in the mouse despite the fact that the cross section is made at a level cranial to that of the mouse. This is because the section has been made at a location close to the origin of the transverse processes.

The muscles (the four solid red masses) and tendons are present in this illustration because of the vertebral level of the cross section. The caudal fascia is well represented dorsally and ventrally, also due to the crosssection level.

The inner surface of the vertebral canal shows bony trabeculae.

Skeletal Structures

Figure 59A. Mouse. Skeleton (left lateral aspect).

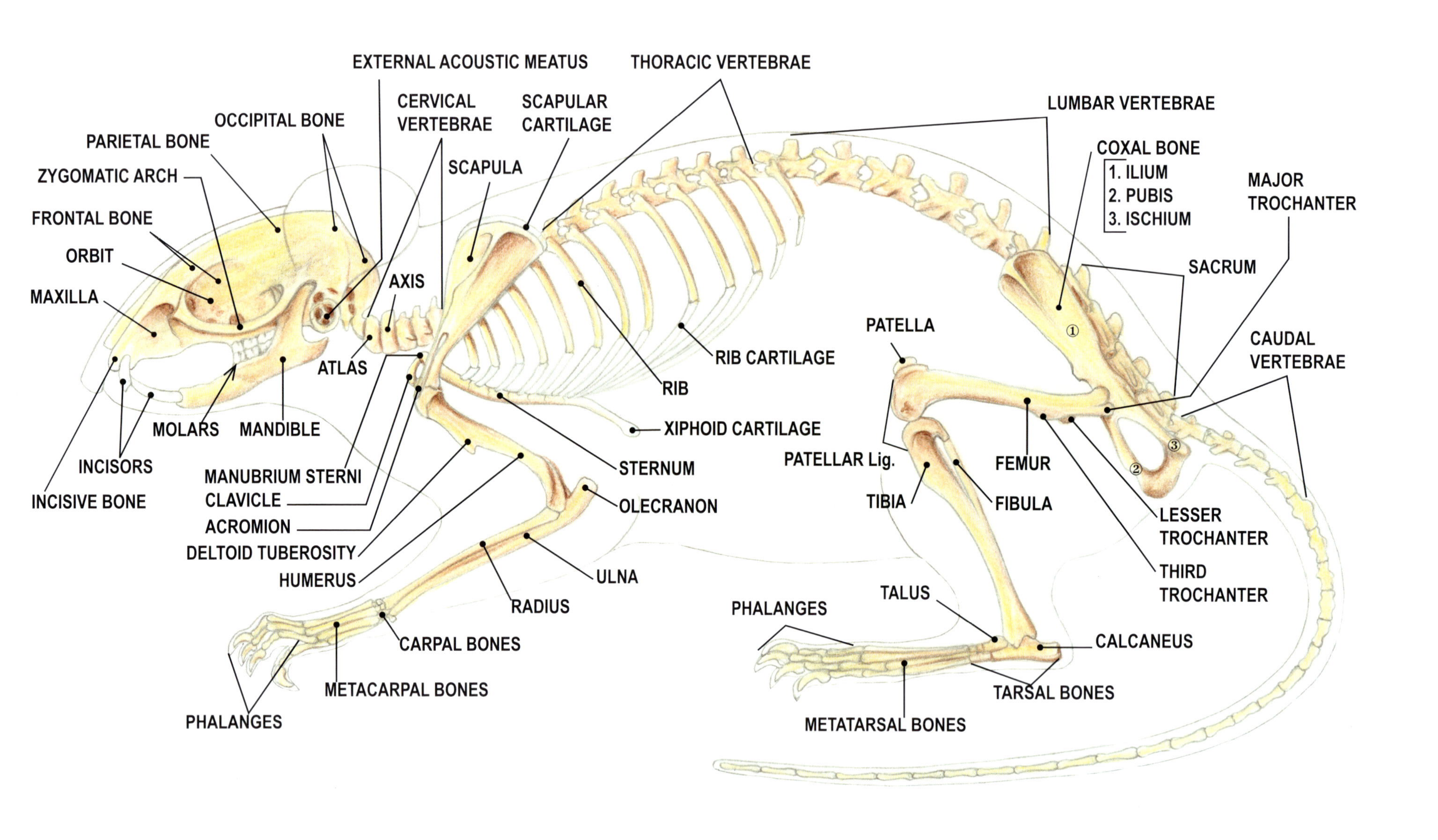

Figure 59B. Rat. Skeleton (left lateral aspect).

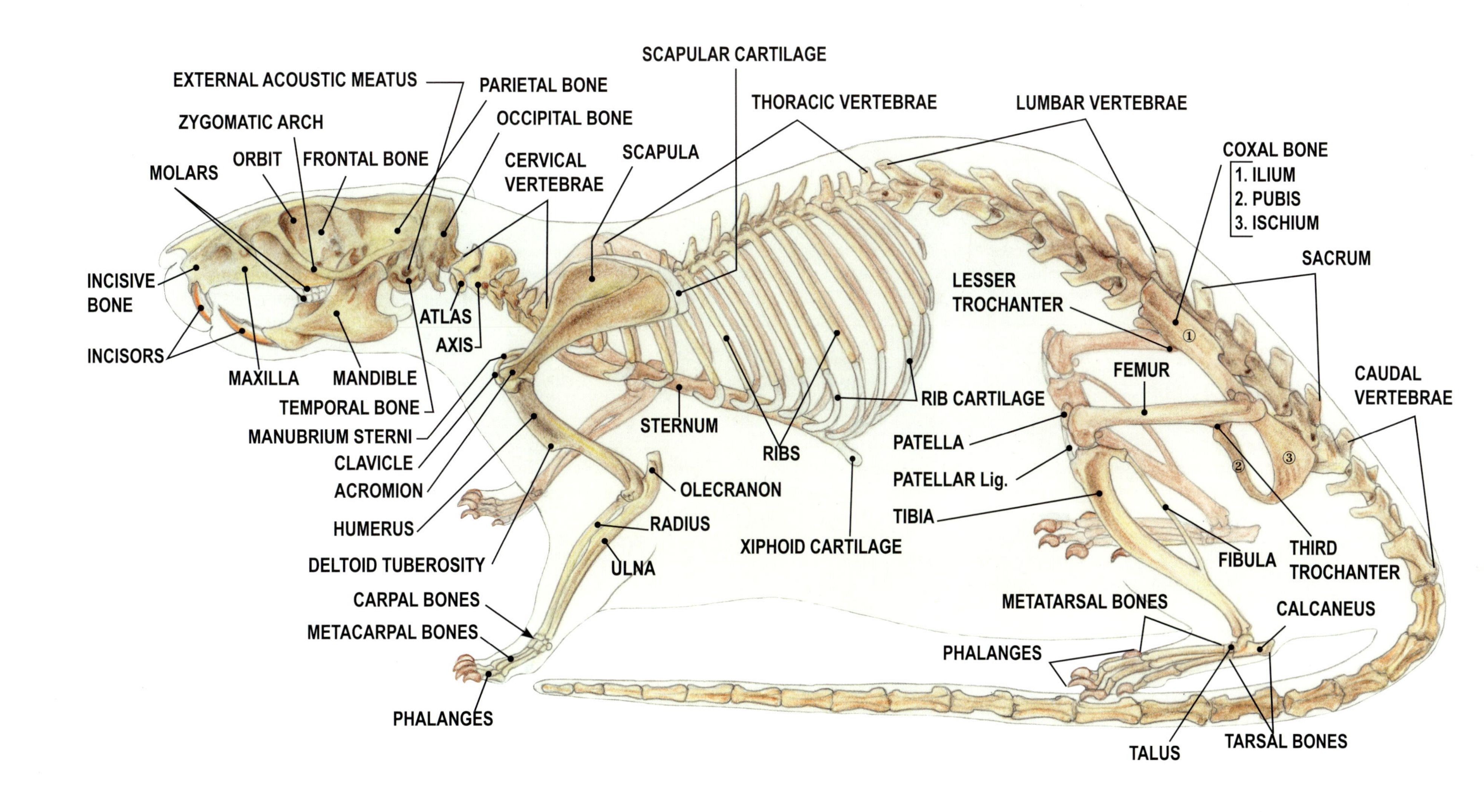

Figure 59A. Mouse. Skeleton (left lateral aspect).

Figure 59B. Rat. Skeleton (left lateral aspect).

Figure 59. Skeleton (left lateral aspect).

The skeleton of the rat shows both left and right structures (limbs and ribs), whereas only the left structures are shown in the mouse. The skeleton of both species shows the same bones, with the same number of vertebrae, sternebrae, carpal and tarsal bones, metacarpal and metatarsal bones, and phalanges in both limbs. In general, the rat has many topographical differences that reflect this species' greater musculature and physical strength. Differences in the skulls are described in Figure 60.

Vertebrae

The rat has a comparatively larger atlas and axis (located in the cervical region) than the mouse. In the mouse, the cervical vertebrae are similar in shape and size. The spinous and transverse processes in the thoracic, lumbar, sacral, and proximal caudal vertebrae are elongated in the rat and diminutive in the mouse.

Ribs and Sternum

In the mouse, the rib cartilages are angled more cranially than those of the rat. The points of attachment of the cartilages on the sternum are also more cranial, and the sternum is comparatively shorter in the mouse. The rat has a deeper chest cavity dorsoventrally than the mouse.

The mouse has a more pronounced xiphoid process than the rat.

Thoracic Limbs

Only subtle differences exist in the shape of these limb bones.

Pelvic Limbs

The patellar ligament is shown in the rat, not in the mouse. In the rat, the fibula is comparatively longer and more separated from the tibia than in the mouse.

Figure 60A. Mouse. Skull and details of teeth (left lateral aspect).

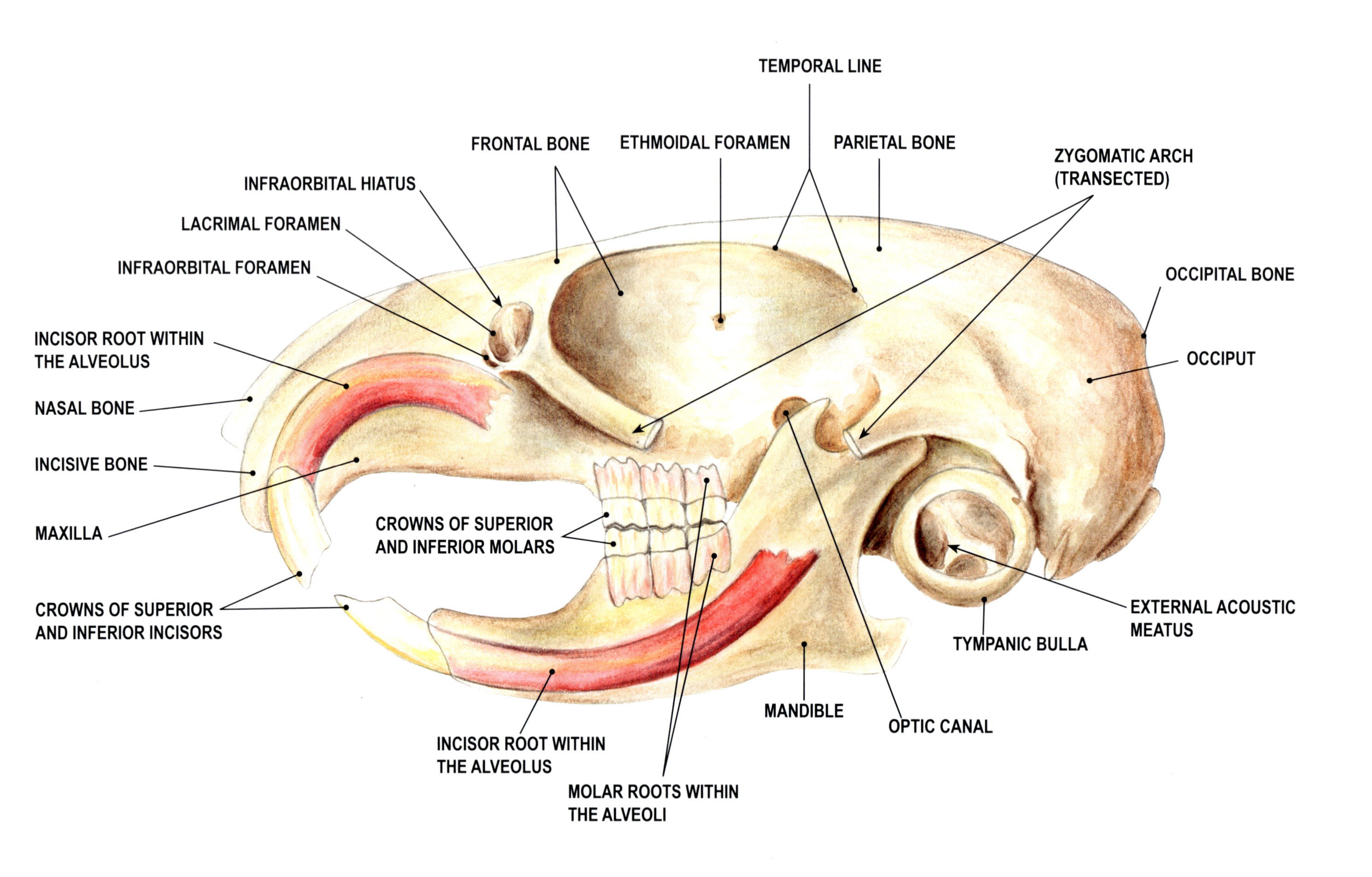

Figure 60B. Rat. Skull and details of teeth (left lateral aspect).

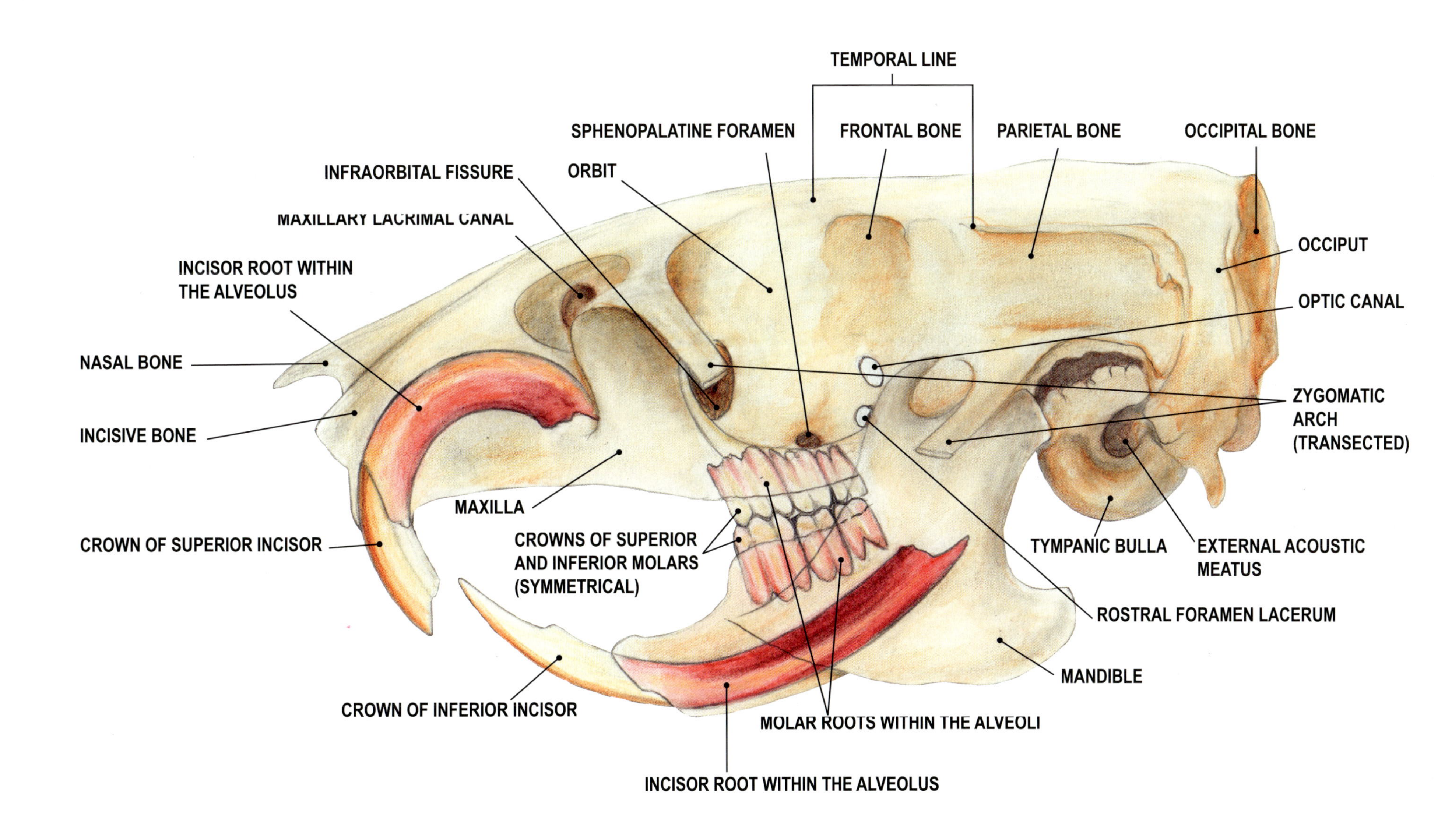

Figure 60A. Mouse. Skull and details of teeth (left lateral aspect).

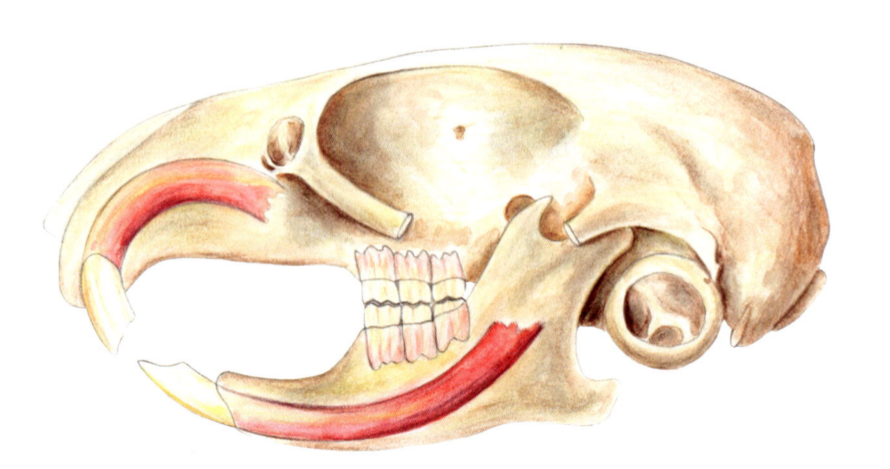

Figure 60B. Rat. Skull and details of teeth (left lateral aspect).

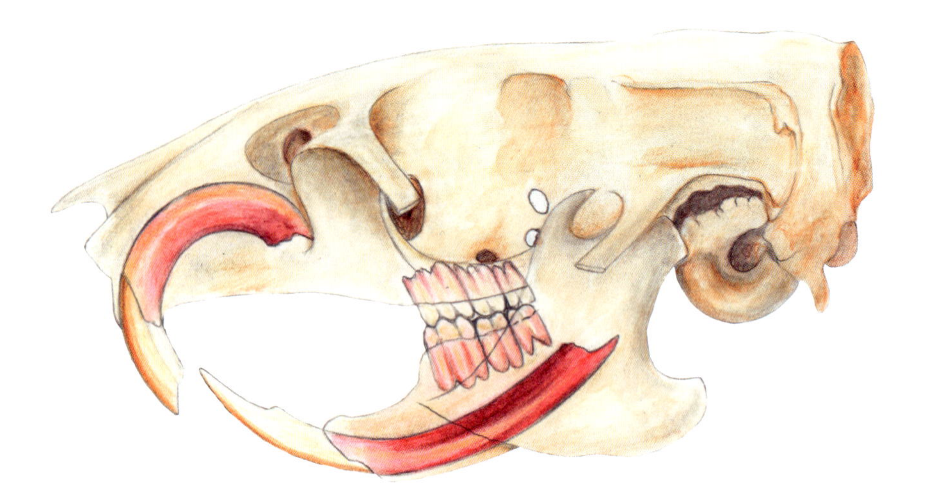

Figure 60. Skull and details of teeth (left lateral aspect).

In both figures, the zygomatic arches are cut for exposing the skull foramina related to the orbital veins, which are illustrated in Figure 18. The teeth are illustrated inside of the alveoli (bony sockets), after the lamina dura (the lateral wall of the alveoli) was removed.

The mouse skull is proportionately longer than that of the rat; the incisor teeth are therefore less curved. Note the curvature of these teeth in the rat. The differences in the position and extent of the teeth are shown and labeled. In any individual mouse or rat, the length of incisors with normal apposition is related to factors of diet and age, which affect the wear of the occlusal surface and therefore the tooth length. The molars have similar shapes in the two species.

The nasal and incisive bones are curved rostrally in the mouse, less so in the rat.

The temporal line of the skull is more pronounced in the rat than in the mouse, reflecting the more powerful temporal muscle of the rat.

The orbit is more circumscribed in the mouse, but wider in the rat.

The zygomatic arch is more ventrally oriented in the rat.

The external acoustic meatus is wider in the mouse, than in the rat.

The occipital bone is rounded in the mouse, and abruptly ends as a vertical wall in the rat.

Bibliography

1. **Albe-Fessard DG, Libouban S, Stutinsky F.** 1971. Atlas stéréotaxique du diencéphale du rat blanc. Paris (France): Centre National de la Recherche Scientifique.
2. **Chiasson RB.** 1988. Laboratory anatomy of the white rat. Dubuque (IA): William C Brown.
3. **Cook MJ.** 1965. The anatomy of the laboratory mouse. New York (NY): Academic Press.
4. **De Groo J.** 1963. The rat forebrain in stereotaxic coordinates. Amsterdam (the Netherlands): Noord-Hollansche UM.
5. **Dong HW.** 2008. Allen reference atlas: a digital color brain atlas of the C57Black/6J male mouse. Hoboken (NJ): Wiley.
6. **Franklin KBJ, Paxinos G.** 2007. The mouse brain in stereotaxic coordinates 3rd ed. New York (NY): Academic Press.
7. **Greene EC.** 1968. Anatomy of the rat. New York (NY): Hafner Publishing Company.
8. **Hebel R, Stromberg MW.** 1986. Anatomy and embryology of the laboratory rat. Wörthsee (Germany): BioMed Verlag.
9. **Iwaki T.** 2001. A color atlas of sectional anatomy of the mouse. Tokyo (Japan): Braintree Scientific.
10. **Lehmann A.** 1974. Atlas stéréotaxique du cerveau de la souris. Paris (France): Centre National de la Recherche Scientifique.
11. **Montemurro DG, Dukelow RH.** 1972. A stereotaxic atlas of the diencephalon and related structures of the mouse. Mount Kisco (NY): Futura Publishing Company.
12. **Olds RJ, Olds JR.** 1979. A color atlas of the rat: dissection guide. New York (NY): Wiley.
13. **Pellegrino LJ, Cushman AJ.** 1979. A stereotaxic atlas of the rat brain. New York (NY): Plenum Press.
14. **Popesko P, Rajtová V, Horák J.** 2003. A colour atlas of anatomy of small laboratory animals, vol 2: rat-mouse-hamster. Philadelphia (PA): WB Saunders.
15. **Raabe O, Yeh H, Schum G.** 1971. Tracheobronchial geometry: human, dog, rat, hamster. A compilation of selected data from the project respiratory tract deposition models. Washington (DC): US Government Printing Office.
16. **Sealander JA, Hoffman CE.** 1967. Laboratory manual of elementary mammalian anatomy: with emphasis on the rat. Minneapolis (MN): Burgess Publishing Company.
17. **Sidman RL, Angevine JB Jr, Pierce ET.** 1971. Atlas of the mouse brain and spinal cord. Cambridge (MA): Harvard University Press.
18. **Silverman S, Tell L.** 2005. Radiology of rodents, rabbits, and ferrets: an atlas of normal anatomy and positioning. Burlington (MA): Elsevier
19. **Smith RS, John SWM, Nishina PM, Sundberg JP, editors.** 2001. Systematic evaluation of the mouse eye: anatomy, pathology, and biomethods (research methods for mutant mice). Boca Raton (FL): CRC Press.
20. **Suckow MA, Weisbroth SH, Franklin CL.** 2006. The laboratory rat, 2nd ed. New York (NY): Academic Press.
21. **Turner CW, Gomez ET.** 1934. The experimental development of the mammary gland. I. The male and female albino mouse. II. The male and female guinea pig. Columbia (MO): University of Missouri.
22. **Turner CW, Schultze AB.** 1931. A study of the causes of the normal development of the mammary glands of the albino rat. Columbia (MO): University of Missouri.
23. **Zilles KJ.** 1985. The cortex of the rat: a stereotaxic atlas. Berlin (Germany): Springer-Verlag.

References

1. **Ben Rhouma K, Sakly M.** 1994. Involution of rat thymus: characterization of cytoplasmic glucocorticoid receptors, evidence of glucocorticoid resistant dexamethasone receptor-positive cells. Arch Int Physiol Biochim Biophys **102:**97–102.
2. **Constantinescu GM, Hillebrand A, Simoens P, editors.** In press. Illustrated veterinary anatomical nomenclature, 3rd ed. Stuttgart (Germany): MVS Medizinverlage Stuttgart.
3. **Cooke PS, Young PF, Cunha GR.** 1987. Androgen dependence of growth and epithelial morphogenesis in neonatal mouse bulbourethral glands. Endocrinology **121:**2153–2160.
4. **Cunliffe-Beamer TL, Feldman DB.** 1976. Vaginal septa in mice: incidence, inheritance, and effect on reproductive performance. Lab Anim Sci **26:**895–898.
5. **Gearhart S, Kalishman J, Melikyan H, Mason C, Kohn DF.** 2004. Increased incidence of vaginal septum in C57BL/6J mice since 1976. Comp Med **54:**418–421.
6. **Habel R, Stromberg MW.** 1976. Anatomy of the laboratory rat. Baltimore (MD): Williams and Wilkins.
7. **International Committee on Veterinary Gross Anatomical Nomenclature, World Association of Veterinary Anatomists.** [Internet]. 2005. Nomina anatomica veterinaria, 5th ed. [Cited 26 Aug 2010]. Available at: www.wava-amav.org/Downloads/nav_2005.pdf
8. **Maronpot RR, Boorman GA, Gaul BW, editors.** 1999. Pathology of the mouse: reference atlas. St Louis (MO): Cache River Press.
9. **Murakami R, Mizuno T.** 1984. Histogenesis of the os penis and os clitoridis in rats. Dev Growth Differ **26**:419–426.
10. **Paxinos G, Watson C.** 2005. The rat brain in stereotaxic coordinates, 5th ed. New York (NY): Academic Press.
11. **Scremin O.** 2004. Cerebral vascular system, p 1165–1199. In: Paxinos G, editor. The rat nervous system, 3rd ed. Burlington (MA): Elsevier.
12. **Shier JGM.** 1984. Studies on the inheritance of vaginal septa in mice, a trait with low penetrance. J Reprod Fertil **70:**333–339.
13. **Sundberg JP, Brown KS.** 1994. Imperforate vagina and mucometra in inbred laboratory mice. Lab Anim Sci **44:**380–382.
14. **Van den Broeck W, Derore A, Simoens P.** 2006. Anatomy and nomenclature of murine lymph nodes: Descriptive study and nomenclatory standardization in BALB/cAnNCrl mice. J Immunol Methods **312**:12-19.

Index

Acoustic meatus, external 244, 245, 248, 249
Adipose capsule of kidney 186
Ampulla coli of cecum 152, 153
Angular notch, of stomach 172, 173
Arteries – *See also* Brain, Female genital apparatus, Heart, *and* Male genital apparatus
- Adrenal A. 218
- Aorta 94, 95, 124, 125, 128, 129, 132, 133, 136, 137, 140, 142, 143, 146, 147, 164, 165, 176, 177, 218, 219, 223, 224
- Axillary A. 94, 125, 136, 146
- Brachiocephalic trunk 94, 95, 124, 125, 128, 129, 132, 133, 136, 137, 140, 146, 147
- Carotid A., common 94, 95, 125, 129, 132, 133, 137, 140, 146, 147
- Carotid A., external 95
- Carotid A., internal 74, 75
- Caudal Aa., lateral, median 238, 239
- Cecocolic A. 164
- Celiac A. 146, 147
- Cervical A., superficial 146
- Colic A. 177, 223
- Costocervical trunk 136
- Epigastric A. 223
- Facial A. 94, 95
- Femoral A. 232, 233
- Hepatic A. 160
- Iliac Aa. 146, 147, 177
- Iliolumbar A. 147
- Internal thoracic A. 136, 146
- Labyrinth A. 74, 75
- Maxillary A. 94
- Mesenteric A., caudal 146, 147, 177, 223
- Pudendal A., external 222, 223
- Pulmonary Aa., trunk 94, 124, 125, 128, 129, 132, 133, 136, 137, 140, 146, 147
- Rectal A., cranial 177, 223
- Renal A. 146, 147, 218, 219, 223
- Saphenous A. 232, 233
- Subclavian A. 94, 95, 124, 125, 128, 129, 132, 133, 136, 137, 140, 146, 147
- Vertebral A. 74, 75, 136, 146

Arytenoid cartilage 86
Atlas 86, 87, 244, 245
Axis 86, 244, 245
Basihyoid 86, 87
Body Regions
- Abdominal mammary R. 4
- Abdominal R. 2, 3
- Acromial R. 2
- Anal R. 3, 4
- Antebrachial R. 2, 3
- Auricular R. 2, 5
- Axillary fossa 3
- Axillary mammary R. 4
- Axillary R. 3
- Brachial R. 2, 3
- Brachiocephalic R. 2
- Buccal R. 3, 5
- Calcanean R. 2
- Cardiac R. 2
- Carpal R. 2, 3
- Caudal thoracic mammary R. 4
- Chin R. 3
- Clavicle 3
- Common calcanean tendon R. 2
- Costal arch 2, 3
- Costal R. 2, 3
- Cranial thoracic mammary R. 4
- Crural R. 2
- Elbow joint R. 2, 3
- Frontal R. 5
- Hip joint R. 2
- Hypochondriac R. 2, 3
- Infraorbital R. 5
- Infraspinatus R. 2
- Inguinal mammary R. 4
- Inguinal R. 3
- Intermandibular R. 3
- Interscapular R. 2
- Ischiatic tuberosity R. 2
- Jugular fossa 2, 3, 6
- Jugular groove 2, 3, 6
- Laryngeal R. 2, 3
- Lumbar R. 2
- Mandibular R. 3, 5
- Masseteric R. 2, 3, 5
- Maxillary R. 5
- Mental R. 5
- Metacarpal R. 2, 3
- Metatarsal R. 2
- Nasal R. 5
- Neck R. 2
- Occipital R. 2, 5
- Olecranon R. 2, 3
- Oral R. 3, 5
- Orbital R. 5
- Paralumbar fossa 2
- Parietal R. 5
- Parotid R. 2
- Patellar R. 2
- Perineal R. 4
- Phalangeal R. 2, 3
- Pharyngeal R. 2

Popliteal R. 2
Preputial R. 3
Prescapular R. 2
Pubic R. 3
Root of the tail R. 2
Sacral R. 2
Scapular cartilage R. 2
Scapular R. 2
Scrotal R. 2, 3
Shoulder joint R. 2
Sternal R. 2, 3
Sternocephalic R. 2, 3
Sternohyoideus R. 3
Stifle R. 2
Subhyoid R. 3
Supraorbital R. 5
Supraspinatus R. 2
Tail R. 2
Tarsal R. 2
Temporal R. 5
Temporomandibular joint R. 5
Thigh R. 2
Thoracic mammary R., caudal, cranial 4
Thoracic vertebrae R. 2
Tricipital R. 2, 3
Trochanteric R. 2
Tuber coxae R. 2
Umbilical R. 3
Urogenital R. 4
Xiphoid R. 3
Zygomatic R. 5
Bones of skeleton 244, 245
Bones of skull 248, 249
Brain – *Includes arteries and veins*
Arbor vitae 82, 83
Basilar A. 74, 75
Cerebellar A., Aa., caudal 74, 75
Cerebellar A., rostral 70, 71, 74, 75
Cerebellar V., dorsal, ventral 70, 71
Cerebellomedullar cistern 82, 83
Cerebellum 70, 71, 74, 75, 78, 79, 82, 83, 86, 87
Cerebral A., caudal 70, 74, 75
Cerebral A., middle 70, 71, 74, 75, 78, 79
Cerebral A., rostral 70, 71, 74, 75
Cerebral arterial circle 74, 75
Cerebral epiphysis 70, 71, 82, 83, 86, 87
Cerebral V., Vv., dorsal 70, 71
Cerebrum 70, 71, 74, 75, 78, 79, 82, 83, 86, 87
Colliculi 70, 71, 82, 83, 86
Commissures 82, 83
Communicating A., caudal 74, 75
Corpus callosum 82, 83, 86, 87
Culmen 70, 71
Declive 70, 71
Flocculus 74, 75
Folium vermis 70, 71
Fornix 82, 83
Habenula 82
Hypophysis 78, 79, 82, 83, 86, 87
Interthalamic adhesion 82, 83, 86, 87
Lamina quadrigemina 87
Lobulus ansiformis 70, 71
Lobulus paramedianus 70, 71
Lobulus quadrangularis 70, 71
Lobulus simplex 70, 71
Mamillary body 74, 75, 82, 83
Medulla oblongata 70, 71, 74, 75, 78, 79, 82, 83, 86, 87
Mesencephalic aqueduct 82, 83
Olfactory bulb 70, 71, 74, 75, 78, 79, 80, 82, 83
Olfactory peduncle 74, 75
Olfactory tracts 74, 75
Olfactory tubercle 74, 75
Optic chiasm 74, 75
Paraflocculus, dorsalis, ventralis 74, 75
Peduncle, cerebral 86, 87
Pineal recess 82, 83
Piriform lobe 74, 75, 82, 83
Pons 74, 75, 78, 79, 82, 83, 86, 87
Pontine A. 74, 75
Pyramid 74, 75
Pyramis vermis 70, 71, 74, 75
Sagittal sinus, dorsal 70, 71
Splenium 82, 83
Sulcus, median 70, 71
Transverse sinus 70, 71
Trapezoid body 74, 75
Tuber cinereum 74, 75, 82, 83
Tuber vermis 70, 71
Uvula vermis 70, 71
Ventricles 82, 83, 86, 87
C7 vertebra 124, 125, 128, 129
Calcanean tendon, common 228, 229, 232, 233
Canthus, medial, lateral; of eye 58, 59
Cardiac notch 172, 173
Cardiac orifice 172, 173
Caudal vertebrae 238, 239, 244, 245
Central canal of spinal cord 82, 83
Clavicle 6, 94, 244, 245
Coxal bone 198, 199, 244, 245
Cricoid cartilage 86, 87, 142, 143
Digestive apparatus
Anus 14, 15, 18, 19, 24, 25, 42, 43, 182, 183, 214, 215
Anus, site of 28, 29, 38, 39, 165
Bile duct 168, 169

Cecum 147, 152, 153, 156, 160, 161, 164, 165, 182, 183, 214, 215, 223
Colon 152, 153, 156, 157, 160, 161, 164, 165, 177, 182, 183, 186, 187, 214, 215, 218, 219, 222, 223
Duodenum 156, 157, 160, 161, 164, 165, 168, 169, 172, 173
Esophagus 86, 87, 136, 137, 142, 143, 168, 169, 172, 173
Gallbladder 156, 164, 168
Hepatic duct, common 160, 168
Ileum 152, 153, 160, 161, 164, 165, 183, 223
Jejunum 152, 153, 156, 157, 160, 161, 164, 165, 182, 183, 214, 215, 223
Lip 86, 87
Liver 146, 147, 152, 153, 156, 157, 160, 161, 164, 165, 168, 169, 187, 219
Oropharynx 86, 87
Palate, hard, soft 86, 87, 90, 91
Pancreas 152, 153, 156, 157, 160, 161, 164, 165, 168, 169, 182, 214, 219
Rectum 153, 154, 164, 165, 182, 183, 214, 215, 218, 219, 223
Stomach 146, 147, 152, 153, 157, 160, 161, 164, 165, 168, 169, 172, 173
Teeth 86, 87, 244, 245, 248, 249
Tongue 86, 87, 90, 91
Duodenal bulb 172, 173
Ethmoidal foramen 248
Ethmoid bone, endoturbinates 86, 87
Eyelid 59
Fat 95, 164, 165, 182, 183, 186, 187, 190, 191, 194, 195, 218, 219
Female genital apparatus – *Includes arteries, gland, muscle, and veins*
Cervix 176 (in mouse), 219 (in rat)
Clitoris 14, 15, 18, 19, 24, 25, 28, 29, 42, 43, 214
Ovarian A., V. 147
Ovarian V. 223
Ovary 153, 157, 164, 176, 214, 215, 218, 219, 222, 223
Preputial gland 176, 214, 215, 219
Salpinx 153, 157, 164, 176, 214, 215, 218, 219, 222, 223
Ureter 176, 177, 218, 219
Urethra 214, 215, 218, 219
Urethralis M. 215
Urethral ostium, external 14, 15, 18, 19, 22, 23, 26, 27, 42, 43, 215, 218, 219
Urinary bladder 153, 157, 160, 161, 164, 214, 215, 218, 219, 222, 223
Uterine A., V. 222, 223
Uterine horn 153, 157, 164, 176, 177, 214, 215, 218, 219, 222, 223
Uterine body, body of uterus 214, 215, 222, 223
Vagina 176, 177, 214, 215, 218, 219
Fold of skin over diastema 90, 91
Foot pads 15, 18, 25, 46, 47
Foramen lacerum, rostral 67, 249
Fornix of prepuce 208, 209
Fornix of vagina 176
Gastric groove 172
Glands and ducts – *See also* Female genital apparatus *and* Male genital apparatus
Adrenal gland 218, 219
Harderian gland 62, 63
Lacrimal gland, with duct 62, 63
Mammary glands 28, 29, 42, 52, 53
Parathyroid gland, external 142, 143
Salivary glands 62, 63, 94, 95
Thymus 94, 100, 101, 104, 105, 108, 109, 112, 113, 116, 117, 120, 121, 142, 143
Thyroid gland 142, 143
Heart – *Includes arteries and vein*
Apex 94, 124, 125, 128, 129, 132, 133, 136, 137, 140
Atrium, left, right 128, 129, 132, 133, 136, 137, 140
Auricle, left, right 94, 95, 100, 101, 104, 105, 108, 109, 112, 113, 117, 118, 121, 122, 124, 125, 128, 129, 132, 133, 136, 137, 140, 143, 144, 146, 147
Cardiac V., great 128, 129, 132, 133, 136
Chordae tendineae 140
Conus arteriosus 124, 125, 132, 133, 136
Coronary sinus 128, 129
Interventricular A., paraconal 124, 132, 133, 136, 146
Interventricular A., subsinuosal 128, 132
Musculi papillares parvi 140
Musculus papillaris magnus, subarteriosus, subatrialis, subauricularis 140
Pectinate muscles 140
Trabeculae carneae 140
Trabecula septomarginalis 140
Valves 140
Ventricle, left, right 94, 95, 100, 101, 104, 105, 108, 109, 112, 113, 116, 117, 120, 121, 124, 125, 128, 129, 132, 133, 136, 137, 140, 146, 147
Ileocecal fold 164, 165
Incisive bone 86, 87, 244, 245, 248, 249
Incisive papilla 86, 87, 90, 91
Infraorbital hiatus, fissure 248, 249
Ischial arch 198, 199
Ischiopubic symphysis 182, 183, 215
Kidney 146, 147, 152, 153, 156, 157, 160, 161, 164, 182, 183, 186, 187, 214, 215, 218, 219, 223
Lacrimal canal, foramen 248, 249
Laryngopharynx 86, 87
Ligaments
Arteriosum, Lig. 133
Interdeferent Lig. 202
Lig. of tail of epididymis 183, 187, 191, 195
Ovarian suspensory Lig. 214, 215
Patellar Lig. 244, 245
Penis suspensory Lig. 187, 198, 199
Pulmonary Lig. 100, 101, 104, 105, 109, 113
Round Lig., of uterus 161, 214, 215, 222, 223
Sternopericardic Lig. 100, 101, 104, 105, 108, 109, 112, 113
Tracheal annular Lig. 142, 143
Urinary bladder Ligg., lateral, median 182, 183, 214, 215, 222, 223
Lymph nodes
Aortic lumbar Lnn. 176, 177, 218, 219
Cecal Ln. 165
Cervical Lnn., caudal, cranial 143, 144
Colic Lnn. 165
Gastroduodenopancreatic Ln. 165
Iliac Lnn. 176, 177, 219
Jejunal Lnn. 165

Mandibular Ln. 63, 94, 95
Mediastinal Lnn. 142, 143
Mesenteric Ln. 165, 176, 177, 218
Sacral Ln. 219
Sternal Ln. 143
Tracheobronchial Ln., Lnn. 142, 143
Male genital apparatus - *Includes arteries, glands, muscles, and vein*
Ampulla of ductus deferens 202
Ampullary gland 186, 187, 202
Bulbocavernosus M. 187, 208, 209
Bulbospongiosus M. 198, 199, 205
Bulbourethral diverticulum 182, 186, 202-204
Bulbourethral gland 182, 183, 186, 187, 198, 199, 202, 205
Bulbourethral gland, duct of 203-205
Bulbourethralis M. 182
Coagulating gland 165, 182, 183, 186, 187, 202
Ductus deferens 165, 182, 183, 186, 187, 191, 195, 202
Epididymis 182, 183, 186, 187, 190, 192, 194, 196
Ischiocavernosus M. 186, 187, 198, 199, 202, 205
Os penis 198, 199, 208, 209
Pampiniform plexus 191, 195
Penis 186, 187, 198, 199, 202, 208, 209
Penis A., dorsal 198, 199
Plate, fibrocartilaginous, of urethra 202, 203
Prepuce 14, 15, 18-20, 24, 25, 28, 29, 38, 39, 186, 187, 199, 208, 209
Preputial gland 182, 183, 186, 187, 199
Prostate gland 165, 182, 183, 186, 187, 202
Scrotal site, in juvenile 14, 15, 19
Scrotum 18, 20, 24, 25, 28, 29, 34, 38, 39
Spermatic cord 190, 191, 194, 195
Testicle 182, 183, 186, 187, 190, 191, 194, 195
Testicular A., V. 147, 183, 186, 187, 190, 191, 194, 195
Ureter 183, 187, 202
Urethra 182, 183, 198, 199, 202-205, 208, 209
Urethralis M. 183, 186
Urethral isthmus 203-205
Urethral recess 203-205
Urinary bladder 152, 165, 182, 183, 186, 187, 198, 199, 202
Venous plexus of penis 199
Vesicular gland 165, 182, 183, 186, 187, 202
Mandible 86, 87, 90, 91, 94, 244, 245, 248, 249
Manubrium sterni 6, 94, 104, 105, 112, 113, 120, 121, 124, 125, 128, 129, 244, 245
Margo plicatus 152, 153, 160, 161, 172, 173
Mesepididymis 187
Mesocolon 182, 215
Mesoductus deferens 186, 191, 195
Mesofuniculus 191, 195
Mesometrium 214, 215, 222, 223
Mesorchium 183, 191, 195
Mesorectum 182, 214
Mesosalpinx 214, 215
Mesovarium 214, 215
Milk spot of neonates 10, 11
Muscles – *See also* Female genital apparatus, Heart, *and* Male genital apparatus
Adductor longus M. 233
Biceps femoris M. 228, 229
Brachiocephalicus M. 94, 95
Buccinator M. 91
Cleidobrachialis M. 6, 94
Coccygeus M. 183
Cranial tibial M. 228, 229, 232, 233
Descending pectoral M. 6, 7
Digastricus M. 94, 95
Digital extensor M., long, lateral; and tendon of 228, 229
Digital flexor M., superficial 233
Digital flexor M., superficial, tendon of 232
Fibularis longus M. 228, 229
Flexor hallucis longus M. 228, 233
Flexor pedis longus M., tendon of 232, 233
Gastrocnemius M. 228, 229, 332, 333
Geniohyoideus M. 86, 87
Gluteal M., superficial 228, 229
Gracilis M. 232, 233
Hyoepiglotticus M. 86, 87
Intertransversarius M., and tendons of 238, 239
Longus capitis M. 86, 87
Longus colli M. 86
Masseter M. 90, 91, 94, 95
Mylohyoideus M. 86, 87
Omohyoideus M. 95
Pectineus M. 232, 233
Pterygoid M., medial 90, 91
Quadriceps femoris M. 229
Rectus abdominis M. 222, 223
Rectus capitis M., lateralis 86, 87
Rectus capitis M., ventralis 87
Rectus femoris M. 228, 233
Sacrocaudal Mm., and tendons of 238, 239
Sartorius M. 232, 233
Semimembranosus M. 232, 233
Semitendinosus M. 228, 229, 232, 233
Sternocephalicus M. 94
Sternohyoideus M. 6, 86, 87, 94, 95, 143
Sternomastoideus M. 6
Sternothyroideus M. 94, 95, 143
Tensor fasciae latae M. 228, 229, 232, 233
Tibial M., caudal 232, 233
Tibial M., cranial 228, 229, 232, 233
Vastus medialis M. 232, 233
Nasal concha, dorsal, ventral 86, 87
Nasopharynx 86, 87, 92, 93
Nerves
Cranial nerves 74, 75, 78, 79, 82, 83
Femoral N. 232

Fibular N. ... 228, 229
Saphenous N. ... 232, 233
Sciatic N. ... 228, 229
Sural N., caudal, cutaneous ... 228
Tibial N. ... 228, 229
Omentum, greater ... 169
Optic canal, foramen ... 248, 249
Phalanges ... 244, 245
Philtrum ... 90, 91
Pyloric antrum ... 172, 173
Pylorus ... 172, 173
Respiratory apparatus
Bronchus ... 142, 143
Choana ... 86, 87
Diaphragm ... 100, 101, 104, 105, 108, 109, 112, 113
Epiglottis ... 86, 87, 90, 91, 142
Larynx ... 86, 87, 90, 91, 142
Lung ... 100, 101, 104, 105, 108, 109, 112, 113, 116, 117, 120, 121, 142, 143
Nostrils ... 90, 91
Trachea ... 142, 143
Tracheal cartilages ... 86, 87, 142, 143
Rib cartilage ... 244, 245
Ribs ... 104, 105, 112, 113, 120, 121, 124, 125, 128, 129, 244, 245
Saccus cecus ... 152, 153, 160, 161, 169, 172, 173
Scapular cartilage ... 244, 245
Sphenopalatine foramen ... 67, 249
Spinal cord ... 78, 79, 82, 83
Spleen ... 146, 147, 152, 153, 160, 161, 164, 165, 168, 169, 219
Sternum ... 124, 125, 128, 129, 244, 245
T1 vertebra ... 124, 125, 128, 129
Temporal line ... 63, 67, 248, 249
Thyroid cartilage ... 86, 87, 142, 143
Trochanters, lesser, major, third ... 244, 245
Tympanic bulla ... 63, 248, 249
Umbilicus ... 19, 20
Vaginal fornix ... 176
Vaginal ostium ... 14, 15, 18-20, 24, 25, 28, 29, 42, 43, 215, 218, 219
Vaginal ring ... 214, 215, 222, 223
Vaginal tunic ... 191, 195
Veins – *See also* Brain, Female genital apparatus, Heart, *and* Male genital apparatus
Adrenal V. ... 147
Angularis oculi V. ... 66, 67
Axillary V. ... 94
Azygous V., left ... 124, 125, 128, 129, 132, 133, 136, 137
Caudal Vv., dorsolateral, lateral, median, ventrolateral ... 238, 239
Cephalic V. ... 6, 94, 97
Colic V. ... 223
Facial V. ... 94, 95
Femoral V. ... 229, 232, 233
Iliac V., common ... 146, 147, 177
Iliolumbar V. ... 147
Infraorbital V. ... 67
Ischiatic V. ... 228, 229
Jugular V., anterior ... 95
Jugular V., external ... 6, 94, 95
Jugular V., internal ... 95
Jugular V., posterior ... 95
Maxillary V. ... 94, 95
Ophthalmic V. ... 67
Palpebral V., inferior ... 66
Phrenic V., cranial ... 101, 105, 109, 113
Portal V. ... 146, 147
Pudendal V., external ... 222, 223
Pulmonary Vv. ... 124, 125, 128, 129, 132, 133, 136, 140, 146
Rectal V., cranial ... 223
Renal V. ... 146, 147, 218, 219, 223
Retroorbital venous plexus ... 66, 67
Retroorbital venous sinus ... 66
Saphenous V., lateral ... 228, 229
Saphenous V., medial ... 232, 235
Sphenopalatine V. ... 67
Spinal V., dorsomedian ... 71
Spinal V., ventral ... 74, 75
Suprahepatic V. ... 146, 147
Supraorbital V. ... 66
Temporal V., superficial ... 66
Vena cava, caudal ... 109, 113, 116, 117, 120, 121, 124, 125, 128, 129, 132, 133, 136, 137, 140, 142, 143, 146, 147
Vena cava, cranial ... 94, 95, 124, 125, 128, 129, 132, 133, 136, 137, 140, 146, 147
Vibrissae ... 58, 59
Vocal cord ... 90, 91
Xiphoid cartilage ... 104, 105, 112, 113, 120, 121, 124, 125, 128, 129, 164, 244, 245
Xiphoid process ... 124, 125, 128, 129, 164
Zygomatic arch ... 63, 67, 244, 245, 248, 249